*New Horizons for Early
Modern European Scholarship*

New Horizons for Early Modern European Scholarship

EDITED BY
ANN BLAIR *and* NICHOLAS POPPER

Johns Hopkins University Press
Baltimore

This publication is made possible in part with support from the
Barr Ferree Foundation Fund for Publications, Department of Art and Archaeology,
Princeton University.

Johns Hopkins University Press
2715 North Charles Street
Baltimore, Maryland 21218-4363
www.press.jhu.edu

Library of Congress Cataloging-in-Publication Data
Names: Blair, Ann, 1961– editor. | Popper, Nicholas, 1977– editor.
Title: New horizons for early modern European scholarship / edited by Ann Blair
and Nicholas Popper.
Description: Baltimore : Johns Hopkins University Press, [2021] | Includes
bibliographical references and index.
Identifiers: LCCN 2020033861 | ISBN 9781421440934 (hardcover) |
ISBN 9781421440941 (ebook)
Subjects: LCSH: Europe—History—1492-1648—Historiography. |
Europe—History—1648-1715—Historiography. |
Europe—Civilization—Historiography. | Europe—Intellectual
life—Historiography. | Historians—Europe.
Classification: LCC D206 .N49 2021 | DDC 001.2094/09031—dc23
LC record available at https://lccn.loc.gov/2020033861

A catalog record for this book is available from the British Library.

Special discounts are available for bulk purchases of this book. For more information,
please contact Special Sales at specialsales@jh.edu.

Johns Hopkins University Press uses environmentally friendly book materials,
including recycled text paper that is composed of at least 30 percent post-consumer
waste, whenever possible.

For Louise Grafton (1941–2019)
"Give her of the fruit of her hands;
and let her own works praise her in the gates."

CONTENTS

*New Horizons for Early
Modern European Scholarship*

Introduction

NICHOLAS POPPER AND ANN BLAIR

Not all that long ago, early modern European historians justifiably perceived themselves as both center and cutting edge of the historical profession. Works like Keith Thomas's *Religion and the Decline of Magic: Studies of Popular Beliefs in the Sixteenth- and Seventeenth-Century England* (1971), Christopher Hill's *The World Turned Upside Down: Radical Ideas during the English Revolution* (1972), Carlo Ginzburg's *The Cheese and the Worms* (1976; Eng. trans., 1980), Natalie Zemon Davis's *The Return of Martin Guerre* (1983), and Robert Darnton's *The Great Cat Massacre: And Other Episodes of French Cultural History* (1984) became staples of undergraduate and graduate syllabi, read not only by specialists in early modern Europe but by historians across chronological and geographical boundaries and by the wider public as well.[1]

The field's flourishing rested on several overlapping elements. For professional historians in the Anglo-American world, early modern Europeanists had long been known for methodological innovation, whether imported by mid-century European émigrés such as Hans Baron, Paul Oskar Kristeller, and Felix Gilbert or developed in Europe by scholars such as Fernand Braudel, Eugenio Garin, and Emmanuel Le Roy Ladurie, whose works were made easily accessible through translation and abridgment. Early modernists were among the first to absorb the French Annales School's commitment to integrating insights from other social sciences, especially anthropology but also economics, sociology, and psychology. Others, especially Quentin Skinner and J. G. A. Pocock at Cambridge, pushed interdisciplinarity in new directions by integrating history with political philosophy to form what became known as the history of political thought. Still others were inspired by the quantitative approach of "cliometrics." Many shared a Marxian-derived focus on history "from below" to challenge the dominance of high political history.[2] The influence of these methods spread through journals devoted to them, such as *Past & Present* and *History Workshop Journal*, as well as through field-shaking monographs.

The most prominent of these golden age early modern histories were rooted in social history of the 1960s, but they developed as new methodologies in the 1970s and 1980s that crystallized under the rubrics of cultural history and microhistory. They were united in several meaningful ways: while they resembled conventional social histories in granting ordinary men and women significance and attention previously reserved for kings and statesmen, they differed in analyzing texts rather than accumulating data, in interrogating structures of perception rather than deriving universal laws of social change from quantitative phenomena, and in emphasizing narrative rather than adopting a stance of scientific objectivity. At the same time, they continued to participate in the kinds of big-picture debates—concerning, for example, the causes of the English Civil War and French Revolution, secularization, and the European crisis of the seventeenth century—that had animated previous generations. Historians throughout the profession embraced these new models, redirecting them with energy and vibrancy toward a wide range of other contexts, from nineteenth-century Vienna to the Aztec Empire.[3] Scholars working at the peak of cultural history reenvisioned statecraft and masques, friendship and money, families and civic parades as opportunities to study how individuals mediated the cultural, social, economic, and political structures pressuring them. Scholars in other disciplines, too, adopted these historical methods in ways that both transformed their own fields and suggested new directions for historical study.[4]

Several of the features that so attracted historians—especially the emphasis on narrative—helped further elicit unusual public interest in early modern European histories. This widespread popular appeal was abetted by the vision they propounded of Europe between roughly 1450 and 1800: one of a world precariously balanced between change and continuity, developing novel contours as early modern Europe's inhabitants generated and confronted political, demographic, economic, and technological innovations. As these historians showed, early modern Europeans struggled to comprehend and assimilate the vortices of change swirling around the venerable institutions and interpretive practices they had inherited from the ancient and medieval worlds. In the process they developed epistemologies, languages of social description, and publics that resonated with the sensibilities of late modern audiences. Indeed, one might argue that groundbreaking cultural historians and microhistorians gravitated toward this period precisely because it was particularly susceptible to the tools they prized, which highlighted the symbolic and linguistic tensions that accompanied gradual structural change. Early modern Europe, in this telling, was both familiar and foreign, recapitulating earlier periods through, for example, the intellectual inheritance of classical antiquity and medieval piety, while simultaneously reconfiguring and transmitting them for the modern age. The early modern served as a mirror to the present, offering an ideal vantage point from which to question and examine assumptions of our own time.

The argument that Europe in these centuries was the seedbed of modernity was hardly pioneered in the late twentieth century. The claim had a long tradition, exemplified by Jacob Burckhardt's assertion that the Renaissance's revitalization of antiquity was in fact a point of rupture, inaugurating the birth of the individual, secularization, and the transformation of dynastic realms into states.[5] Perhaps no less paradigmatic was the Weberian assertion that the Reformation was ultimately responsible for replacing a world of communal superstition with one of rational, secular, economically minded individuals.[6] And there were other avenues to the claim that Europe modernized over these centuries. A trifecta of technological innovations originally heralded by Renaissance humanists emblematized the alterations of the age: the compass, representing European exploration, also explained the resultant influx of commodities and capital; gunpowder, which accounted for the heightened capacity for coordinated violence by centralizing nation-states, also propelled Europe's domination of civilizations whose wealth and sophistication had previously outstripped it; the printing press drove religious dislocation and democratized access to texts, while also facilitating the rapid dissemination of all kinds of knowledge through the Reformation, Scientific Revolution, and Enlightenment.[7] These Whiggish visions created a sense that over the period from roughly 1450 to the end of the eighteenth century—bracketed by the invention of the printing press or Columbus's voyage on one end, and Enlightenment and French Revolution on the other—modern subjectivity, institutions, and social structures came into being.

The works of Ginzburg, Davis, Darnton, Thomas, and others capitalized on the assumed importance of early modern Europe. At the same time their revelation of cultural logics suffused by tensions, eccentricities, and contradictions undermined the tidy Whiggish narratives that lent early modern Europe much of its prestige. In subsequent years, a series of critiques have made early modern Europe more alien by appropriately challenging the blindnesses and assumptions implicit in many earlier, grand visions of the period. It no longer is acceptable, for example, to portray Europe's globalization as a process driven solely by internal factors while treating the rest of the world's inhabitants as raw material inertly awaiting domination (in much the same way that European colonists and subjugators had exploited them). Nor is it viable to see the Reformation as portending an age of unabashed secularization. And Joan Kelly's provocative "Did Women Have a Renaissance?"—and the emphatic negativity of her answer—revealed the need for new chronologies and narratives that centered on historical actors other than a small cadre of elite white men.[8] The elements of emergent modernity detectable in these years now seem contingent and precarious rather than inexorable, universal, and irreversible.

In part because of this heightened awareness of the complexity and local variation of Europe in these years, many historians determined to examine the birth of the modern have gravitated away from the sixteenth and seventeenth centuries and instead toward the emergence of capitalism, modern notions of race, reconstructed gender roles, and dominant European empires in the eighteenth and early nineteenth centuries. Historians have ascribed elevated importance to the eighteenth century in recent years, but predominantly by linking it to future developments rather than by portraying it as a culmination of the preceding ones. It seems safe to say that early modern Europe as conventionally conceived no longer enjoys its former position of professional privilege.

Though increasingly viewed as just one among an expanding array of subfields of history defined by time and place, the field of early modern European history has continued to gain exciting new historiographical dimensions in the past three decades. This volume revisits traditional topics in early modern European intellectual and cultural history, including humanism, contact with other cultures, revolutionary historical and scientific sensibilities, and the early Enlightenment, in light of new emphases on globalization, religion, and the practices and material forms of knowledge that each resonate especially with preoccupations current in our world today. These essays draw on both the foundational studies of the "glory years" (1960s–1980s) and the recent "turns" of various kinds to explore new frontiers of the field, in particular its unsettled borders, whether geographical, temporal, or thematic.[9] Early modern Europe's claim to significance now largely resides in its role as an engine of globalization. To be sure, the early modern globalization depicted by historians in the early twenty-first century differs markedly from versions articulated by past generations, who in essence viewed the world as quarry for control by empires created by imperial European nation-states. Most significantly, Asians, Africans, and Americans are now—at least ideally—invested with agency previously reserved for Europeans, and historians aspire to treat their societies, cultures, communities, and knowledge symmetrically while still foregrounding their frequent exploitation and oppression—not to mention genocide in the Americas.

Investigations attentive to the global scope of early modern life have revealed that, contrary to the rigid cultural and social boundaries often unreflexively assumed in earlier histories, early modern Europe was permeable and porous, shot through with peoples, commodities, beliefs, ideas, images, flora, fauna, and much else originating all over the world—and that these forces acted with previously undetected or understated agency in the process of globalization. Earlier portrayals of the Ottoman Empire in its powerful confrontation with a decentralized Europe made no mention of the continual trade and diplomatic relations between the two or the heavily European population of the former (except for the Janissaries). A regime of horrific enslavement, for example, now appears a precondition of Europe's rise, while intellectual transformations like the Scientific Revolution owed considerable debts to previously overlooked participants such as enslaved African and Indigenous healers, Jewish sailors, and aristocratic women, as well as to the circulation of material culture previously considered exotic.[10] Early modern Europe now appears imperfectly characterized by the often parochial visions propounded by the dominant cultural and social histories of the end of the twentieth century, for it appears constituted less by villages in which families dwelt unmoving for centuries as broader forces seeped in than by individuals maneuvering through a vibrant and animated world of movement, kinesis, exchange, and turbulence generated both within and without Europe.[11] Indeed, Europe's flux and volatility increasingly appears a consequence of both its own incessant dynamism and its inexorable interactions with the rest of the world.

For those of us who, in the early twenty-first century, see our contemporary world as sharing this multiplicity, early modern Europe holds a particular appeal and relevance as an opportunity to see the lenses behind the kaleidoscope, to dissolve monolithic big narratives into jarring shocks of the particular where meaning is vested not in a static past but in its fascinating waves of receptions, where paradox and reason, contradiction and unity entangle. The process of modernization in these years resists characterization as monolithic, unified, and singular; it appears to have been a highly contingent process assuming distinct and precarious forms across cultural, social, and political geographies. The outcome was not a coherent assemblage of alike subjects, institutions, and mentalities, but rather a series of relentless experiments in subjectivity, politics, and intellectual apparatuses—a world of multiple competing modernities engaged in relentless transmission and conflict. Early modern Europe can no longer be described as divesting the medieval from the modern; rather it should be seen as brimming with contingent, fractal ideologies, social organizations, governments, religious beliefs, and cultural codes. Competition to control these underdetermined forms occurred between empires no less than within towns.

The collective body of recent scholarship that has revealed the stunning multiplicity of early modernity has been less concerned with developing meth-odological innovations of the sort that commanded such widespread fascina-tion for early modern cultural histories and microhistories in the 1970s and 1980s. But the focus on early modern European contacts with other societies and the deep, textured connectedness of the world after 1492 has suggested that the dynamics of exchange were no less multivalent than the myriad meth-ods early modern Europeans used to navigate more local irruptions of differ-ence and the unfamiliar. Accordingly, such work suggests the need for close, sustained attention to the practices that structured how individuals procured and assimilated texts, artifacts, information, images, and bodies. Practices of interaction and interdependence constitute the new site demanding enhanced methodological attention.

To be sure, there exists a massive scholarship on encounter and the shock of the new. But such work has tended to focus more on the exercise of power within such contacts, whether through its unilateral articulation of inequalities or through—as more recent scholarship emphasizes—processes of negotiation. In the latter, the worthy goal of restoring agency to the colonized or historically marginalized and underrepresented has resulted in a focus on demonstrating the act of negotiation without closely examining the instruments and methods by which mediation was structured. The overarching power dynamics should always be conspicuous in such studies. But historians' characterizations of power relations in such encounters are often overdetermined by modern assumptions that presume European dominance (or that, in a revisionist vein, refute it) rather than viewed as a field of contingent possibility.

Recent work, ranging from the history of go-betweens to that of archiving, attends increasingly to those contingent, mediated negotiations by focusing on the people and practices involved, and on the ways that the sources we use were generated.[12] Alongside studies of economic and political power, this historiog-raphy focuses on understanding the structure of cultural transmissions and contact, of reception and exchange. These studies examine often overlooked strands of interaction that drove the precarious warp of change and continuity in the patterns of power as well as culture. In particular, relatively new fields such as book history and the history of knowledge have devoted special atten-tion to early modern Europe even as they have introduced methods applicable to other contexts. Both of these new subfields draw on the revisionist instincts of the past generation of early modernists while also reflecting ongoing twenty-first-century preoccupations with the mediation of information, the porousness of knowledge, and the uneven topography of expertise. Accordingly, as the fol-lowing essays show, these historiographies offer powerful means by which to understand the relentless mediations, fluctuations, transmissions, entanglements, divestments, and subordinations of the early modern world. And they may simultaneously offer to other fields the sort of rich vein for methodological mining offered in the 1970s and 1980s by cultural history and microhistory.

The term "book history" dates from its original focus, in Lucien Febvre and Henri-Jean Martin's 1958 *The Coming of the Book*, on the spread and impact of printing in early modern Europe.[13] In the 1980s it garnered attention as a distinctive field that had developed at the intersection of historical interests in studying mentalities (for example, through the production and circulation of cheap print), literary questions (most prominently concerning reception), and the long-standing expertise of book professionals developed by bibliographers, librarians, dealers, and collectors with a special focus on rare and valuable books and manuscripts that had previously been integrated into academic research only rarely.[14] In the decades since, book history has experienced a rapid expansion in the range of participating disciplines, of questions posed and methods applied, and of periods and places investigated. At this point, anywhere there is text, there can be book history, and in this broad conception, the approach examines forms other than the codex (from the clay tablet to the tweet), and media beyond the printed book, including orality, manuscript, and digital media and their varied interrelations.

The recent surge of interest in book history and related fields such as media studies and the history of information has no doubt been inspired in part by our awareness of powerful and ongoing changes in today's media ecology. We might well call these changes "revolutionary" in principle, except that in practice we also notice how fitful, slow, and uneven they are (an illuminating parallel to broader revisions of early modern Europe's ostensible process of modernization). Alongside the massive growth of digital publications of many kinds, more printed books are being produced every year than ever before, and some digital media resemble oral and manuscript more than print publication in how they operate. The urge to compare the innovations of our period with other examples of the arrival of a new technology has fueled a renewed interest in the impact of printing in early modern Europe, even though other examples are equally fascinating (for example, the introduction of paper in ninth-century Islam).[15] As in other areas of historiography of early modern Europe, recent work in book history has fractured earlier accounts of inevitable modernization through the spread of printing, by emphasizing, for example, the persistence of manuscript and orality as modes of publication, or the association (perceived both at the time and by recent historians) of print capitalism with cutthroat profiteering rather than admirable truth-seeking. Understanding early modern Europe as a complex media society thus provides a promising avenue for capturing its processes of transformation; at the same time, it captures a vital dimension of its role as precursor to the modern world.

Recent work on the "history of knowledge" has introduced a promising umbrella term that encompasses recent converging trends in cultural and intellectual history. The history of knowledge attends to methods and practices of making knowledge, in addition to the concepts and ideas more traditionally associated with intellectual history and the history of science. In casting the net wide to encompass "knowledge" rather than specific fields, this approach strives to avoid using predetermined (e.g., modern) definitions of disciplines in order to study the connections and combinations of different topics made by the historical actors. And those actors include a broader array of people and places than the canonical figures and institutions traditionally emphasized in intellectual history. "Knowledge" can thus encompass not only all the sciences and humanities but also expertise, savvy, and tacit know-how generated and exchanged outside disciplinary structures, for example, in domestic spaces, marketplaces, artisanal workshops, printing houses, travel, and secretariats, or through cross-cultural contacts. The history of knowledge builds on developments under way since the 1980s especially in history of science and book history, but it has also drawn inspiration from sociology, cognitive anthropology, and postcolonial studies, among other fields.[16]

A history of knowledge approach can apply to any time and place, but it has blossomed especially in the study of early modern European history, where its impact can be compared with that of microhistory some thirty years ago. New people and places became involved in European knowledge making between 1450 and 1750 thanks to the impact of printing, the rising status of vernacular languages, the growth of court patronage and of cities, increased trade across the globe, and a general broadening of the locales for intellectual work and interaction—including the new printing houses. Equally significant is the greater survival rate of sources from this period than from earlier ones, sources through which to piece together practices of reading and writing, observing and analyzing, communicating and collaborating. Because of the durability and wide availability of paper in this period (fueled by the spread of printing, which required paper to be economically viable), we have working notes and documents that generally do not survive from earlier contexts that used different media for short-term writing (wax tablets, or parchment fragments) and long-term record keeping (fragile papyrus rolls or durable but expensive parchment). Equally essential to the survival of personal papers such as diaries, letters, notes, and working papers was the effort on the part of both the record makers and their heirs to preserve them. Rarely have personal papers survived in families all the way since the early modern period; immediate descendants often played a crucial role in the survival of papers in preventing their immediate destruction, but the collections that survive were generally placed in institutions that have endured more or less intact since then (such as libraries and museums, learned societies, religious orders, and government archives of various kinds).[17]

Annotations also survive in printed books—signs of reading and ownership and other unrelated pieces of writing—generally transmitted unintentionally along with the book itself. Although individual and institutional collectors in previous generations preferred pristine books and therefore washed off annotations to increase the book's value, the reverse is true today. Annotated books are now highly valued for the light they shed on the interests and reading methods—both those specific to individual readers and those widely shared among the educated.[18] The growing number of studies of annotated books has helped historians determine which contents and methods of working were widespread in a given context (for example, as diffused through dictation in a classroom setting) and which were idiosyncratic (like the judgmental exclamations of an Isaac Casaubon). Recent digitization projects have facilitated the study of annotations by making them more widely available and easier to compare across different copies and contexts.[19]

Historians of early modern Europe also benefit from having access to troves of letters that survive in diplomatic records, personal papers, and collections that were printed at the time. While some of these have long been studied for insight into individual thinkers, in recent decades scholars have sought to uncover patterns of exchange across national and confessional barriers, social and intellectual habits, or conventions of politeness and writing among the wide variety of educated correspondents who formed the "Republic of Letters." Alongside multivolume editions equipped with careful indexing, digital methods have also been developed to help visualize and study these letters in new ways. Thanks to these new approaches to sources that are comparatively abundant, early modern Europe has proved a fecund proving ground for the history of knowledge and knowledge practices.

It is challenging to assess the extent to which early modern patterns were unique to that period or place, or when they are more visible evidence of a history of practices extending back for centuries or dispersed more broadly throughout knowledge-making systems generated across the world. Comparative work in this vein would help disclose specificities in particular time-place contexts. And questions need not have answers to be valuable incitements to new kinds of research and analysis. One of the virtues of the history of knowledge is that it neatly captures and amplifies how expertise, learning, and intellectual authority might coil through any number of spaces, bodies, practices, forms, media, and the like. It encourages historians to unearth and make sense of previously obscure or obscured testimonies of learning, to view materials such as account books, ships journals, oral testimony, archaeological sites, images, and material culture not as transparent evidence of the past but as the objects of inquiry themselves. A history of knowledge perspective encourages scholars to expand their scope of analysis beyond texts to cultural products of all sorts and to see them as hybrid statements embodying distinctive lineages that articulate social dimensions of knowledge, as sheaths that cinched and clenched tangled networks of influence.

The ten essays in this volume each approach an area of early modern European cultural or intellectual history in order to showcase recent work and future directions that build bridges between various subfields of what has become a large historiography of this particular context. We highlight connections between disciplines, types of sources, time periods, and places. At the same time, the essays offer resources for thinking through early modern Europe's connections to its predecessor and successor societies and to those with which it sustained new or newly developed contacts. They propose models for examining how individuals operated under conditions of media and knowledge pluralism, models that could be applied to the adaptation of native food-provisioning systems or the work of Anglo-American fort builders in the Caribbean, to the narratives elaborated by Indigenous chroniclers at the cusp of the Age of Exploration or the bustling of commercial newsrooms in the Age of Revolutions, to the compilation of medieval cartularies or the racialization of medical knowledge. Indeed, as surveys of recent monographs and ongoing dissertations across the profession indicate, such models can also be applied to the dominant themes of race, globalization, and political economy so urgent and immediate in our contemporary world.[20]

The potential value of the methods outlined here lies in their broad-based applicability: the possibilities they raise for analyzing practices of exchange and reception both within internal European interactions and in Europe's imperial and more putatively exotic encounters. More broadly, they suggest an analytic grammar for interrogating the strategies and practices used by individuals poised at the hub of dynamic worlds. Drawing on early modern European precedents does not derogate from the urgent attention to other societies and geographies that have proliferated in the past several decades, nor does it seek to reprovincialize the global in the service of the European. Rather, we seek to highlight the flexibility and fertility of practice- and reception-based methodologies that have germinated and matured in early modern European history and may be of interest to other chronological and geographical fields. We hope that a focus on the transformations in the knowledge practices in early modern Europe will highlight the specificities of this context but also foreground the pluralism and contingency of the world we now inhabit. The selections made by the editors and authors among many other possible themes only reinforce our belief that early modern Europe harbors many opportunities for continued methodological innovation.

NOTES

1. K. Thomas 1971; Hill 1972; Ginzburg (1976) 1980; Davis 1983; Darnton 1984.

2. Stone 1979; Novick 1988; Burke 2015a.

3. Schorske 1979; Clendinnen 1991.

4. Most prominent was New Historicism's rise in English literature, which fed back into historical study through works such as Greenblatt 1980.

5. Burckhardt (1860) 1990.

6. Weber 1930.

7. McNeill 1989; Eisenstein 1979.

8. J. Kelly 1977. For a recent discussion of the historiographical background to this article, see Davis 2013.

9. See also *American Historical Review* Forum 2012.

10. See, for example, the places and people included in Park and Daston 2006, part 2: "Personae and Sites of Natural Knowledge"; Rankin 2013; Zaken 2010; Parrish 2006; Cook 2007; Schiebinger 2017.

11. See, for example, Davis 2006; Ruderman 2010; Spence 1984; Carlebach 2011.

12. Schaffer et al. 2009; E. Rothman 2009; Premo 2017; K. Burns 2010; Fuentes 2016.

13. Febvre and Martin (1958) 1976.

14. Darnton 1982, 2007.

15. Bloom 2001.

16. Burke 2015b; Jacob 2007–11.

17. M. Hunter (1998) pioneered the study of surviving collections of papers. See, more recently, Corens, Peters, and Walsham 2016, 2018; Keller et al. 2018; Head 2019.

18. Sherman 2008.

19. Online projects include Annotated Books Online and the Archaeology of Reading. Studies of annotated books are legion at this point; for a seminal model, see Jardine and Grafton 1990; on Casaubon, see Grafton and Weinberg with Hamilton 2011.

20. See, for example, A. Curran 2011; Seth 2018; Hogarth 2017; McCormick 2009; Reinert 2011; Warsh 2018; Strang 2018; Eacott 2016; Lennox 2017; S. Davies 2016; Dubcovsky 2016.

PART ONE

CHRONOLOGICAL HORIZONS

CHAPTER ONE

Humanism between Middle Ages and Renaissance

ELIZABETH MCCAHILL

Yet again, humanism threatens to break through the levees crafted to manage and channel it. In part, this is due to developments within modern humanism, which has long confused efforts to classify earlier movements of the same name.[1] But scholars of pre-Enlightenment Europe are also expressing some discomfort with the neat definition provided by Paul Oskar Kristeller, a definition that has served for the past seventy-five years as a valuable foundation for consensus, especially among Anglo-Americans. Rejecting nineteenth- and twentieth-century accretions to the term, Kristeller demonstrated that Petrarch and his followers used the phrase *studia humanitatis* to denote a precise group of disciplines: grammar, rhetoric, history, poetry, and moral philosophy.[2] Kristeller's humanist was a professional rhetorician, the heir of the medieval *dictatores*, and he was distinguished from scholastic thinkers by his lack of interest in systematic philosophy. In the long arc of competition between rhetoric and philosophy, according to Kristeller, the rise of humanism represented a triumph for rhetoric.[3]

Kristeller's definition aptly describes a group of fifteenth-century intellectuals who documented their allegiance to ancient rhetoric in classicizing texts and who, usually as teachers or secretaries, earned a living based on their prowess in Greek and Latin. But the predecessors and successors of fifteenth-century humanists challenge the parameters Kristeller offered, and, in doing so, they raise important questions about the place of humanism in European intellectual history. Before Petrarch, there were writers who used some of the texts and methods of later humanists but who were not professional rhetoricians. Conversely, by the early sixteenth century, few of Europe's most influential intellectuals were simply, or even primarily, humanists in Kristeller's mode, although almost all well-educated Europeans were versed in the *studia humanitatis*. With humanism's triumph, it ceased to be a primary form of identification and became a scholarly technique, used to greater or lesser extents depending on the intellectual task at hand. Given this trajectory, should historians present humanism as a major turning point in European intellectual history or as a somewhat distinctive approach to disciplines that played an important role in ancient, medieval, early modern, and modern Europe?

In *Lineages of European Political Thought: Explorations along the Medieval Modern Divide*, Cary Nederman criticizes the tendency to describe modern political theory either as a direct outgrowth of medieval ideas or as a radical rupture from them.[4] Analyzing the field of Renaissance intellectual history, Christopher Celenza identifies a similar disjunction between synchronic and diachronic approaches to Renaissance scholarship and urges scholars of humanism to combine these perspectives.[5] The present chapter follows the lead of Nederman and Celenza, arguing that studies of humanism between 1350 and 1550 will be more faithful to the intellectuals studied and more illuminating about the broader culture of the period if they consider both change and continuity.[6] Recognizing the role of humanism in this period requires appreciation of its interaction with other scholarly practices. It demands that the intellectual life of the fourteenth, fifteenth, and sixteenth centuries be located firmly within social, political, economic, and religious as well as intellectual contexts. In order to achieve these goals, those interested in humanism need to consider a broader range of subjects. Rich studies of the era's polymaths have explored and explained the period's intellectual breakthroughs.[7] A few scholars have begun to trace the presence of humanism among doctors, ambassadors, teachers, and merchants, individuals who left no scholarly treatises but offer other testimony of their interest in classical study.[8] But there is less study of scholars who published a few works, interacted with a range of colleagues, and included humanism as one part of their tool kit. To understand humanism in the fifteenth and sixteenth centuries, it is essential to look at these middling figures, as well as the era's superstars.

Case Study from the Court of Leo X

In 1513 Giovanni de' Medici, the second son of Lorenzo il Magnifico, ascended to the throne of St. Peter and took the name Leo X. Ever since his election, Leo

and his curia have been celebrated (and derided) as the culmination and quin-
tessence of the Italian Renaissance.[9] A passionate musical enthusiast, the pope
devoted special care and attention to the musicians in his employ.[10] He appointed Raphael as chief architect of St. Peter's, while also commissioning him
to draw a map of ancient Rome, design tapestries for the Sistine Chapel, deco-
rate the Vatican Loggie, and complete the Vatican Stanze.[11] As one would expect from the former student of Poliziano, Leo also promoted classical scholarship; above all, he revived the Sapienza, Rome's University. He sponsored the
foundation of a Greek institute and a Greek printing press, which produced,
among other texts, a massive Greek-Latin dictionary. The printer Aldus Manu-
tius dedicated his 1513 Greek edition of the complete works of Plato to Leo.
The pope's classical interests were not confined to ancient Greece. Scholars,
hoping that the Medici pontiff would prove to be a new Maecenas, celebrated
his election with a flood of adulatory Latin verse. The strict Ciceronian Pietro
Bembo became one of Leo's two personal secretaries, and informal humanist
sodalities flourished.[12]

Yet, if Leonine Rome sounds like a humanist paradise, the texts dedicated
to Leo point to a more complex intellectual scene and to a patron with wide-
ranging interests. For example, Christoforo Marcello's *De sanitate animae*, a
Neoplatonic dialogue dedicated to Leo, demonstrates Marcello's interest in the
studies of Ficino.[13] However, Marcello's magnum opus, the *Universalis de anima
traditionis opus* (Venice, 1508; dedicated to Julius II), evinces his deeper engagement with Paduan debates about Averroes. Other works that Leo received—
particularly theological ones—fall squarely within scholastic conventions. At
the pope's request, in 1515 Tommaso De Vio (Cajetan) penned a Thomistic
discussion of the immaculate conception of the Virgin Mary.[14] Filippo Donato
dedicated his father Girolamo's work on the procession of the Holy Spirit to
Leo, claiming that the pope's interest in learning was restoring the faith of all
scholars.[15] Other dedicated works were more eclectic. The pope received a 697-
folio treatise on moral and theological subjects (including books on the mechanical arts, the pleasure of the body, the liberal arts, and the Assumption
of the Virgin), a call for the popular origins of political sovereignty, a massive
geography, and several studies of calendar reform.[16] This list is representative
rather than exhaustive, but it provides some indication of the range of intellectual currents that pervaded Leonine Rome.

That Christian and classical elements combined in the culture of Renaissance Rome is hardly news.[17] Ingrid Rowland has deftly demonstrated that close
perusal of texts that challenge the stereotype of Rome as a bastion of Ciceroni-
anism lead to a richer understanding of the city's intellectual life.[18] However,
the text that I most fully discuss here, the *De re publica christiana* (1518), comes
not from an Egidio of Viterbo or Athanasius Kircher but a less well-known

scholar, Pietro Galatino. Galatino, an Observant Franciscan, came to Rome in the 1480s, where he remained, almost without interruption, until his death around 1539. He served as a professor of philosophy, theology, and Greek at the University of Rome, held the office of apostolic penitentiary of St. Peter's, and was chaplain to several cardinals.[19] He was one of the first Italians to embark on the study of Ethiopic.[20] Galatino helped to pioneer study of the Jewish mystical writings of the Kabbalah and wrote in defense of the much more skilled German Hebraist Johannes Reuchlin.[21] He was also drawn to the tradition of Franciscan prophetic writings, especially the real and alleged works of Joachim of Fiore.[22] By 1525, Galatino had convinced himself that he was the Angelic Pastor predicted in late fifteenth-century apocalyptic texts; thus, he was destined to oversee the destruction of the Turks and the spread of Christianity and peace throughout the world.[23]

Galatino's intellectual interests were even more eclectic than this overview of his Hebraic and Joachimite thought suggests. In 1518, he wrote the short treatise entitled *De re publica christiana*, which offers a vision not of the end of times but of religious reform.[24] The first and much the longest section of *De re publica christiana* promotes a traditional hierarchical vision of Christendom by comparing the "Christian republic" to the human body. According to Galatino, the pope is the head and the cardinals are the heart. Bishops represent the eyes, priests the ears, and deacons the nose. Members of the religious orders form the mouth. The emperor is the neck, the kings of Spain and France are the shoulders, while the kings of Poland, Bohemia, Sicily, and England are the arms. Lesser princes and nobles are the hands. Orators and secretaries serve as the breast. Oddly, the Church seems to consist only of an upper body.

Galatino makes no claims to originality in his body metaphor. Like John of Salisbury, he refers to the pseudo-Plutarchan *Institutio* of Trajan for the idea that the prince is the head of the republic. Thus, he continues the tradition of using the term "republic" as a shorthand for good government of any sort, regardless of a particular government's power structure.[25] Galatino also cites numerous Biblical passages, including Paul's description of the Church as the body of Christ.[26] Like other papalists before him, Galatino speaks of both Christ and the pope as the head of the Church.[27] Galatino's emphasis on complementarity and on the organic nature of communities accords with better-known works of medieval political theory and with other authors who drew on John of Salisbury.[28] But he goes to much greater lengths than John to integrate secular and religious authorities, making the pope and cardinals not just a soul but the embodiment of some of the bodies' most vital, and tangible, members.[29] Comparison with Nicholas of Cusa's *Catholic Concordance* is instructive. Like Galatino, Nicholas emphasizes the importance of collaboration between the parts of the body, but in his work, written in 1433–34, secular and religious hierarchies are parallel but distinct.[30] This schema not only accorded with Cu-sanus's belief in a well-ordered universe; it also promoted his defense of the Council of Basel against the attacks of Pope Eugenius IV and his role as repre-sentative to the bishop of Trier.[31] When Galatino was writing, Cusanus's vision of parallel authorities was still familiar; it had been used by members of the Council of Pisa to argue that Julius II was subject to its judgment.[32] But con-ciliarism does not seem to be Galatino's target. Instead, he emphasizes a single hierarchy that combines both secular and religious figures under the leader-ship of the pope. Ideally, this hierarchy would work together for practical ends, most especially church reform and crusade, those perennial goals of the medi-eval papacy.

So why include this case study in a discussion of humanism? Based on the above description, *De re publica christiana* seems more like a late revival of R. W. Southern's scholastic humanism than a piece that fits into Kristeller's disci-plinary definition.[33] However, as the little treatise proceeds, Galatino begins to flaunt his knowledge of grammar and rhetoric. His discussion of bishops as eyes demonstrates his commitment to allegory and also to etymology:

> First indeed [bishops are the Church's eyes] because the eyes are accustomed to
> be called lights either since light emanates from them, according to Plato's judg-
> ment, or since, as Aristotle believes, they pour forth the light they have received
> from without to make things visible. From this it can easily be perceived that
> bishops of every sort ought not only to make things clear to others by the light
> of virtues, but in truth ought always to keep their eyes open, as the names given
> to them clearly indicate. These names are *episcopus, pontifex, praesul*, and *an-
> tistes. Episcopus* is the Greek equivalent to superintendent or scout because bish-
> ops ought to watch over and be devoted to the flock entrusted to them, and they
> ought to look at and investigate the mores and vices of the people under them.[34]

Galatino goes on to explain that the bishop is called *pontifex* because his exam-ple should provide a bridge to help his flock reach heaven. He is called *praesul* because he should be superior in integrity and learning. Finally, the term *an-tistes* is fitting, as the bishop must oppose heretics. Having made the most of his various terms for bishop, Galatino then returns to the eye analogy. Just as sight is, of all the senses, the one nearest to the spirit, so bishops ought to be nearest to God. According to Galatino, "oculus" comes from *occultum*, or secret, and bishops should hide their good works unless revealing them glorifies God. Like eyes, which are lidded, bishops should be hidden from the judgment of the secular arm. Here, Galatino praises Constantine for establishing the indepen-dence of clerical justice.

Galatino's intertwining of philology and allegory is bizarre to those who associate humanist philology with accuracy and sensitivity to the historicity of language.[35] The fanciful etymology for *oculus* is as important to his discussion of a bishop's role as the translation of the Greek term *episcopus*. Both Galatino's allegories and his etymologies derive from his vision of the bishop's duties. Instead of moving from an etymological root to a word's classical meaning, Galatino starts with a conception of what bishops should do, and then uses his linguistic knowledge to provide supporting evidence.

Clearly, Galatino did not adopt the philological rigor of some of his contemporaries in his study of language. But his etymological practice has venerable roots of its own. In *On Dialectic*, Augustine argues that "the origin of a word is contained either in the similarity of things and sounds, in the similarity of things themselves, in their proximity, or in their contrariety."[36] Citing the authority of Cicero, Augustine mocks the Stoics' insistence that the origin of all words can be determined; instead, he emphasizes the subjective nature of etymology, comparing study of the origin of a word to the interpretation of dreams. Does *verbum* come from *verberando*? From *vero*? From *verum boando*? According to Augustine, it is up to the reader to decide.

If Galatino does not adopt the new philology of Poliziano, he also does not offer a novel vision of the episcopal office. Quattrocento works as disparate as the reform decrees of the Council of Constance, Alberti's *Pontifex*, and the sermons of Savonarola all argued that bishops should be learned, illuminate the truth, watch over their flocks, help humbler Christians, oppose heretics, glorify God, and be judged by fellow bishops, not secular authorities. And, from at least the time of Gregory VII (1073–1085), similar directives had issued both from within the curia and from church councils.[37] Other parts of *De re publica* also combine humanist flourishes with traditional messages. Toward the end of the work, Galatino condemns contemporary piety and calls for reform. Following a popular humanist technique, he has the Church herself deliver a speech in which she begs Leo, her bridegroom and "one true hope," to save her. Both the ills the Church identifies and the prescriptions she offers reprise the reform literature of the preceding four centuries. At a time when Leo X was abandoning the directives of the Fifth Lateran Council, Galatino made an impassioned plea to carry out its program of reform.[38] In doing so, he echoed not just a long tradition of reform rhetoric but, more particularly, the themes of the speeches that had opened the various sessions of the Council.[39]

Some humanists probably concurred with Casaubon's assessment of Galatino's scholarship: "Rubbish, rubbish, rubbish."[40] But Galatino's work was popular throughout Europe; the *De re publica* was published in 1518 along with Galatino's text in defense of Reuchlin, *De arcanis Catholicae veritatis*. In 1612, the *De arcanis Catholicae veritatis* was republished along with Reuchlin's *De arte cabalistica*. Thus, instead of being condemned for failing to live up to new humanist standards, Galatino's work was associated with more innovative scholarship.

What does *De re publica christiana* suggest about the role of humanism in early sixteenth-century Rome and early modern Europe more generally? First, it serves as a reminder that, behind the vanguard of humanist innovators, there was a whole host of scholars who used ancient texts and languages in an erratic, uncritical fashion.[41] Second, this short text offers an evocative example of the range of intellectual traditions with which sixteenth-century authors could and did engage. Galatino employed techniques and arguments used by philologists, linguists, papalists, conciliarists, reformers, and proponents of crusade, as well as humanists, Hebraists, and prophets. He provides almost comical proof of the claim made above, that by the early sixteenth century almost no one in early modern Europe could be simply, or even primarily, a Kristellerian humanist.[42] Third, *De re publica christiana* shows that, to make any sense of what Galatino was trying to accomplish, it is necessary to consider his work from many perspectives, some of them typically considered medieval and others characteristic of humanism. Finally, if a synchronic approach to Galatino's work is essential, so too is a diachronic one. Galatino's eclecticism suggests that the triumph of humanism at the papal court was not as total or clear-cut as humanists claimed, most particularly in their longing evocations of Leo's pontificate after the sack of 1527.[43] This text deserves attention not because of the originality of its ideas but because of its ability to illuminate the cultural and intellectual hybridity of Leonine Rome.

Placing Galatino in a Scholarly Context

What would or did Kristeller make of Galatino's little treatise? Although many of those who use Kristeller's definition of humanism have focused on humanists as creators of a new intellectual milieu, he himself was keenly sensitive to and interested in the interplay between humanism and earlier intellectual traditions. *Renaissance Thoughts and Its Sources*, probably Kristeller's best-known work, consists of five sections: "Renaissance Thought and Classical Antiquity"; "Renaissance Thought and the Middle Ages"; Renaissance Thought and Byzantine Learning"; "Renaissance Concepts of Man"; and "Philosophy and Rhetoric from Antiquity to the Renaissance." In short, the work is premised on intellectual continuities, or at least connections, across long periods of time, and Kristeller's efforts as a bibliographer supported this wide-ranging vision of humanism.[44] An inveterate manuscript enthusiast, Kristeller combed the libraries of Europe, compiling his findings into the massive, six-volume catalog *Iter Italicum*, which includes references to thirty manuscripts that contain works by Galatino.[45]

In Anglo-American circles, Ronald Witt did more than any other scholar to develop Kristeller's emphasis on intellectual continuities between Renaissance humanism and medieval rhetoric.[46] His last two books deftly trace a series of prequels to the Petrarchan story of the *studia humanitatis*.[47] In doing so, they delineate a host of rich avenues for research, but unfortunately, few English-speaking scholars have pursued them. This may be due partially to the difficulties of commencing work on little-known Latin manuscript sources, in spite of their increasing availability in the digital age, but it also reflects the pressures of the academic job market. European scholars of humanism are more likely to pursue this time-consuming philological labor than Anglo-Americans, as evidenced in journals like *Humanistica Lovaniensia*, *Italia Medioevale e Umanistica*, and *Rinascimento*. Leo S. Olschki (the publisher of *Rinascimento*) continues to produce excellent editions and studies of an ever-wider range of medieval and Renaissance texts, in both Latin and Italian. The association and publishing enterprise Roma nel Rinascimento has been particularly successful at combining philology and social history, situating humanist subjects within their urban contexts.[48]

But for Anglo-American scholars, the pull of modernity is strong, in part because of the need to justify ourselves to search committees and deans. One might be able to give a successful job talk on Galatino's contribution to modern notions of the state; in most American history departments, analyzing his ties to Augustine and John of Salisbury would elicit less enthusiasm. This reflects not just the presentist perspective of much historical research in North America but also a more particular tradition of associating the Renaissance with the birth of modernity. Petrarch began this narrative, and his humanist successors followed him in self-aggrandizing claims of radical innovation.[49] During the Enlightenment, the philosophes described the humanists as their intellectual forefathers, and in the nineteenth century, *The Civilization of the Renaissance in Italy* offered a particularly vivid and influential portrait of the age. Jacob Burckhardt insisted that not just humanism but every facet of fourteenth-, fifteenth-, and sixteenth-century life broke with the stultifying culture of the Middle Ages. A century later, Hans Baron's *The Crisis of the Early Italian Renaissance* (1955) offered a new version of the nexus between the Renaissance and modernity, one shaped by and appealing in the context of Cold War America.[50] Like Burckhardt, Baron saw the Italian Renaissance as an innovative period that contributed in essential ways to the liberal values of modern Europe; more particularly, he saw republican Florence's 1402 struggle with Milanese tyranny as a prototype for America's struggle against Soviet totalitarianism.[51] Baron's work has been challenged from a variety of perspectives.[52] Yet, if few scholars now accept his image of Renaissance Florence, Baron's focus on the humanists' political engagement has encouraged study of their public roles in a variety of different regimes. The fact that most Renaissance states were oligarchies or monarchies rather than republics shaped, rather than squelched, this participation.[53] Galatino is just one example of a Renaissance scholar who used the political legacy of ancient Rome in surprising ways.

Kristeller and Baron are examples that Celenza uses to clarify the split between synchronic and diachronic approaches to the study of humanism.[54] As defined by Celenza, the synchronic approach "attempts to take a kind of cultural snapshot of a given epoch" while seeking connections between the era under study and earlier eras. Conversely, the diachronic approach "moves through time and seeks what is important in a given epoch for the future."[55] As noted above, Celenza encourages scholars to combine these approaches. In a variety of ways, the works of Kristeller and Baron, like that of Burckhardt, provide models for new approaches to familiar questions, some of which combine synchronic and diachronic methods.

Burckhardt divided his work into chapters with titles that emphasized bold and systemic innovation: "The State as a Work of Art"; "The Discovery of the Individual"; "The Revival of Antiquity"; "The Discovery of the World and of Man"; "Society and Festivals"; "Morality and Religion."[56] These titles offer insight into Burckhardt's understanding of modernity, which was more complex and ambivalent than certain oft-quoted passages from *The Civilization* suggest.[57] The postmodern turn in literary studies has, depending on one's perspective, shorn Burckhardt's work of its Romantic biases or revealed the ambivalence already present in his original description of individualism. In *Renaissance Self Fashioning* (1980), Stephen Greenblatt offered six case studies from the English Renaissance, beginning with Thomas More. He argued that More exhibited "the invention of a disturbingly unfamiliar form of consciousness, tense, ironic, witty, posed between engagement and attachment and, above all, fully aware of his own status as an invention."[58] For Greenblatt, Renaissance individualism was a painful and ceaseless effort to craft the self in response to the agonistic environment of sixteenth-century London and the Tudor court. Greenblatt's More possesses a version of Burckhardtian individualism, but it is a psychological burden, not a source of liberation.[59] More recent explorations of the ethical thinking of Alberti, Poggio, and other Italian humanists make similar claims, mapping a moral philosophy far less modern, less coherent, and less optimistic than that posited by Baron.[60]

At the same time that Greenblatt emphasized the personal challenges of self-fashioning, he also encouraged modern scholars to be mindful of the political, professional, and religious contexts in which humanists wrote their texts. An increased interest in situating famous figures within the broader social, religious, political, and technological environments that shaped their work has been apparent in the past several decades of scholarship on Erasmus.[61] In *Eras-*

mus, Man of Letters, Lisa Jardine parsed the relationship between Erasmus's elaborate textual self-fashioning and his use of the new medium of print.[62] Marjorie Boyle, Christine Christ-von Wedel, Hilmar Pabel, and others have traced the intricate interplay of humanism and theology in Erasmus's writings as he sought to forge his own path, overlapping with but also deviating from Protestant and Catholic orthodoxies.[63]

In short, in the past forty years, there has been steady, though perhaps insufficiently appreciated, effort to build on Burckhardt's vision and to present Renaissance humanism as a vital factor in the political, cultural, and social life of Europe prior to 1550. It remains difficult, however, to determine how significant humanists' ideas were outside their immediate networks and which of their practices and priorities reached beyond a narrow group of specialists. Recent intellectual biographies of these specialists, along with burgeoning interest in book history, have encouraged an important shift in studies of humanism, a shift to the study of practices and institutions. Kristeller's oeuvre hovers in the background of the study of practices, and Baron's influence can still be felt in the study of institutions. But these newer foci are less likely to impose modern disciplinary and political assumptions onto the scholarship of the fourteenth, fifteenth, and sixteenth centuries. They provide important opportunities for the consideration of scholars such as Galatino who were not major intellectual players but nevertheless represented a vital part of the European intellectual landscape. Further study of institutions and practices promises vital insights into the roles of humanism in early modern society.

Institutions and Practices

Schools and universities were the institutions that humanists fought most vociferously to conquer. By the early to mid-sixteenth century, humanist curricula had a prominent role in Italian and northern European schools and universities, as well as in private teaching.[64] But, as all teachers know, there are gaps between the goals of curricula, the realities of what happens in the classroom, and the material that a student takes from a particular class. In *From Humanism to the Humanities*, Grafton and Jardine argued that students were more likely to leave the classroom befogged by a plethora of rhetorical detail than inculcated with prudence, magnanimity, and *pietas*. What students learned in humanist classrooms was to be effective, successful diplomats and secretaries, supporters of the political and religious establishment.[65]

Over the past two decades, studies of specific institutions and localities have complicated Grafton and Jardine's (avowedly impressionistic) overview, detailing the ways in which funding, confessional loyalties, municipal engagement, individual personalities, and a host of other factors combined to determine the quantity and quality of humanist education in particular contexts.[66] Robert Black has provided an important corrective to the narrative of humanism's triumph in Italian schools, demonstrating continuities in the teaching of Latin between the twelfth and fifteenth centuries and proving that moral education had little place in humanist schoolrooms.[67] Analyses of textbooks used outside Italy have clarified how teachers taught classical rhetoric and helped their students to retain and use it.[68] Given the great disparities in early modern educational practices, more studies of schools, universities, professors, students, and educational materials are needed before it will be possible to write a compelling sequel to *From Humanism to the Humanities*. At present, it seems that important elements of change and continuity coexisted in the schools of early modern Europe.

Studies of humanism in other institutional settings have not kept pace with the work on schools and universities. In spite of the number of humanists who worked as secretaries and in spite (or because) of the example set by Baron, historians have paid relatively little attention to the role humanists played in reshaping courtly culture.[69] Classical scholars who worked for princes and popes heralded their own achievements in creating a more classicizing milieu, but the evidence for their cultural and political influence is often equivocal. For example, the humanist pope Nicholas V, once touted as the archetypal patron of classical scholarship, has more recently been portrayed as unsympathetic to the humanists in his employ, especially when their interests turned toward reviving Roman republicanism.[70] As discussed above, even in the Rome of Leo X, humanism was only one intellectual current among many. The role that it played in this and other courts requires further study. Art historians and those interested in ceremony have been responsible for much of the analysis of classical scholarship at Renaissance courts.[71] As diplomatic history has become a broader and more culturally focused discipline, its practitioners have tended to gloss over the role of humanism.[72] Important recent studies argue that it was only in the later sixteenth century that humanist scholars established practices of political guidance and diplomatic exchange that fundamentally altered the dynamics and dialogues of late medieval courts.[73] But much more work needs to be done to understand the ethos of these rich cultural centers, which Norbert Elias, quoting Erasmus, described as the nurseries of refined conduct.[74] Other institutions, such as hospitals, academies, and printing houses, likewise provide fertile arenas for study of the role of humanism.

Another body of scholarship that emphasizes both continuity and change is the study of humanist practices, especially practices of reading and note taking. Scaliger's scientific approach to philology built on the innovations of several generations of Italian and French scholars.[75] Poggio rediscovered the text of

Lucretius in 1417, but Ada Palmer has argued that it was rediscovered in a more fundamental way in the 1550s, when reading practices changed thanks to the ready availability of classical sources.[76] The hard work of fifteenth-century editors, translators, and commentators, as well as the refinement of printing technology and the painstaking work of correctors and typesetters, made it possible for sixteenth-century scholars to roam through classical forests without having to stop to shore up each individual tree.[77] Changes in understanding of intellectual property, use of manuscripts, editing practices, reading habits, textual fixity, and other supposed characteristics of print culture did not emerge overnight, but by the late sixteenth century, book culture was very different than it had been for Petrarch and his friends.[78] By the later sixteenth and seventeenth centuries, scholars of all types hung on for dear life in the face of massive "information overload," which encouraged the publication of new types of texts and in turn necessitated the development of new scholarly practices.[79]

A move to the study of practices has demonstrated that humanism stretched its tentacles, or at least its techniques, into varied and surprising parts of European culture. Humanists were integral to pre-Reformation efforts to correct hagiography and deepen patristic scholarship.[80] The Reformation spread humanist textual practices, especially in the realm of Biblical exegesis and preaching, and humanist methodology also shaped Counter-Reformation efforts to reform religious education and the cult of the saints.[81] Far from constituting repudiations of classical scholarship, the achievements of the Scientific Revolution depended on humanistic modes of reading and writing.[82] Even those, including women, who did not attend humanist schools still participated in the ever growing *res publica litterarum*.[83] In short, recent scholarship seeking to map the impact of humanism demonstrates that, by 1550, it had a substantial role in political, religious, scientific, and domestic spheres. The arrival of humanism may not have constituted a rupture in early modern life, but it did introduce significant changes to the types of texts in circulation, the extent of their availability, and the tools with which readers approached their reading material.

One problem with the study of practices—and, more generally, with most of the scholarship on humanism—is that it focuses on those who left written records. Innovative works by Brian Maxson and Sarah Ross offer models for future study that encompass broader swathes of the population. In *The Humanist World of Renaissance Florence*, Maxson argues that modern scholars have focused too much on literary humanists, those who composed classicizing texts, and not enough on social humanists, those who may not have been writers but nevertheless participated in the new culture.[84] He carefully documents the connections between these social humanists, situating their intellectual interests among the complex ties that bound together elite Florentines. In *Everyday Re-*

naissances: The Quest for Cultural Legitimacy in Venice, Ross uses wills and book inventories to identify the surprisingly diverse array of sixteenth-century Venetians who expressed commitment to, if not an expertise in, classical scholarship.[85] Eschewing distinctions between elite and popular culture, she argues that cultural legitimacy was a widely shared Venetian goal, and one that was often dearest to those who have not previously been recognized as part of the *res publica litterarum.*

These burgeoning areas of research are thrilling to all those who study European humanism. Nevertheless, the more the field of study grows, the more questions reemerge about how to define humanism in a way that is inclusive but still retains enough specificity and content as to be useful. As historians shift their focus from the solitary scholar in his library to the harassed undergraduate, the overworked chancery official, the reformed preacher, the experimental philosopher, the female letter writer, and the doctor book collector (and thus from texts to practices), we run the risk of describing everyone in early modern Europe as a humanist. And even in a volume that celebrates the exciting future of intellectual history, that is going a bit too far.

If the study of practices threatens to make humanism omnipresent, it also may encourage another, more dangerous tendency. If we look only for certain practices, we may not take time to contextualize them fully. To describe Galatino merely as a bad etymologist would flatten and oversimplify his project. The challenge now is to provide the same type of rich, variegated studies of eccentric, marginal, or "ordinary" figures that have already been fashioned for many of the age's polymaths. We need to consider the practices "ordinary" scholars employed while also situating them within a variety of institutions. Understanding Galatino, for example, will require further study of his engagement with the Sapienza and the Franciscan order as well as his role in the complex court and bureaucracy that was the papal curia. More generally, study of figures like Galatino demands a multifaceted approach to the range of traditions that combined to make up medieval and Renaissance intellectual life. It also requires consideration of the ways in which the members of institutions interacted and the dynamics between institutions and the societies of which they were a part. Modern scholars will have to pick and choose those works that best capture the ambience of the particular milieu they are investigating. With the publication of the I Tatti editions, those who teach Italian humanism have more opportunities than ever before to engage students with primary sources. To encourage these students to pursue scholarly research on humanism, however, it is also essential to provide them with innovative studies that demonstrate the variegated ways in which humanism permeated the culture of Europe both during and after the early modern period.

NOTES

1. Copson and Grayling 2015. Cf. Yoran 2007, 326–44.

2. Kristeller 1944–45, 346–74. See also Campana 1946, 60–73.

3. For reflections on the influence of Kristeller's definition after fifty years, see Grafton 1998, 118–21.

4. Nederman 2009.

5. Celenza 2004.

6. This chapter focuses on humanism prior to 1550. On "late humanism," see chapter 3, by Bulman.

7. Grafton 1983–93; Grafton and Weinberg with Hamilton 2011.

8. Maxson 2013; S. Ross 2016.

9. Gouwens 2012.

10. Cummings 2012.

11. Jones and Penny 1983.

12. Gaisser 1999.

13. Biblioteca Apostolica Vaticana, Vatican City (hereafter BAV), MS Vat. Lat. 3647. On Marcello, see Palumbo 2007.

14. Cajetan 1575. De Vio was made a cardinal two years later. For discussion of humanism's influence on Cajetan, see A. Jenkins and Preston 2007, 152–56.

15. BAV, MS Vat. Lat. 4326.

16. BAV, MS Vat. Lat. 3971; Salomoni 1955; BAV, MS Vat. Lat. 3844; BAV, MS Vat. Lat. 7046, 8226. On calendar reform literature, see Nothaft 2018.

17. Stinger 1998.

18. Rowland 1998, 2000. See also Chiabò, Ronzani, and Vitale 2014.

19. Colombero 1982.

20. For a useful overview of scholarship on Galatino, see Wilkinson 2007, 58–61.

21. Price 2011, 163–92. For Galatino's work on the Talmud, see J. Cohen 1982.

22. Reeves 1993, 234–38, 441–49.

23. Rusconi 1992.

24. Perrone 1973. The original text is found in BAV, MS Vat. Lat. 5578, f. 86–106v. For an edited version, see Perrone 1973, 609–32.

25. Coleman 1997.

26. Romans 12:5.

27. Kantorowicz 1981, 194–206.

28. Nederman 2009, 63–80.

29. Lachaud 2015, 436–38.

30. Nicholas of Cusa 1991, 215–16.

31. Izbicki 2006.

32. Izbicki 1999.

33. Southern 1995–2001. For an interesting analysis, which also offers proposals for future study of scholasticism, see Otten 2001, 275–89.

34. "Et primum quidem quia oculi lumina vocari solent, sive quod ex eis lumen (iuxta Platonis sententiam) exterius emanet, sive quod lucem extrinsecus (ut Aristoteles sentit) accaeptam, usui proponendo refundant. Ex quibus facile percipitur Episcopos cuiuscumque generis, non solum virtutum luce alios illustrare debere, verum etiam oculos ad viden-

dum semper apertos tenere, quod nomina eis attributa aperte indicant. Quae sunt Episco-pus, Pontifex, Praesul et Antistes. Episcopus namque idem graece est quod superintendens, seu speculator latine, eo quod et super gregem sibi creditum semper intendere atque invig-ilare debeat, et populorum infra se positorum mores ac vitia iugiter speculari atque in-spicere." Perrone 1973, 615.

35. Lewis and Short agree with Galatino's interpretation of the origin of the term "pon-tifex," but they are more hesitant about the bridge connection. *Pontifex*: "Doubtless from pons-facio; but the original meaning is obscure."

36. Augustine 2009, 348.

37. Rusconi 1992, 175.

38. Perrone 1973, 548–49.

39. Minnich 1969.

40. For Casaubon's dismissal of Galatino's *De arcanis Catholicae veritatis*, see Grafton and Weinberg with Hamilton 2011, 37–42.

41. Grafton 2007b; Grafton 1990a.

42. For more on the connection between humanism and philosophy, see Celenza 2017 and Kraye's chapter in this volume. On the connection between humanism and science, see the chapters by Popper and Bulman in this volume.

43. Gouwens 1998.

44. Monfasani 2006a.

45. For an important effort to continue this work, see Union académique internationale 1960–.

46. See also Monfasani 2006b.

47. Witt 2003, 2011.

48. See the works of Anna Modigliani and the conference volumes on the pontificates of Martin V, Sixtus IV, Julius II, and Leo X.

49. On Petrarch, see Ascoli and Falkeid 2015. On later humanists as self-promoters, see Amato 2014, 59–80; and Rummel 1995.

50. See also Baron 1988. For more on America and the Renaissance, see Molho 1998.

51. A. Brown 1990, 441–48.

52. Witt et al. 1996, 107–44; Hankins 2000. See also Jurdjevic 2014.

53. Findlen, Fontaine, and Osheim 2003; N. Baker and Maxson 2015.

54. Garin is Celenza's main example of a diachronic thinker; Baron is his secondary example. Celenza 2004, 36–39.

55. Celenza 2004, 28–29.

56. For some compelling readings of Burckhardt, see Kerrigan and Braden 1989, 3–35; Starn 1998, 122–24; and Caferro 2011.

57. Ruehl 2015, 58–104.

58. Greenblatt 1980.

59. J. Martin 2004.

60. Garin's work (1972) on Alberti served as a harbinger of this change. See also Fubini 2003; Calzona 2009.

61. For one of the best-known examples of this trend, see Biagioli 1993.

62. Jardine 1993. See also Naquin 2013. Cf. Vessey 2017, 23–44.

63. Boyle 1981; Christ-von Wedel 2013; Pabel 2008.

64. Grendler 2002; Overfield 1984; De Ridder-Symoens 1996.

65. See, more recently, Enterline 2012.

66. Amos 2015, 25–45; A. Ross 2015; Van Liere 2003.

67. Black 2001.

68. Coroleu 2014; Campi et al. 2008.

69. Exceptions include D'Elia 2009, 2016; and McCahill 2013.

70. Westfall 1974. Cf. Modigliani 2013; D'Elia 2007.

71. The predominance of scholarship by art historians is apparent in the journal *Court Studies*. On royal ceremonial, see Bertelli 2001; Mulryne, Watanabe-O'Kelly, and Shewring 2004; and Visceglia 2009.

72. See, for example, Adams and Cox 2011.

73. Millstone 2016; Popper 2012.

74. Elias 2000, 66.

75. Grafton 1983–93.

76. Palmer 2014.

77. Botley 2004; Pade 2005. For an illuminating example of the practices of two lesser-known translators and commentators, see Butcher, Czortek, and Martelli 2017; and Grafton 2011b.

78. Johns 2010; McKitterick 2003; Pettegree 2010. See, more generally, Darnton 2007.

79. Rosenberg 2003; Blair 2003; Ogilvie 2003; Sheehan 2003; Yeo 2003; Blair 2010; Goeing 2017.

80. Backus 2004; Collins 2008; Frazier 2005.

81. Olds 2015. For more on the intersection between humanism and sacred history, see Van Liere, Ditchfield, and Louthan 2012.

82. Exemplary studies include Grafton 1991a; Ogilvie 2006; Siraisi 2013; Raphael 2017.

83. Furey 2006; S. Ross 2009.

84. Maxson 2013.

85. S. Ross 2016.

CHAPTER TWO

From Renaissance to Enlightenment

WILLIAM J. BULMAN

Sometime in the later seventeenth century, we have been told, the intellectually innovative elites of Europe discarded their institutional and intellectual inheritance. Struck by the moral and intellectual bankruptcy of their world, which had been gruesomely revealed by the Wars of Religion, they traded a thought environment dominated by clerics, universities, princely patronage, and humanist pedantry for a brave new world of lay men of letters, public spheres, secular mores, and new philosophies. This world, somehow created ex nihilo, formed the foundation for an array of new sciences.[1]

For more than three decades now, this view has been under sustained attack. We now know that the basic concepts, norms, concerns, and practices long associated with the Enlightenment were by no means wholly novel, confined to philosophy, or wedded to secularism and emancipation. The scholars who have offered these insights have mostly operated in a revisionist mode, subdividing and complicating the Enlightenment beyond recognition, stressing continuity over change.[2] This work has been immensely productive, but in the long run, these revisionist stances are no more tenable or sustainable than their opposites. The challenge for future work in the field is to accommodate both the familiarity and the undeniable turbulence of the late seventeenth and early eighteenth centuries in a compelling account of the transition from Renaissance and Reformation to Enlightenment.[3] The linchpin of such an account is a sound description of the Enlightenment's early stages.

Against the long-standing characterization of the Enlightenment as a set of specific propositions, intellectual systems, or political programs, recent work has pointed to the conclusion that Enlightenment is most fruitfully understood as the attempt to articulate, defend, disseminate, and implement ideas in accordance with a particular set of imperatives and by means of a distinctive set of practices and institutions.[4] These imperatives were products of the era of globalization and religious war that stretched from the Lutheran Reformation to the Peace of Westphalia. By 1648, the practices of Europe's empires, churches, and confessional states had led to doctrinal discord, confessional violence, domestic turmoil, and an awareness of the planet's bewildering social and religious diversity.[5] Many elites became preoccupied by the conviction that religious and public life needed to be reordered to prevent religious zeal from destroying civil peace. Accordingly, they asked: What forms of intellectual, social, religious, and political organization could secure such an order?[6] This question dominated the early Enlightenment, which stretched from about 1650 to 1715.[7] As the lens-grinding philosopher Baruch Spinoza put it in 1670, "Religious worship and pious conduct must be accommodated to the peace and interests of the state."[8]

The second imperative of Enlightenment dictated how this concern was to be addressed in practice. By 1650, elites had become more acutely aware than ever before that their own religious commitments (or lack thereof) constituted a choice among many available forms of religion (and irreligion), all of which could be embraced by sane and intelligent (if erring) people. This relativizing awareness did not require anyone to relinquish personal theological commitments or to cease striving to persuade or even coerce others. It meant only that elites could no longer assume that others would normally make the same theological assumptions that they made.[9] They had to act with an appreciation that in some sense, as John Locke wrote in 1689, "every one is orthodox to himself."[10] This meant that matters of public concern should be debated on metaphysically and epistemologically minimal premises that people of widely varying types and degrees of belief could be expected to accept.[11] This imperative accounts for the familiar turn in Enlightened argument away from the theological, the systematic, the providential, and the revealed, and toward the useful, the natural, the rational, the probable, the civil, the moral, the peaceful, the cosmopolitan, and the terrestrial.

The early Enlightenment's double imperative of civil peace and secularized argument was encapsulated in a single question that guided early Enlightenment practice: How could plans for ending civil and religious strife be defended, evaluated, and implemented in a manner that would elicit acceptance from such a diverse population?[12] Enlightened answers to this question typically bracketed confessional theology. The Huguenot minister and professor Jacques Bernard, for example, sought to defend the terrestrial utility of religion by presenting "a truth which would not have been less of a truth, had there been no religion." Starting from natural, historical, and rational—as opposed to specifically Christian—premises, Bernard outlined a set of basic verities and attendant practices to show that religion effectively produced earthly peace and happiness by teaching duty.[13] Arguments like this one hardly put an end to conflict about foundational assumptions or public policies, of course. Political and religious struggle endured, but its nature was transformed. It was now a struggle between Enlightenment and its opponents, and among rival forms of Enlightenment. Other Enlightenment writing countered arguments like Bernard's with more radical claims, asserting the utter superfluity of all religion and theism. As Spinoza put it, "The divine law which makes men truly happy and teaches the true life, is universal to all men" and "innate to the human mind."[14] Or as the French philosopher Pierre Bayle insisted, even the acceptance of mere natural religion was unnecessary for the inculcation of ethics.[15]

As this range of solutions to the problem of civil peace and secularity suggests, Enlightenment could certainly take the form of a principled deism, "indifferentism," libertinism, or tolerationism, but it did not always lead to these positions. Because the questions that guided Enlightened argument primarily concerned order, security, and prosperity, answers could be intolerant, religious, authoritarian, and communitarian just as easily as they could be liberal, secularist, egalitarian, or individualist.[16] The content, dissemination, and implementation of Enlightened speech was also channeled through a diverse array of practices and institutions. These included inherited and invented scholarly methods (both humanistic and nonhumanistic), learned disciplines, literary genres, rhetorical techniques, voluntary associations, and reading publics, but also universities, churches, governments, and empires.[17] Enlightenment could thus be pursued in a variety of settings and on a variety of geographical scales, from the local to the international. The fact that specific people, institutions, ideas, and practices were vehicles for Enlightenment does not imply that they were Enlightened in toto or in essence; this is why we can speak of many people and institutions as Enlightened even when they retained traditional theological and intellectual commitments. This fact does imply, though, that it can be misleading to specify an Enlightenment pantheon. Many practitioners of Enlightenment spent much of their time engaging in non-Enlightened actions that either had no bearing on the problem of civil peace or took place in a realm of theological consensus. The Enlightenment, on this reading, is less a framework for studying intellectual, social, religious, or political history in isolation than it is a lens on their interrelationships.[18]

This perspective allows us to take a prism to the age of lights, simultaneously appreciating its unity without ignoring its diversity, and giving "Enlightenment" meaningful content without exaggerating its specificity. It also allows for a nuanced assessment of continuity and change that improves upon existing accounts of the early Enlightenment by clarifying, in particular, the early Enlightenment's relationship to the Renaissance. To do so, we first need to determine to what extent the practices of sixteenth- and early seventeenth-century humanism provoked the distinctive imperatives of the early Enlightenment. Second, we need to investigate how the very perpetuation of Renaissance practices in the service of non-Enlightened programs prior to the late seventeenth century generated new or altered practices that were later used in the service of Enlightenment imperatives. Finally, we need to identify the extent to which the practices of the early Enlightenment inherited from late humanism were transformed in the early stages of their new deployment.[19]

The Renaissance Sources of Enlightened Imperatives

The application of Renaissance humanism to religious questions was partly responsible for the conflicts and awareness of diversity that provoked the mindset characteristic of the early Enlightenment. Renaissance humanism was originally a repertoire of practices for the discovery, authentication, interpretation, and creative use of an expanding range of ancient texts.[20] The recovery and application of ancient methods and models of criticism and public action in early modern circumstances regularly led to unexpected novelty. Novelty led in turn to instability and conflict.

Perhaps the most spectacular example of this dynamic in the history of the Renaissance is the humanists' attempts to use ancient tools in the service of Christianity. The most consequential attempt of this sort was the effort, best exemplified early on in the work of Desiderius Erasmus, to use philological learning to restore the text of the Bible to its original purity and clarity. At the same time, theologians slowly turned away from scholastic approaches, increasingly using philological tools to ascertain the doctrinal significance of the Old and New Testaments. Well into the seventeenth century the most important humanists believed that, as Joseph Scaliger allegedly put it, "there is no other source of disagreements in religion than the ignorance of grammar."[21] In saying this, Scaliger was not making an irenic, ecumenical, or secularizing gesture; he meant, instead, that a properly critical understanding of the Bible would bring all its students to a single theological truth—and very likely, he must have assumed, to his own Calvinism.

Yet, over the course of the sixteenth and seventeenth centuries, in their attempts to capture the single, true content and meaning of the holy text, the humanists achieved the opposite of what they intended. Their efforts were generally inconclusive, divergent, and, from Martin Luther onward, confessionally inflected. As a result, their work led not to harmony among Christians but to mounting hermeneutical discord, heated polemic, and violence, both between and within the main confessional groupings. It proved impossible for Christians to agree on either the meaning or even the content of the Biblical text. Competing claims to divine truth proliferated as never before, both because the humanists' hermeneutical enterprise was inherently inconclusive and because humanist practices were increasingly drawn into the service of dogmatic certainty, and rarely employed in a completely open-ended fashion. Taking cues from its most learned inhabitants, Christendom splintered into innumerable churches, sects, and factions.[22] Further compounding this division, the philology of the later sixteenth and early seventeenth centuries was a weapon eventually wielded by deists and atheists in the study of the Bible and the wider ancient world. These challenges to Christian orthodoxy and authority were not primarily the result of an external assault from new philosophies but rather the outgrowth of a Christian culture of late humanist criticism.[23]

By the mid-sixteenth century, furthermore, a second terrain of learned confessional conflict, closely connected to Biblical scholarship, had emerged. These were the new histories of true and false religion, ancient and modern. Each history was predicated upon a particular confessional or heterodox stance on the presence and meaning of divine revelation in Scripture, and many historical studies of ancient religion were used, in turn, to bolster those stances. Here again the humanists occupied center stage, compounding the conflicts that had begun with variant interpretations of God's Word. These histories took a variety of forms. Rival ecclesiastical histories, from the Lutheran team effort of the *Magdeburg Centuries* to Cesare Baronio's Catholic response, the *Annales Ecclesiastici*, presented visions of true and false churches and worship from antiquity to the present.[24] More politically themed narratives of Christianity that mixed church history with the dominant ancient understanding of history as past politics did much the same. The geographical range of these histories of religion extended beyond Europe and included a burgeoning historiography of religious error and corruption in the wider world. Antiquarian treatises examined pagan idolatry, Judaism, early Islam, and other ancient traditions in order better to understand the ubiquity of religious error and corruption. Similar studies of contemporary religions, inspired by the evangelical impulses of the Catholic Reformation and later Protestant missionary movements, explored the practices of modern pagans, Jews, and Muslims from Asia to America.[25] All strands of this historical enterprise produced staggeringly detailed records of religious diversity, which was mostly depicted as diabolically inspired error and corruption that could eventually be defeated by the light of the Gospel. Before too long this flowering discourse of idolatry, superstition, infidelity, and error had been applied to nearly every cult and sect known to Europeans, including every form of European Christianity.[26]

In these two ways—the proliferation of variant understandings of divine revelation and the production of a global catalog of religious diversity—some of the central achievements of the Renaissance humanists directly produced the global spectacle of religious pluralism and violence that provoked the Enlightenment search for civil peace under conditions of pluralism.

The Renaissance Sources of Enlightenment Practices

The perpetuation of Renaissance humanism also furnished many of the practices that in the late seventeenth and eighteenth centuries were adopted in the service of Enlightened imperatives. The fact that Enlightenment was pursued in reaction to doctrinal discord did not lead it to jettison the practices that led to that discord. The historical and philological techniques of late humanism were particularly important. Once the necessity of bracketing doctrinal conflict in the consideration of worldly, public matters had gained traction, the record of late humanist erudition, unmoored from confessionalized claims to dogmatic certainty, became an indispensable resource.

The secularization and historicization of the Biblical text was not, for the most part, accomplished by an external, rationalist assault led by Cartesians, Spinozists, and Socinians. It was a process internal to the humanist tradition, in many cases carried out unwittingly by scholars of undoubted piety, but also by impious figures like Spinoza, whose Biblical criticism was mostly a rather straightforward radicalization of the criticism of rabbis and late humanist Christians.[27] The heterodox exegetes of the mid-seventeenth century thus relied as heavily upon work done by their predecessors as their more orthodox counterparts did. The perpetuation and refinement of older methods came in many forms, such as the divergent responses of the French Oratorian Richard Simon and the Dutch Cocceian minister Fredrik van Leenhof (1647–1712) to the impossibility of recovering the pristine text of the Old Testament. Whereas Simon argued that this fact should encourage Christians to converge on the Vulgate and Catholic tradition as a clear guide to daily Christian practice while scholarship continued unimpeded, Leenhof argued as a Protestant apologist that the difficulty of reconstructing the Old Testament text and its meaning meant that Christians ought to converge on a minimalist, Christian natural religion that conduced to both social peace and salvation and had no input from philosophy.[28]

The philological labors directed at the Bible were only one example of a wide variety of historically oriented practices that were immensely consequential in the late sixteenth and early seventeenth centuries. History arguably became the master discipline of late humanism, and historical practices were late humanism's primary instruments.[29] Around 1550, Europeans began to conceive of a style of history that was more mindful of method and more expansive in time, space, and topic. Some spoke of *historia* in a fuller, ancient sense, as a sort of empiricism, or a systematic form of causal inquiry that proceeded by induction and could absorb information that was human or natural, political or religious, with no geographical, cultural, or temporal limits. Others demanded that historians pay closer attention to the procedures they used to evaluate historical testimony, and that they draw on the full range of artifacts and testimony appropriate to their task. This entailed, among other things, the merger of the antiquarian and literary-historical traditions previously kept separate in the study of ancient Greece and Rome; the rapid extension of historical inquiry to the globe, to the present and recent past, and to the natural world; and an increased investment in travel and collection of the material remains of the past. Late Renaissance histories glittered with geographical, topical, chronological, and evidential breadth and slowly shifted focus from the provision of examples of virtue and vice to causal, comparative, and philosophically oriented analysis.[30]

These developments in historical inquiry, while first abetting the visualization of global diversity, also set the stage for the slow emergence of the "new sciences" and "new philosophy" so strikingly evident by the end of the seventeenth century. These novel assemblages of intellectual practices were uncommon and were present only in outline before they were appropriated to Enlightened imperatives, but their emergence prior to the Enlightenment is crucial. The centrality of historical inquiry to these developments is clear if we consider changes in the study of nature, religion, and politics and describe the origins of the new philosophy, the anthropology of religion, and the science of politics—all realms of practice that often supported Enlightenment projects. These areas of inquiry were not separate disciplines with distinctive methods; instead, each of them featured the application of the same practices.

We can begin with the study of nature. Sixteenth-century natural philosophers were normally deeply immersed in late humanist culture and its characteristic pursuits. Natural philosophy, natural history, mathematics, astronomy, and medicine were all repeatedly transformed by humanist practices, and in general, the study of nature was humanistic in orientation. The rediscovery, publication, editing, interpretation, and historical contextualization of relevant ancient texts as the proper basis of future natural knowledge was itself a force for change. Further, by dismantling medieval scholasticism and thereby enabling the new systematic philosophical projects of the seventeenth century, the general Renaissance approach to nature facilitated the later work of the new philosophers.[31]

But more specific links to the transformation of early modern natural inquiry in the seventeenth century emerge if we consider the historical bent of late humanist inquiry. In the sixteenth century, nature was fully within the remit of *historia*. The emphasis on copious collection within natural historical communities was nearly identical to that practiced by the antiquarians and other late humanists. Experience—both textual and physical, and interlocked with philological erudition—was just as central to the practice of students of nature as it was to the practice of their counterparts studying religion, history, and politics.[32] Specific approaches to the natural disciplines were legitimated by late humanist histories of the ancient sciences. Humanist tools were used as part of a wide critical repertoire, and novelties were commonly understood as recoveries of ancient knowledge.[33]

Indeed, this culture of history produced late Renaissance blueprints for natural inquiry that were embraced and elaborated after 1650. One of the foremost—devised earlier in the century but enthusiastically embraced in subsequent years—was Francis Bacon's inductive schema for natural philosophy, which distinguished itself from Renaissance natural history and natural philosophy because it was a transposition, abstraction, and systematization of developments in late Renaissance historiography. Philosophers, Bacon argued, should consider the regularities emerging from the collection of natural and textual particulars, engage in inductive reasoning, and thereby assess traditional accounts of causation on the basis of experience.[34] In addition, many branches of the new philosophy, frequently marshalled for Enlightened ends, were indebted to ancient philosophy as understood through the lens of historical scholarship. The links made between the ancient and the new were paradigmatic of late humanist historical legitimation of innovation by means of the appropriation of ancient texts. Gassendi, for instance, built his refutation of Aristotle and his own Epicurean atomism on a humanist history of philosophy. This continued to be a widespread practice in the later seventeenth century, even if it was opposed by Descartes and others.[35] Even the most unlikely figures of the early Enlightenment, including outspoken critics of humanist learning such as Hobbes, owed a significant debt to humanist anti-Aristotelians and their historical and philological treatments of the true Aristotle.[36] The study of mathematics, too, was partly rooted in late humanist erudition, the close study and editing of ancient philosophical texts, and the historiography of ancient science.[37] When approached from this angle, it becomes clear that the "new science," which is often posed as a radical intellectual rupture, resulted in part from the circulation of humanist practices.

The study of history proper was more obviously the engine of counsel, conflict, and polemic among elites in the late sixteenth and early seventeenth centuries. In this role it gave rise to the new religious and political sciences, in addition to the natural, that later supported Enlightenment projects. When applied to the political realm, the increasingly causal, comparative, and philosophical bent of historical study in the late sixteenth and early seventeenth centuries yielded a "politic" style of analysis—often referred to as "reason of state" or "Tacitism"—and accompanying prudent conduct. It was meant to support the establishment, maintenance, and advancement of state power in a manner consistent with the Christian ethics and competitive demands of an age of religious war. The ancient paganism to be found in the primary sources for this emergent tradition was no barrier to a fundamentally pious use of them. The intent of widely read figures such as Giovanno Botero and Justus Lipsius was to guide efficacious action on behalf of the state by relying upon experience (obtained through reading history, traveling, and attending to affairs of state), which through comparison yielded predictive knowledge of political regularities.[38]

This late humanist mode of analysis remained important throughout the seventeenth century, exemplified, for instance, by the Spanish diplomat Diego de Saavedra Fajardo's frequently translated *Empresas Politicas* (1640) and the Jesuit Baltasar Gracián's extremely popular *Oraculo manual y Arte de prudencia* (1647).[39] In the early Enlightenment, politic commentary became an ordinary phenomenon. Its characteristic practices were directed beyond the traditional imperative of maintaining state power to the overlapping but wider and often countervailing imperative of civil peace. Accordingly, these practices were shared by men working outside and often against absolutist states, and in new institutional realms. This could take many forms, from the subversive editorial work of the Gracián translator Abraham-Nicolas Amelot de la Houssaye (1634–1706) to the digestion of earlier politic humanism, including Amelot's, in Bayle's *Critical and Historical Dictionary* (1697).[40]

Enlightened Transformations of Renaissance Practices

Lurking behind these changes was a fundamental epistemological problem posed by religious war and metaphysical diversity: the apparent difficulty of even establishing any commonly accepted truths or responsibly absorbing an ever-increasing deluge of books, information, and opinion.[41] This problem encouraged many figures to use their late humanist educations as a foundation for developing reliable practices upon which the intellectual basis for a novel social and political ordering might plausibly be constructed.[42] Contrary to Immanuel Kant's famous slogan of 1784, they dared *not* to know what was beyond the limits of human ability.[43] Instead they charted (and contested) a middle way between the Scylla and Charybdis of Pyrrhonian skepticism and metaphysical dogmatism.

Part of the foundation for this new political and social ordering was a series of mechanisms for the communication, dissemination, and deployment of knowledge. The media through which Enlightened knowledge circulated included practices and institutions both new and old that underwent important changes in the late seventeenth and eighteenth centuries. This was one of the arenas in which the early Enlightenment most clearly drew upon Renaissance traditions. The Renaissance Republic of Letters was transformed into the Enlightenment Republic of Letters and a broader constellation of Enlightenment publics. The Renaissance Republic of Letters was primarily a Latinate network of learned correspondence, travel, and local institutions that crossed many of the religious, political, and institutional divides of the sixteenth and early seventeenth centuries without managing full independence from them.[44] In the late seventeenth and early eighteenth centuries, the Republic expanded in all dimensions—linguistic, geographical, cultural, and institutional. Bolstered by learned periodicals, salons, and academies, it facilitated the interaction of scholars with men and women of letters as well as a wider, educated readership.[45] Reconstructed as a broadened network of learned exchange and a constellation of Enlightenment publics, it was equipped to translate scholarly debate and the creation of knowledge in accordance with the ultimately public, practical imperatives of the Enlightenment. It was also in itself a setting for Enlightenment to the extent that it facilitated intellectual sociability under conditions of civil peace by bracketing the ongoing political and religious divisions of the day.

It was in this emergent environment that practitioners of Enlightenment began to work out the epistemological foundations of a post-Westphalian order beyond the borders of their imagined republic. The Pyrrhonian skepticism of the late seventeenth and eighteenth centuries was a reaction to civilizational diversity that threatened to undermine Enlightenment. It too had been made possible by late humanism and, in particular, by the publication of the works of Sextus Empiricus in the 1560s and the skeptical currents of that period.[46] Pyrrhonian texts, transposed into the present and beyond their original remit of philosophy, made it possible to understand the profusion of variant opinions, experiences, and traditions and the sheer explosion of books and information in the seventeenth century as evidence of the impossibility of making statements about the world—and by extension, improving the world—with any certainty. Like the resurgent dogmatism so common in the later seventeenth century, skepticism was a negative practice against which the so-called Age of Reason was directed and constructed. Enlightened commentators tended to agree that skepticism led not to peace but to barbarity, moral degradation, irreligion, and chaos.[47]

There were few viable, strictly philosophical responses offered to skepticism before the later eighteenth century. Instead the responses tended to be utilitarian ones. Minimal and purportedly shared standards for reliable knowledge in an environment of permanent pluralism and doubt were developed across all realms of inquiry. In novel articulations of some of the hallmarks of late humanist logic and rhetoric, moral certainty, probability, and verisimilitude were prized over metaphysical certainty. A via media between absolute certainty and absolute doubt came to predominate. Many chose "practicable reason" (empirical observation for practical purposes) over the "speculative reason" of Leibniz's *Monadology* (1714) and other metaphysically inclined investigations in philosophy and theology. "Practicable reason," as Jean le Rond D'Alembert would later call it in the "Discours préliminaire" of the *Encylopédie*, became the standard for knowledge in natural philosophy, historiography, and theology. Practicable reason made possible the production of reliable knowledge on the topics on which it was deemed imperative for Europeans to be acting, and it created a basis for assent to knowledge claims that could itself gain wide acceptance by those who were struggling with practical problems. After the 1650s, this arrangement was generally accepted, as opposed to debated. Metaphysical philosophy, for instance, continued to be practiced in the universities, but it ceased to be of central importance to public life.[48]

This general response to the problem of certainty was evident in each of the historically oriented realms of inquiry that had been central to the late Renaissance. These commitments and projects gradually changed the nature of Biblical criticism, for instance, in a movement from *critica sacra* to *critique*. While the originality and consistency of Jean Le Clerc's textual criticism has been exaggerated, in the interest of creating a minimal Arminian theology, it did clear away what Le Clerc saw as the philosophical, typological, and prophetic baggage that had corrupted Protestant exegesis. In the process, he also furnished some recognizably novel practices, moving beyond the confessional and partly ahistorical exegesis of late humanist Protestants and the philosophical inflections of the Augustinian tradition. His most important maneuvers included the disentanglement of linguistic and intellectual similarities among texts, especially the Old and New Testaments, which disrupted the typological tendencies of his predecessors.[49]

Historical research and writing—produced by both learned travelers and scholarly homebodies—was at the core of the early Enlightenment response to skepticism and the demands of civil peace under conditions of pluralism.[50] Early Enlightenment historians often derided their humanist predecessors but in fact built directly upon late Renaissance erudition. They developed rigorous procedures for examining official manuscripts, coins, monuments, inscriptions, vases, statues, and reliefs, and some of them helped invent the double narrative of footnoted histories. Simon, Bayle, Le Clerc, Jacob Perizonius, Jean Mabillon, Bernard de Montfaucon, Scipione Maffei, Jean Hardouin, George Hickes, and Richard Bentley offered new rules for source criticism that were more focused on achieving regularity in the evaluation of testimony than the historical theory and practice of the late Renaissance had ever been.[51]

A similar set of developments can be seen in the study of nature. Many natural philosophers eluded skeptical challenges by incorporating skepticism into their experimentalism and empiricism. Marin Mersenne, Pierre Gassendi, and other mid-seventeenth-century natural philosophers had used Pyrrhonism tactically to challenge Aristotelianism and Cartesianism while grounding their own knowledge claims in a form of mitigated skepticism drawn from humanistic, probabilistic forms of logic. Humanist philology also underpinned their arguments about the use of language in philosophy.[52] Their successors in the late seventeenth and early eighteenth centuries largely followed suit: humanist rhetoric and dialectic provided ample resources for the post-Pyrrhonian probabilism and mitigated skepticism of the "new science" and Enlightenment.[53] Scientific facts produced through experiment and other means were authenticated in ways that were extremely similar to methods in history.[54]

In fact, experimentalism itself was commonly legitimated not by actual experiments but by the critical examination and narration of textual testimonies according to humanist historical methods. Eventually this created a schema for a science proceeding on the basis of experimentation.[55] In any case, the Royal Society hardly acknowledged any distinction between human and natural history, especially when it came to the increasingly common use of travel writing in the investigation of both human and natural facts.[56] On the receiving end of natural learning, late-seventeenth-century virtuosi adapted late humanist traditions of note taking in order to secure, organize, and access particulars, both read and observed, in an environment of informational deluge.[57] Galileo, for instance, despite his attacks on bookish learning, was largely digested even by prominent proponents of the new science as a contributor to traditional, partly humanistic, partly scholastic natural philosophy focused on central Aristotelian topics.[58]

It was by means of these familiar literary practices that new natural learning was disseminated and eventually attained intellectual hegemony. Its strictly empirical practices relied upon the same probabilism and utilitarianism to be found in the generation of historical and political forms of knowledge. Science exhibited perhaps more than any other realm of inquiry a strong association between knowledge and the public interest, and the corollary notion that knowledge was based on probability and particulars.[59] Like their historian contemporaries, exponents of experimental philosophy such as Robert Boyle and Johann Christoph Sturm (1635–1703) explicitly argued that a consensual definition of its criteria for truth was a means of avoiding controversy.[60] Beyond the confrontation with skepticism, early modern disciplines were considered to be realms that might build stable, generally accepted truths that could eventually lay a foundation for civil peace and improvement.

At the same time as these post-skeptical ways of knowing developed in response to the imperatives of Enlightenment, new practices of inquiry, inscription, and publication were built upon them. Enlightened answers to the question "Why should I read the Bible?" for example, prompted a series of new scholarly and translation practices that eventually produced the "cultural Bible," a text (or more precisely, a variety of texts) created and valued independently of its confessional or theological significance.[61] More generally, the "new sciences" of nature, religion, and politics became solidified, expanded, and repurposed versions of the causal, inductive, empirical, and comparative forms of inquiry based upon experience and authenticated facts that had first emerged in the late sixteenth and early seventeenth centuries. History also provided a conceptual underpinning to the Enlightenment by demarcating a realm of the "natural" that became a theologically neutral space for inquiries into human betterment. On all these fronts, the resources of late humanism were crucial starting points.

The study of the natural world, like the study of the Bible, did not become a terrain of Enlightenment simply because Descartes, Spinoza, and others strictly distinguished philosophy from theology or because physics was used to question the book of Genesis. In fact, the book of nature largely retained its significance as a form of revelation, but that revelation was no longer discovered by studying nature with recourse to the Bible, but rather by piecing together nature's own order and structure. This development was made possible in large part by means of advances in humanist Biblical criticism that gradually rendered the meaning or authoritative status of many passages relating to natural history so unclear that they could no longer serve as a foundation or guide for studying the book of nature. Weeks after a great comet was seen in 1664 across Europe, for instance, the Utrecht professor of history and rhetoric Johannes Graevius (1632–1703) asked whether comets should primarily be considered signs of God's wrath, as Reformed orthodoxy would dictate, when the Bible never explicitly referred to them as such. In fact, he wondered, was this notion actually a remnant of pagan superstition? Were comets more obviously evidence of God's omnipotence, and therefore no fertile ground for confessional polemics? These sorts of questions, emerging from humanist practices, enabled an independent, non-Biblical understanding of both nature and revelation to emerge and guide human conduct. In other words, scientific inquiry could in theory now become a consensual, probable (and in some cases, demonstrable) basis for moral and civil improvement.[62]

The form of this new style of natural investigation that eventually came to dominate was experimental philosophy. There is of course a sense in which the case of English natural philosophy better fits this discussion of early Enlightenment science than do the continental versions, which for some time remained less empirical, experimental, probabilistic, and utilitarian. But the experimental style of scientific exploration certainly rivaled the mathematical sciences in its importance to Enlightenment on the continent in the eighteenth century. As the study of nature secured its independence from Biblical theology as a form of inquiry, it created a realm for studying nature in a manner that took on only minimal theological commitments. This positioned students of nature to contribute to the stabilization and improvement of state and civil society in an Enlightened manner, without invoking their theological commitments. Natural inquiry, over the course of the late seventeenth and eighteenth centuries, thereby became central to Enlightened governance and empire.[63]

In this environment, the realm of "nature" could be constructed independently with recourse to deductive or inductive practices. Whether construed theistically or atheistically, it provided a means of bracketing theological discord in the construction of schemes for civil peace. This in fact transformed the underlying visions of such a peace. In the late Renaissance, "politic" analysis was still largely particularist in the manner in which it accumulated and presented political knowledge. But as the authority of nature was increasingly used to furnish schemes for civil peace, these schemes assumed a more universalist character. In this more "scientific" or "philosophical" guise, politic analysis became a recognizably different practice. This shift began in radical fashion, of course, with Thomas Hobbes's attempt to invent a post-rhetorical, post-experiential form of civil science in *Leviathan* (1651) and other works. But Enlightened styles of political analysis still deeply rooted in history were more common.[64]

Even when presented as a form of natural jurisprudence or civil philosophy, early Enlightenment political knowledge was founded in probable, historical particulars and a rejection of previous, scholastic versions of such inquiries. This generalization would apply both to John Locke and to important German figures such as Christian Thomasius and Samuel Pufendorf. Late humanist historiography and philology were also brought into the service of natural law theory in the form of "histories of morality" built upon early humanist models. This natural jurisprudence was then used as the basis for the training of civil servants. In Germany above all, where reason of state and humanistic forms of political jurisprudence became deeply embedded in the universities by the mid-seventeenth century, a historically inflected civil philosophy thrived. It explicitly aimed at secularizing the state and heightening its power in absolutist fashion in order to bracket theological difference from governance and avoid religious war.[65]

Similar developments occurred in the study of religion. Here the movement from late humanism was arguably even more dramatic, since politic analysis that bracketed theological commentary had hardly even been applied in detail to religion itself before the later seventeenth century. The nature of religious analysis was transformed once the imperative of civil peace encouraged reflections on both religious corruption and religion done right that were not primarily conducted with an eye to the vindication of doctrinal propositions or grounded in theological explanations of error. In late-seventeenth-century religious discourse, understandings of natural religion and historically rooted understandings of religion as a technology of power took center stage. Civil and natural religions were the early Enlightenment's typical answers to the problem of holy violence. These cults of truth and peace were occasionally described in philosophical treatises, but most of the time, they appeared in stories about past departures from natural religion. These stories tended not to be happy ones, since the easiest way for historians, antiquarians, exegetes, and travel writers to identify a pure and orderly religion was to describe what it was not. These writers slowly cobbled together a global history of religious imposture, intrigue, and ignorance and turned their full attention to the earthly causes of error and corruption. They defended schemes for civil peace on the basis of deistical or atheistical premises, bracketing the theological propositions that had led to religious war in the past. But by doing so, they by no means committed themselves to politically or religiously emancipatory positions.[66] Here again the early Enlightenment quite clearly drew on late humanist politic analysis—and, in particular, the Jesuits' reflections on Catholicism as the ultimate civil religion— but with different purposes and cautions in mind.[67]

From Renaissance to Enlightenment Now

The emerging account of the relationship between Renaissance and Enlightenment sketched here offers an alternative to the traditional and still-dominant understanding of the Enlightenment as a philosophical movement of liberal secularism. It combines elements of the traditional interpretation with an awareness of how profoundly both religious belief and Renaissance practices triggered, enabled, and partly constituted the Enlightenment while undergoing considerable change themselves. Far more research remains to be done in execution of this scholarly agenda, especially on the pivotal moment of the Enlightenment's emergence in the later seventeenth century, a topic on which scholarship is still overwhelmingly conducted within liberal-secularist or revisionist paradigms. Future work must pay much closer attention to practices and institutions and to the problematic category of "philosophy," with an eye to both the powerful role of humanistic culture and the subtle relationships between scholasticism and early Enlightenment philosophy. Fine-grain studies focused on the turning point of the early Enlightenment (which I have addressed here only in general terms) can provide the precise account of practical transformation and continuity that we require.[68]

The account provided in this chapter of the relationship between the world of the Renaissance and Reformation and the world of the Enlightenment is important not simply because it is more empirically accurate, conceptually consistent, and ideologically open-ended. It is also important because it provides lessons for the academy and the public in an age of culture wars and related confusion about the value of humanistic inquiry and its relationship to modernity. All parties to the domestic and transnational culture wars of the twenty-first century accept the idea that the Enlightenment created an inherently secularist, liberal, and scientific civilization in the West, one inherently opposed to religious and humanistic perspectives, for better or worse. They structure their positions accordingly, arguing for or against the Enlightenment project and its successors in the nineteenth, twentieth, and twenty-first centuries. A thoroughly revised understanding of the relationship between the world of the Renaissance and Reformation and the world of the Enlightenment offers a way to dispel this series of false dilemmas and to explore the possibilities of a more accurate understanding of the world's Enlightenment inheritance.[69]

The story told here is also one about how early modern societies adapted fecund cultural resources primarily rooted in the humanities to confront the instabilities and crises of truth-telling, political subjectivity, and social stability wrought by globalization and pluralism. The early Enlightenment, in particular, was a moment in which humanism and science flourished in tandem and could not do without each other, a moment in which history stood as the linchpin of an entire intellectual universe and its relevance to public life, and a moment in which the dangers of crude scientism and system making were well understood. It was a moment in which the value of humanistic inquiry could neither be dismissed nor smugly taken for granted by its adherents; instead, its value had to be articulated and proven through its application to profound worldly problems, not to mention otherworldly ones. This, of course, had been demanded of humanism from the beginning. It had gained a foothold in European society because of its perceived social utility, not because of its transparent superiority; and its relationship to ethical conduct was (and remains) intrinsically indeterminate and had therefore to be defended and actively constructed.[70] We face very similar challenges today. The problems and the proper responses to them are of course different in content, but fundamentally alike in character and structure.

NOTES

1. Gay 1966–69; Hazard (1935) 2013; Israel 2006.

2. The best major revisionist work is Pocock 1999–2015.

3. For further discussion and citation of relevant work, see Bulman 2015; Bulman and Ingram 2016; and Bulman 2016. The primary focus of these works is on the relationship between Reformation and Enlightenment, but portions of this essay are drawn from them.

4. Pocock 1999–2015, vol. 1; Sheehan 2005; Gierl 1997; D. Goodman 1994; Hunter 2007.

5. Israel 2006, 63; Hazard (1935) 2013, 3–28.

6. Bulman 2015, xii.

7. Bulman and Ingram 2016, 12–13. This is to combine the chronology of Hazard's *crise* (Hazard [1935] 2013) and Israel's "crisis" or "prelude to the early Enlightenment" (Israel 2001, 14).

8. Spinoza 2007, 238 (19.2). The later Enlightenment, as Robertson (2005, 28–31) has argued, translated the early Enlightenment imperative of civil peace into broader, positive terms as "human betterment in this world" and "sociability" under conditions of secularity. The present essay focuses on the early period in order to capture the transition between Renaissance and Enlightenment and to accommodate space constraints, but the argument is compatible with Robertson's and other accounts of the later eighteenth century.

9. C. Taylor 2007, 3–4, 12–14, 19–20, 192–94.

10. Locke (1689) 2010, 7–8.

11. Stout 2004, 92–117.

12. Bulman 2015, xiii.

13. Bernard 1714, 147.

14. Spinoza 2007, 68 (5.1).

15. Bayle 1705, 519–22.

16. Koselleck 1988; I. Hunter 2007; Malcolm 2002, 535–45.

17. Sheehan 2005; Bulman 2016.

18. Bulman 2015, xiii–xiv.

19. This chapter thus primarily explores the extent to which the Enlightenment was made possible by late Renaissance humanism, and the extent to which humanistic practices changed as a result of their involvement in the Enlightenment. Because of space constraints it necessarily excludes or sidelines other elements of what would be a comprehensive alternative to traditional interpretations of the Enlightenment, its consequences for Renaissance practices, and its relationship to other traditions that preceded it. Such elements include the links between Enlightenment philosophy and scholastic philosophy; the simple perpetuation of traditional humanism in Enlightened and non-Enlightened contexts in the late seventeenth and eighteenth centuries; and the extinction of certain humanist practices in the eighteenth century.

20. On definitions, see Kraye's chapter in this volume.

21. Hardy 2017, 1, quoting Des Maizeaux 1740, 2:96.

22. Rummel 2000; Shalev 2012; Shuger 1994; De Jonge 1975, 65–109; Grafton 1983–94.

23. Laplanche 1986; Muller 2003; van Miert et al. 2017; Touber 2018; van Miert 2018.

24. Ditchfield 1995; Lyon 2003; Cochrane 1981, 445–78; Van Liere, Ditchfield, and Louthan 2012.

25. Mulsow 2005; P. Miller 2001; Loop 2013; MacCormack 1991.

26. Levitin 2012.

27. Grafton 2017; Malcolm 2002, 383–431.

28. Nellen and Steenbakkers 2017, 37; Touber 2017, 342–45; Touber 2016, 171–73.

29. Pomata and Siraisi 2005; Popper 2012.

30. Gilbert 1965; Pocock 1987; Momigliano 1990, 132–52; Grafton 2007b; Stagl 1995; Rubiés 2000; Hodgen 1964; Tinguely 2000. See also Clark's chapter in this volume.

31. Grafton 1990b; Cochrane 1976; Grafton and Siraisi 1999; Ogilvie 2006; Siraisi 1997. See also Popper's chapter in this volume.

32. Shapiro 1979; Siraisi 2007; Siraisi 2012; Pomata and Siraisi 2005; Blair 1997; Seifert 1976.

33. Kelley 1997; Popper 2012, 294–95, including citations; Lüthy 2000; Grafton 1997c; Siraisi 2007; C. Martin 2014.

34. Popper 2011; Popper 2012, esp. 291; Pumfrey 1991; Findlen 1997.

35. Joy 1987; Levitin 2015, 230–446; Shelford 2007, 119–24.

36. Leijenhorst 2002; Paganini 2007; Sorell 2000; Shelford 2007, 164, 183.

37. Goulding 2010; Feingold 1990; Guicciardini 2009, esp. 81, 291–327; Dear 1995, 216–43.

38. Kahn 1994; Burke 1991; Malcolm 2010; Senellart 1989; Viroli 2005; Millstone 2014, 2016.

39. J. Snyder 2009, 133–41.

40. Soll 2005.

41. Blair 2010, esp. 57–59.

42. Shelford 2007, esp. 7.

43. K. Baker 2001.

44. Waquet 1989, 1993; Bots and Waquet 1997; Dibon 1990; Hardy 2017; Furey 2006.

45. B. Curran 2012; Caradonna 2012; Bulman 2016; Robertson 2005; Siskin and Warner 2010; Van Horn Melton 2001; Brockliss 2002; Marshall 2006; Goldgar 1995; Mulsow 2007; Shelford 2007; Bevilacqua 2018; Hamilton 2005.

46. On this and the more general phenomenon of which it was a part, see Kraye's chapter in this volume.

47. For an example, see Clark 2017, esp. 393.

48. Matytsin 2016 (D'Alembert is cited on 273); Shapiro 1993; K. Baker 1994, 114–17; Keller 2015; I. Hunter 2001.

49. Hardy 2017, 391–98.

50. Bulman 2015; Pocock 1999–2015; Hochstrasser 2000; Champion 1992; Bödeker and Iggers 1986; Völkel 1987.

51. Grafton 1997a; Momigliano 1990, 54–79; Levine 1991, 267–418; Barret-Kriegel 1993; Völkel 1987, 68, 99–202; Borghero 1983; Israel 2006.

52. Dear 1991, 9–47, 170–200; Joy 1987, 165–94; Blackwell 2001.

53. Shapiro 1991, 50–52.

54. Serjeantson 1998; Dear 1985; Shapin 1984; Shapin and Schaffer 1985.

55. Popper 2016.

56. Gascoigne 2009.

57. Yeo 2014.

58. Raphael 2017.

59. Keller 2015.

60. Shapin and Schaffer 1985; Hochstrasser 2000, 24–25.

61. Sheehan 2005.

62. Jorink 2010 (on Graevius, 155, citing Graevius 1681); Poole 2010.

63. Drayton 2000; Bleichmar 2012.

64. See also Clark's chapter in this volume.

65. Keller 2015; I. Hunter 2001, 2007; Koselleck 1988; Malcolm 2010; Tuck 2012; Kraynack 1990.

66. Bulman 2015, ch. 4; Rubiés 2006; Stroumsa 2010; Bevilacqua 2018.

67. Höpfl 2004, 112–39.
68. See, for example, the work of Mulsow (2007, 2012).
69. Bulman and Ingram 2016.
70. Grafton and Jardine 1986.

PART TWO

GEOGRAPHICAL HORIZONS

CHAPTER THREE

New Worlds, New Texts

Rewriting the Book of Nature

DANIELA BLEICHMAR

The early modern period was the first truly global era, a time of unprecedented voyages and encounters. People, things, and ideas moved farther and more often, and many of those who stayed put nevertheless found their lives touched by what came their way. Travelers and homebodies alike experienced their time as one marked by novelty and difference, and by encounters that forced them to reckon not only with things they had never imagined but also with those they thought they already knew. Writing in the late 1570s about his time in Brazil, the French author Jean de Léry reflected:

> I am not ashamed to confess that since I have been in this land of America, where everything to be seen—the way of life of its inhabitants, the form of the animals, what the earth produces—is so unlike what we have in Europe, Asia, and Africa that it may very well be called a "New World" with respect to us, I
> · have revised the opinion that I formerly had of Pliny and others when they describe foreign lands, because I have seen things as fantastic and prodigious as any of those—once thought incredible—that they mention.[1]

For Léry, Pliny was both an obligatory reference point and a source that, before his voyage, had seemed too incredible to believe; America's difference and novelty made it not only fantastic and prodigious but also the source of a new authority that allowed him to rethink the classics. Encounters transformed people and ideas; they also transformed places. Writing of Amsterdam in the 1630s, René Descartes described the city as overflowing with "all the produce of the Indies and everything rare . . . all the commodities and curiosities one could wish for."[2] The same could be said, at different points, of Mexico City, Lima, and Manila; of Canton and Beijing; of Goa, Agra, and Delhi; of Istanbul and Cairo, among others.

Modern historians have often been less cosmopolitan and open to rethinking their beliefs than the early modern globetrotters and worldmakers they have studied. Much scholarship on early modern Europe, especially that produced in the United States and Europe, has tended to remain stubbornly provincial and particularly so when it comes to the study of ideas, science, and art. The study of humanism and the so-called Scientific Revolution, for instance, have traditionally been conceived as purely European phenomena.[3] More recently, scholars of the early modern period have pursued transregional approaches conceived as connected, crossed, and entangled histories. Such studies propose other centers and investigate contact, collaboration, and exchange between communities and regions. They have shown that people, texts, objects, and images circulated (whether willingly or by force); that these exchanges resulted in transmissions, erasures, and translations; and that they ultimately transformed what was known and deemed worth knowing across the globe. For many, such entanglements define the early modern period.[4] When it comes to the history of knowledge in the early modern Americas and Atlantic world in particular, a wave of scholarship since the 1990s has radically transformed what we know, the questions we ask, and how we go about answering them.[5]

This chapter builds on such scholarship, proposing a transregional approach to the history of the book, the history of scholarship, and the history of science that places European and non-European ideas, traditions, and sources in conversation with one another and highlights the active contributions of the Americas to Western knowledge. I emphasize the centrality of Spanish American and Spanish sources to the history of early modern knowledge; underscore the contributions of Indigenous individuals, categories, experiences, and knowledges; consider images and objects as indispensable historical sources; and focus above all on questions of exchange, circulation, and translation. Cultural encounters and exchanges reshaped scholarly methods and epistemic values. The history of early modern knowledge was much less provincial than it was later imagined to be.

The Early Modern Book of Nature

After 1492, Europeans and Native Americans alike lived in a new era. They had mutually encountered and discovered each other, engaged in battles for political and spiritual dominance, and transformed their cultures, customs, and beliefs by choice and by force. This world was new to everyone—Europeans, Native Americans, Africans, and Asians—and in constant transformative flux. Humans, plants, animals, and pathogens traveled back and forth across the oceans in an exchange that had powerful and, for American Indigenous populations, devastating effects.[6] While the exchange went both ways, it was by no means symmetrical. The consequences were incomparably more destructive for Indigenous Americans and enslaved Africans taken by force to the New World, and the balance of power was much weighted in Europeans' favor. But even if Europeans had the strongest say and the loudest voice, they were certainly not the only ones whose ideas and words contributed to the making of new knowledge and new images. Native Americans and Africans in the Americas had great personal expertise with New World nature and much to say about it.

The knowledge that groups like Native Americans and enslaved Africans deployed in the New World permeated texts produced by Europeans such as the Spaniard Gonzalo Fernández de Oviedo, who authored the earliest work dedicated to the natural history of the Americas written by someone with first-hand experience of the land.[7] Oviedo had spent four decades shuttling back and forth between Europe and the New World, where he held various positions in the Spanish imperial administration. His landmark *Historia general y natural de las Indias* (*General and Natural History of the Indies*; 1535–49) provided lengthy descriptions of the American flora, fauna, and landscape, as well as native customs and history, and for centuries served as one of the most consulted and definitive sources on the New World.

In his work, Oviedo stressed the value of *autopsia*—the lived experience of witnessing with one's eyes.[8] He considered eyewitnessing essential to confirming the credibility of knowledge claims; he pointedly described himself as a "*testigo de vista*" (eyewitness) and wrote that readers valued as "most truthful and authentic" the work of authors who had traveled the world and seen distant regions with their own eyes, basing their written accounts on lived experiences and the things they had witnessed themselves.[9] This primacy placed on first-hand testimony was especially necessary with the New World, Oviedo maintained, for it was home to people, animals, and trees so different from anything known to Europeans that they could be almost impossible to comprehend: in order to be known and understood, the New World needed to be seen. Such direct encounters were of course unavailable to the vast majority of his readers, and accordingly he relied extensively on visual renditions of his subjects, which he saw as mimicking the experience. Such visualizations emerged as a vital means of communicating information about the New World to European audiences.

But for all his emphasis on *autopsia*, Oviedo's descriptions of the flora and fauna were deepened by a profound reliance on local informants who recounted to him their own knowledge and often guided his apprehension of the phenomena he described. In this way he was typical of most Europeans seeking to gain knowledge of New World *naturalia*. For while European travelers repeatedly asserted their authority and expertise on matters concerning the lands they visited, the people with the most experience and knowledge of American nature in the sixteenth century were unquestionably the longtime Indigenous inhabitants of the New World. Accordingly, European authors attempted to mine this expertise and eagerly questioned natives about their uses of plants and animals. And if they often elided the identities of their interlocutors, they frequently reported the names that Indigenous peoples gave to the plants and animals, and sometimes also their knowledge and practices, though often with mixed feelings concerning Indigenous beliefs. In particular, as this chapter shows, emphasis on graphic renderings of natural phenomena constituted a critical point of overlap for European and Indigenous practitioners.

To depict as significant the knowledge and practices of Native American and African practitioners primarily as a fragment of a broader European intellectual enterprise, however, fundamentally mischaracterizes the broader epistemological terrain of the early modern world. Indigenous peoples, enslaved Africans, and maroons were not only informants but also knowledge makers in their own right. They grew plants, prepared medicines, cared for sick people of every ethnicity, drew maps, painted, and wrote books. They also often collaborated with missionaries, who had a keen interest in Indigenous knowledge. And in fact they often seized on European technologies and methods to articulate, codify, and disseminate their own knowledge. This chapter examines texts, observations, and images produced with dominant input from native practitioners in order to illustrate the distinct participation of Indigenous peoples in the production and circulation of New World natural knowledge. These case studies provide unparalleled testimony of native knowledge traditions; highlight the interpenetration of New World and European knowledge; and illustrate some of the ways in which Indigenous practitioners, artists, healers, and others could harness European technologies, media, and practices in the production and circulation of knowledge. Closely attending to how groups of Europeans and natives collaborated to produce texts illuminating New World geography, ethnography, and flora and fauna showcases the prevalence and different registers of exchange, transmission, and adaptation that underlay the structural dynamics of the early modern world.

American Authors: Mexican Indigenous Knowledge in Two Sixteenth-Century Manuscripts

Two manuscript books created in Mexico in the fifty years following the conquest century—the *Codex de la Cruz–Badiano* (1552) and the *Florentine Codex* (ca. 1577)—are prime examples of this world of Indigenous knowledge making. But without question, the dynamics of Indigenous knowledge production were transformed by Native Americans' encounter with Europeans. Both works were expressly created for transmission to Spain and produced at the same institution, the Colegio de la Santa Cruz de Tlatelolco (est. 1536). In this Franciscan school, young men from the Aztec ruling class studied Spanish and Latin grammar, religion, the liberal arts, the writings of the Church Fathers, and the Greek and Roman classics. These manuscripts show the active role that Indigenous authors played in producing and circulating knowledge of American nature within a polyphonic, multivalent world.[10]

The earliest work on American nature created by Indigenous authors after the Spanish conquest of the Americas is an illustrated manuscript entitled *Libellus de medicinalibus Indorum herbis* (*Little Book of Indian Medicinal Herbs*) (see fig. 3.1). It is also known as the *Codex de la Cruz–Badiano* after its two Nahua authors, the physician Martín de la Cruz, who provided information on remedies and healing practices that used plants, animal parts, and minerals, and the translator Juan Badiano, who turned this account from Nahuatl into Latin.[11] They created this "little book"—it measures only 8 by 6 inches (20.6 by 15.2 cm), the size of a modern paperback—in 1552, only three decades after the conquest. The long title of the book describes Martín de la Cruz as "a certain Indian, physician of the College of Santa Cruz, who has no theoretical learning, but is well taught by experience alone." In a note addressed to the reader at the end of the work, Badiano identifies himself as the *interpres* (interpreter or translator), a professor at the Franciscan college at Tlatelolco, and "an Indian by race, a native of Xochimilco."[12] This last comment suggests the endurance of preconquest communal affiliations (the mention of the *altepetl*, or community, of Xochimilco as an identifying mark) and also indicates the multiple currencies that the author's indigeneity could have in both local and transatlantic contexts. Badiano's ethnicity served to establish his linguistic expertise in Nahuatl, to highlight his learning (in Latin, classical sources, and rhetoric and composition), and, by extension, to exalt the work of the Franciscans who had educated him.

The manuscript was commissioned by Francisco de Mendoza—son of the powerful Antonio de Mendoza, the first viceroy of New Spain (r. 1535–50) and a major benefactor to the college—and intended as a gift for the Spanish royal family. It was thus conceived as a luxury object, and a gift meant to please and impress. European readers would have recognized the work as evidence of the successful conversion and education of the Indigenous elite, reflecting well on the Franciscans who ran the college at Tlatelolco and on the viceroy as their patron. In addition, the younger Mendoza might have had a commercial agenda: he had established cultivations of ginger and China root (*Smylax pseudinochina*), both of them profitable plants; and after presenting the codex to the royal family, he negotiated a license to import and export medicinal plants and was named general administrator of mines in New Spain. Through the codex, Mendoza would have presented himself as a benefactor of missionaries and Indians alike, a sponsor of investigations into nature, and a person with access to valuable information about the commercial exploitation of New World nature.[13]

The *Codex de la Cruz–Badiano* highlights the mastery that Indigenous authors, scribes, and artists had acquired of European book conventions, which they combined with native elements. The book consists of a Latin text and 184 botanical drawings on 63 folios, or sheets of European paper. It emerged from a close collaboration among the painters and the scribe.[14] Many of the pages have large images with manuscript annotations underneath describing how each plant is used; some of the pages comprise only drawings accompanied by the plants' names, without further commentary. A reader at the time would have recognized the penmanship—a very competent italic in the "chancery" Italian style popular in learned and polite circles throughout sixteenth-century Europe—as the work of a highly educated person.[15] Despite the great number of illustrations and the importance of the painted image in Nahua knowledge traditions, the artists are not named. The drawings are done in an Indigenous style, using brightly colored organic pigments and a few native glyphs, most prominently the glyph for *tetl*, or stone, and the pictograph for flowing water.

The botanical and medical information comes from Aztec medicine, and, although the majority of the text is in Latin, the plant names are provided in Nahuatl. The content is organized into thirteen chapters, each of them dedicated to the maladies and remedies for a different part of the body. The fact that many recipes include expensive and rare ingredients, such as jade, coral, pearls, exotic plants imported from distant tropical lowlands, and rare animal materials such as bezoar stones, suggests that de la Cruz's patients were predominantly the Indigenous aristocracy or that he chose the most refined remedies because the book was intended for the Spanish king and accordingly was expected to circulate only among the most elite circles.[16] One of the recipes suggests a cure "for the fatigue of those administering the government and holding public office," a very particular malady indeed.[17]

The manuscript contains a wealth of information on native medicines and categories, relating the functioning of the body to the Aztec calendar, astronomy, religious beliefs, and cultural practices. Although it is very much a product of the early colonial world—written in Latin by Christianized Indigenous members of a Franciscan college and created for the viceroy's son to send to the king of Spain—it is little concerned with presenting Aztec knowledge within a Christian framework. Thus, whereas most European authors at the time tracked which plants had uses in Indigenous medicine but dismissed native practices as superstitious or idolatrous, this book provides rare and detailed information concerning Indigenous medical and botanical knowledge.[18] It addresses the preparation and application of medicines to treat multiple ailments, and provides insight into Indigenous ideas of how bodies and remedies operated—while referring to Pliny, as any learned author at the time would have done.[19] One entry, for instance, advises, "A fox's eye is wonderfully beneficial to sore eyes, for which purpose it is to be fastened to the upper arm. And if the eyes are so sore that they seem almost torn out, grind up—in the blood of a goose, woman's milk and spring water—a pearl, a purple crystal, a pinkish oyster, the stone found in the little bird called *molotototl*, the stone *tlahcalhuatzin* and the stone that is in the stomach of the Indian dove; and the liquid thus prepared, instill into the troubled eyes."[20]

Despite the princely audience anticipated for the volume, its scope of treatments extended more broadly. The book includes a chapter addressing women's medicine, a subject almost universally ignored at the time by authors describing the New World or writing about medicine more generally. The text discusses "the medicines for recent parturition; for menstruation; for washing the abdomen of a woman in childbed; tubercles of the breast; medicine to produce lactation." It includes the following passage detailing the complex preparations meant to assist an elite woman during childbirth:

> If a woman has difficulty in labour, in order to deliver and bring forth the foetus with little effort, she should drink a medicine of the bark of the *quauhalahuac* tree and the herb *cihuapahtli* crushed in water, the small stone *eztetl*, and the tail of the small animal named *tlaquatzin* [opossum]. She should carry the herb *tlanextia* in her hand. Also the hairs and bone of an ape, the wing of an eagle, the tree *quetzalauexotl*, the skin of a stag, the bile of a rooster, the bile of a hare and sun-dried onions are to be burned together; to which are to be added salt, the fruit called *nochtli* in our language, and *octli*. The above-mentioned things are to be heated, and she is to be anointed with the juice. She is to eat the cooked flesh of a fox, and an emerald and a very green pearl are to be bound to her shoulders. She may also drink a mixture of the excrement of a kite and goose and the tail of the *tlaquatzin* crushed in sweet native wine. Also the stalks of *xaltomatl*, the tail of the *tlaquatzin* and the leaves of *ciuapahtli* are to be crushed, with the liquor of which the vulva is to be washed. Also grind the tail of the animal *tlaquatzin* in water and the herb *ciuapahtli*, with which juice, infused with a clyster, you wash or purge the abdomen.[21]

While the linguistic translation from Nahuatl to Latin made the Nahua physician's knowledge accessible to European readers, the fact that vital native cultural content—such as native plants and animals—was not assimilated into their terms would have frustrated those readers seeking to execute the recipes prescribed by the text. Perhaps for this reason, upon arrival in Europe, the *Codex de la Cruz–Badiano* seems to have been approached as an exotic curiosity that required an inaccessible cypher to decode, rather than as a learned treatise and a precious, rare source of knowledge about American nature and Indigenous medicine.

In subsequent years, however, Indigenous and European practitioners devised techniques to circumvent this obstacle. Twenty-five years later, another great illustrated manuscript based on Indigenous knowledge was produced in the same Franciscan *colegio* in Tlatelolco: the magisterial *Historia general de las cosas de la Nueva España* (*General History of the Things of New Spain*), finished around 1577 in the midst of a terrible plague that devastated central Mexico and decimated the Indigenous population. Strikingly, the manuscript is primarily referred to as the *Florentine Codex*, after the Italian city where it has been housed since it came into the possession of the Medici family in the 1580s, rather than for any of the attributes of its production.[22] It provides a detailed account of Aztec religion, culture, life, and history and concludes with a long account of the conquest from an Indigenous perspective. Organized into twelve parts—called "books" within the work—that are bound into three volumes and occupy 1,200 pages of handwritten text in Nahuatl and Spanish, it includes approximately 2,500 drawings in ink and color. Each page is arranged in two columns, one for the Nahuatl text and the other for the Spanish, with the drawings appearing in the Spanish column. The Nahuatl and Spanish texts are not direct translations of each other; rather, they are related but independent—and often different—statements on a single topic. One column presents the postconquest Nahua view on a subject, the other the Spanish interpretation of that view. The images can be considered a third text, as they are often not direct illustrations of the material in either of the textual columns but rather pictorial statements providing yet another take on a subject. Sometimes the images illustrate the text; at other times they complement or augment it.[23]

The manuscript represents the culmination of thirty years of labor on the part of the Franciscan friar Bernardino de Sahagún, who directed the project, and of numerous Indigenous interviewers, informants, artists, and scribes. The text acknowledges their participation—though it does not make clear whether they operated voluntarily or under compulsion—naming some of the Nahua "grammarians" and scribes who aided Sahagún. Modern scholars have suggested that as many as twenty-two artists worked on the images in the final document, while seven scribes produced the final handwritten text.[24] Thus, while Sahagún is routinely described as the "author" of the *Florentine Codex*, it was truly the result of a great collaboration that involved many voices and hands. Scholars regard the resulting work as one of the most important and significant sources from sixteenth-century Mexico, a rare bilingual and bicultural treasure that provides unique insights into the postconquest world from both Indigenous and European perspectives.

Indeed, the *Florentine Codex* supplies the most detailed account we have of Nahua culture based on Indigenous sources. Even so, its Nahua elements should not be fetishized as autochthonous. Although the manuscript presents both Spanish and Indigenous perspectives, its Indigenous components do not present a pre-Hispanic vision but rather an early colonial one. The Nahua informants who reported on their past and society to a missionary several decades after the conquest spoke as Christian converts who were keenly aware of European attitudes toward Indigenous culture, attitudes they had absorbed in framing their accounts. As members of a new society, speakers of a new language, and converts to a new religion, they did not live in the world their grandparents inhabited, and they would not have interpreted their history and culture exactly as their grandparents did. The codex remains a precious and rare source that vividly captures, in words and images, postconquest Indigenous culture at a time when it was rapidly changing due to conquest, conversion, and disease. But it also reflects a form of collaboration—one rife with fraught power dynamics, to put it mildly—between European and native technologies, knowledges, and practices.

Nevertheless, the *Florentine Codex* does provide evidence of distinct elements specific to Nahua culture; in particular, a wealth of information on Nahua approaches to the natural world. Plants, animals, minerals, and the landscape are constantly mentioned as central components of spiritual beliefs and practices—a subject of keen interest to Sahagún. The sun and the moon had great religious significance and were connected to the gods; rainbows figured in divinatory practices; caves were infused with historical and mythical meanings. Many gods in the Aztec pantheon were related to animals and the elements, most notably Huitzilopochtli ("Hummingbird of the South" or "Hummingbird of the Left") and Quetzalcoatl ("Feathered Serpent"). Both the gods and the men who impersonated or honored them wore feathers, as shown, for instance, in the depiction of the Huitzilopochtli impersonator during the god's celebration (see fig. 3.2).[25] Impressive animal-inspired uniforms, important luxury objects and signs of prestige in a military empire, identified their wearers as members of the two highest ranks in the army, the jaguar warriors and eagle warriors. Flowers, noted as sources of medicines and pigments, feature most prominently in connection with the deities and the divinatory calendar. They adorned sacred spaces and images, played a significant role in offerings, and were important elements in the attire worn during ritual celebrations, as shown in the illustration.[26] Sahagún and his collaborators mention the ceremonial decoration of human and stone bodies with enormous numbers of flower garlands, beautifully crafted and powerfully perfumed.[27] The natural world was thus infused with religious meanings.

Book eleven, on "earthly things," is the longest and most heavily illustrated of the codex's twelve books. It consists of thirteen chapters on animals, plants, and minerals, "the different animals, the birds, the fishes; and the trees and the herbs; the metals resting in the earth—tin, lead, and still others; and the different stones."[28] It notes many medicinal uses of different plants and animals, recommending, for example, the consumption of opossum tail or an infusion made with it to counteract constipation or provoke labor.[29] Chapter seven, on medicinal herbs, lists and illustrates a whopping 150 different plants; intriguingly, the first 30 are described only in Nahuatl, without a Spanish text (see fig. 3.3). Comprising 253 folios of text that include 965 paintings, book eleven proves an incomparably rich source on early colonial Indigenous ideas about nature—as they were conveyed to, and mediated by, a Franciscan priest.

Created in the Franciscan Colegio de la Santa Cruz de Tlatelolco by Indigenous makers with and for Spanish patrons, the *Codex de la Cruz–Badiano* and the *Florentine Codex* show how quickly and deftly Nahua elites and artisans mastered the tools, genres, and media of European culture and then put them to their own use. In these sophisticated and beautiful manuscript books, two great treasures of the Mexican sixteenth century, Indigenous authors set down, in words and images, their own visions of American nature, knowledge, and culture. But in so doing, they also exposed the ineluctable interconnection between communities of practitioners that was the hallmark of early modern knowledge production, and the tight, taut, and tense connections that wove practitioners from a variety of backgrounds—experiencing highly distinct versions of the same moment—together.

Imperial Science, Local Visions

In the same years that Bernardino de Sahagún and his Indigenous collaborators at Tlatelolco worked on the *Florentine Codex*, another ambitious knowledge-making endeavor brought together Spaniards and Indigenous people in Mexico. This one, however, involved not missionaries but the imperial administration. In 1577, the Spanish Council of the Indies distributed copies of a printed document consisting of fifty questions to town officials throughout New Spain. The goal of the questionnaire was to compile *relaciones geográficas* (geographical accounts), containing information helpful to the imperial administration: a town's name, location, and history, including rulers and wars; its current population, in particular the characteristics of the Indigenous natives; and information regarding infrastructure, geography, landscape, flora, fauna, and minerals.

Many of the questions addressed what we would now call natural resources; for example:

22. Which wild trees and their fruits are commonly found in the district? What are the uses of them and their woods, and to what good are they or could they be put?
24. What are the grains and seed plants, and other garden plants and vegetables that are or have been used as sustenance for the natives?
25. Which were brought from Spain? Does the land yield wheat, barley, wine, or olive oil, and in what quantities? Is there silk or cochineal in the region, and in what quantities?
26. What are the herbs or aromatic plants that the natives use for healing? What are their medicinal or poisonous properties?[30]

Three other questions asked for visual responses in the form of maps showing the layout of towns, ports, and coastlines. This questionnaire was thus an essential instrument for generating, recording, and amassing native knowledge.

Through the questionnaire, the Council of the Indies sought to turn individual imperial administrators into a network of informants. The document's author—the *cosmógrafo-cronista mayor* (main cosmographer-chronicler) Juan López de Velasco—hoped to compile local experiences and knowledge from multiple sites in the Americas and to use that information to create a detailed chronicle or account of the Spanish territories in the Americas, as well as an atlas charting those lands.[31]

Over the following years, many local administrators in New Spain responded to the questionnaire. Ninety-eight written *relaciones* exist, penned by Spanish officials, along with sixty-nine maps. Many imperial officials, lacking the skills to produce a map, delegated the task to a collaborator. One modern scholar has attributed forty-five of the maps (almost two-thirds of the total) to Indigenous creators, and the remaining twenty-four to nonindigenous ones.[32] The maps display enormous variety in terms of their contents and styles, but what they all have in common—whether they were the work of a European or an Indigenous maker—is that they do not adhere to the conventions of specialized sixteenth-century cartography, as López de Velasco would have wanted and needed. Carefully painted in towns across central Mexico and shipped back to Spain, the maps proved useless to the cosmographer's goals of creating a standardized, scientific atlas of the Spanish Americas. Rather, they present extremely local, community-focused perspectives and agendas, imagining their localities as wholes unto themselves rather than small pieces of a larger puzzle.[33] This vision proved impossible to slot into an imperial grid, and López de Velasco failed to produce the desired atlas.

On the other hand, the maps created by Indigenous artists manifest the continuity of native traditions as well as the radical transformations introduced by Europeans, both in what they show and how they depict it. The documents include many elements coming from Indigenous artistic practice, such as the inclusion of pictographs to provide place-names, the use of footprints to indicate a road, and the depiction of rivers and other bodies of water through the use of wavy lines and whirlwinds.[34] They combine these visual aspects with other, new ones, among them the use of European paper and alphabetic writing; the three-dimensional perspectival representation of buildings; and a naturalistic style used for landscape elements such as hills, mountains, and some of the vegetation. That said, the act of map making does not constitute the adoption of a European custom imposed by the Spanish administration but, instead, the continuation of a native practice transformed in the viceregal context. A long-standing practice of using maps characterized imperial administration in Mesoamerica, notably in the Aztec Empire. And Indigenous people continued creating and using maps in the viceregal period, for example, commonly presenting them to authorities as evidence of their rights to ancestral lands.

These maps did not always adhere to Spanish imperial logic, even as they bear traces of the colonial reordering of the landscape and environment. The map of Santiago Atitlán (now in the Sololá department of Guatemala), for example, sketches the town quite minimally through seven buildings and a church around a central plaza and focuses instead on the beautiful and imposing natural setting (see fig. 3.4). The town is perched on the shores of the vast Lake Atitlán, high up in the mountains, surrounded by three imposing volcanoes (known today as San Pedro, Tolimán, and Atitlán). The lake's blue, undulating waters cover the vast majority of the sheet of paper. The volcanos, densely forested, rise majestically to reach the clouds in the sky. The human presence is minuscule in comparison to the natural landscape: small rowboats carrying tiny humans dot the lake's surface, while a few buildings appear around the lake. This breathtakingly beautiful natural setting has been a site of conquest. The small town was recently built as a result of the Spanish policy of *congregación*, the mandatory resettling of Indigenous communities. Its main buildings, labeled *monasterio* and *cabildo* (local council), attest to the imposition of a new religion and government. And the town's very name, Santiago Atitlán, is based on a double imperial presence as it is not provided in the local Indigenous language, Tz'utujil Maya, but rather in both Spanish and Nahuatl: Santiago after Saint James, the patron saint of Spain, and Atitlán, the Nahuatl translation of the Tz'utujil name Chiya', meaning "by the water."[35]

Although they did not meet the imperial cosmographer's expectations or needs, the maps of the relaciones geográficas are precious historical documents that present in-depth views of Spanish America as it was seen and experienced by those who lived there in the 1570s. They show the transformation of Indigenous communities through resettlement into Spanish colonial cities, neatly organized according to the mandated *traza* (grid plan), with straight, intersecting streets surrounding a central town square that housed the local government and the main church. Many other churches can be seen, arranging the city into neighborhoods.[36] The maps often indicate relationships with neighboring communities, through such means as the rendering of the roads that connected towns. Natural landmarks like rivers, lakes, mountains, and vegetation feature prominently. Crucially, the maps present rare surviving examples of local visions of the Spanish American natural and human landscape at a time of rapid transformation. But in the uneasy relationship between their motivations and production, they reveal how these visions were in essence perspectival hybrids born of exchanges between European and native practices, ideologies, and instruments.

New World Knowledge and the Republic of Letters

Indigenous knowledge of the New World continued to play a major role in European natural knowledge over subsequent decades, even as traditions of knowledge were threatened within communities devastated by enslavement, illness, and violence. And the native knowledge could be directed toward multiple ends as it coursed through learned conduits.

A third major effort to investigate Spanish American nature took place in the 1570s in Mexico, one that similarly sought to capitalize on native knowledge. Much as with the *Florentine Codex* and the relaciones geográficas, its goal was to collect information about American nature and transmit it to Europe, though in this case it was stimulated more explicitly by imperial demands.

In 1570, the king of Spain, Philip II, named physician Francisco Hernández *protomédico de las Indias* (general physician of the Indies) and instructed him to travel to New Spain and Peru to investigate the natural history of those regions. This order launched the first-ever state-sponsored European scientific expedition. Hernández, at the time in his mid-fifties, brought together the worlds of medicine, humanistic learning, and Christian scholarship. A prolific author, he had translated Pliny's lengthy *Natural History* into Spanish, with his own annotations; he had also written a long missionary poem on Christian doctrine and commentaries on Aristotle and the classical medical author Galen, and later compiled a historical account of Mexico. For Hernández, the voyage to Mexico provided an opportunity to write a contemporary natural history that would do for the New World what Pliny had done for the Old (much like Gonzalo Fernández de Oviedo had hoped to do).[37]

Hernández arrived in Mexico in 1571 and spent the following six years traveling and researching, before returning to Spain in 1577. He visited markets, botanical gardens, and hospitals to gather information from local experts, the majority of them Indigenous. He spent time in Guaxtepec, where he investigated local plants and experimented with patients in the hospital (see fig. 3.5). And he wrote at breakneck speed, producing many volumes of handwritten Latin text in which he recorded information obtained from Indigenous informants as well as his own observations and experiments.

Hernández also hired Indigenous artists to paint American plants, returning to Spain with hundreds of depictions. He knew well the great importance that such images would have as authoritative evidence coming directly from the New World, as testimony to his own eyewitnessing, and as crucial components of a publication that would allow European audiences to examine American nature with their own eyes. While all of these qualities were true for books on the New World in general, this was particularly the case for scientific publications and even more so for botanical works.[38] It is also highly likely that Hernández's interest in visual materials originated not exclusively from European agendas but also out of an engagement with Indigenous perspectives. Given the circles in which Hernández moved in Mexico and his humanist interest in original documents and scholarly methods, he would have learned of the great significance that drawings had as a source of knowledge for Mexican natives—his stay in Mexico, after all, coincided with the work on the *Florentine Codex* and barely precedes the relaciones geográficas. Any scholar interested in mining Indigenous knowledge at the time would have become aware of the importance of pictorial sources.

Hernández's goal was to compose a thorough, definitive treatise on the natural history of New Spain, drawing on Indigenous knowledge but based on his own interpretation of it. Following humanist philological methods, he believed that Nahuatl-language names for plants and animals held important information about their qualities and uses. He methodically recorded native names and planned to use them as the organizing structure for his publication—an approach strikingly different from the common European one of replacing Indigenous names as a symbolic way of taking possession.

Upon his return to Spain, Hernández set to work on his ambitious book on American natural history, which he never managed to complete. Nonetheless, his work seeded a number of other ventures that sought to absorb Indigenous knowledge into expansive natural histories. Indeed, much to Hernández's dismay and frustration, King Philip II in 1580 instructed an Italian physician, Nardo Antonio Recchi, to wrangle Hernández's outsized project into shape. Recchi combed through the hundreds of pages and created a much shorter manuscript, 241 folios in length, which he described as an orderly arrangement of the materials (*in ordinem digesta*).[39] Recchi not only omitted much material but also reorganized what he did include, rejecting Hernández's idea of ordering the work according to Nahuatl names and ideas and choosing a framework based on European categories. His summary rearrangement proved extremely influential, especially after Hernández's original manuscripts perished in a fire at El Escorial palace in 1671.[40]

Despite the powerful European inflection given to this compendium of New World knowledge, Hernández's research was first published not in Europe but in Mexico City, a printing center since 1539, by Francisco Ximénez.[41] Ximénez's text, however, did not return to Hernández's manuscripts; instead he translated Recchi's Latin text into Spanish and added his own commentaries, as *Quatro libros: De la naturaleza, y virtudes de las plantas, y animales que estan recevidos en el uso de medicina en la Nueva España* (*Four Books: On the Nature and Virtues of the Plants and Animals That Are Accepted in Use in Medicine in New Spain*; 1615) (see fig. 3.6). The book consists of 478 short entries, each describing the physical aspect and medicinal uses of a plant. It prioritized a practical, local focus: the title page promises to explain the method of preparing and administering Mexican remedies, advertising the book as "very useful for all kinds of people who live in farms and towns where there are no physicians or pharmacies." Indeed, this appears to have reflected Ximénez's own interests; a lay brother associated with the Dominican convent in Mexico City, he worked at some point at Tlatelolco's Hospital de la Santa Cruz, where he was in charge of the pharmacy. His goal was not to provide a comprehensive, learned account of all of Mexican nature but rather to allow anyone—that is, anyone who could read Spanish—to identify and use local medicinal plants. And the book appears to have been used exactly in this manner. The copy held in the Huntington Library's collection is annotated with marginal notes in Spanish that would have enabled the reader to quickly find the appropriate remedies for various maladies.[42] But while the *Quatro libros* might have proved useful in Mexico, where the plants could be found from local sellers and identified through their Nahuatl names, European readers had a different experience. Johannes de Laet, a geographer who directed the Dutch West India Company and authored an important work on the Americas, complained in a letter that, "because the author added no illustrations, his book is practically useless to us."[43]

As de Laet's lament suggests, the volume did circulate broadly. But Hernández's inquiries lay at the foundation of other European studies of New World *naturalia* as well. The second publication to result from Recchi's version of Hernández, which appeared in 1651, was both more ambitious and more lavish. It also came with an illustrious pedigree, as it was the work of Europe's first scientific society—the Italian Academy of the Lincei. Soon after Prince Federico

Cesi established the academy in 1603, he purchased Recchi's personal copy of his manuscript. Cesi was determined to publish the magisterial natural history of the New World envisioned by Hernández, an impressive feat that would establish the society's importance within European learned circles. The Lincei recognized the uniqueness of Hernández's contribution: an extensive and detailed study prepared by a medical expert with deep humanistic training, based on many years of firsthand access to native informants and American plants in situ. This was completely different from an attempt to assess the dried or powdered ingredients sold by pharmacists in Europe. It was also different from an examination of the few plant specimens growing in European botanical gardens, as scholars worried that their physical aspect and properties might diverge from those expressed in native soil.

The Lincei commenced work on this publication in the 1610s and progressed slowly, given the amount of material and the desire to include woodcut images of the plants in order to make the work definitive and useful. Cutting hundreds of woodcuts was laborious, time-consuming, and costly. Even worse, Recchi's manuscript contained no images, making it necessary to consult the original drawings that Hernández had brought from Mexico, which remained at the royal library in Philip II's imposing monastery at El Escorial. In 1626, Cassiano dal Pozzo, a member of the Lincei, visited Spain as part of the retinue traveling there with Cardinal Francesco Barberini. Following Cesi's instructions, he examined Hernández's illustrations at El Escorial. Dal Pozzo reported seeing "sixteen bound volumes of illustrations and commentary" on the plants, birds, quadrupeds, reptiles, insects, aquatic creatures, and minerals of New Spain. He praised the works, noting that "the beauty and exactness of the colours of all the illustrations . . . cannot be believed. The strangest images of birds were seen."[44]

During that visit to Madrid, Cardinal Barberini obtained the *Codex de la Cruz–Badiano*, which he took back to Rome. The Lincei studied this Mexican manuscript closely and tried—unsuccessfully—to check it against the Hernández materials. It commissioned a copy of the codex, created by a scribe who transcribed the Latin text and an unnamed European (perhaps Italian) artist who copied the illustrations, grappling with Nahua pictorial conventions while attempting to achieve maximum fidelity to the original drawings (see fig. 3.7).

In 1651, four decades after beginning work on the Hernández materials, the Lincei finally published the *Rerum medicarum Novae Hispaniae thesaurus* (*Treasure of the Medical Things of New Spain*). The book opens with a beautifully engraved title page depicting an architectural structure, with figures at either side (see fig. 3.8). At the top, the coat of arms of the Spanish monarchy and its motto, *Et plus ultra* ("And farther beyond"), indicate the project's origin. The two allegorical depictions of American Indians, minimally clothed in feather skirts and capes, hold a globe and a cornucopia from which American fruits and flowers pour out. Below, four additional figures, similarly clad, grasp and are surrounded by additional plants. The one at far right wears and holds gold jewelry, underscoring the continent's mineral wealth. At far left, a small feline peeks from behind the legs of another figure, gazing attentively beyond the page: it is a lynx, acting as a playful signature for the society that published the book. At the center of the page, a banner held high by two putti unfurls to reveal the book's lengthy title, which invokes the work's entire lineage—Hernández, Recchi, King Philip II, and the Lincei—and dedicates the publication to the Spanish monarch Philip III, whose ambassador to Rome had helped fund the project. Below it, a map shows central Mexico, naming many of its cities and regions.

The publication lives up to its designation as a *thesaurus*, or treasure, with almost one thousand pages of text and hundreds of woodcut illustrations of Mexican plants and animals (see fig. 3.9). All vegetal entries begin with the Nahuatl name for the plant, amounting to the richest botanical glossary of non-European names known at the time. The entries, most of them quite brief, describe the plant's physical makeup, the locations where it could be found (if known), and its Indigenous uses. The Lincei's major publication, the "Mexican treasure" became the most prominent and consulted source on Mexican plants and animals until the nineteenth century. The fascinating story of the circulation and publication of Hernández's materials underscores the continued and complicated relationship between Indigenous and European knowledge and perspectives, between travelers to the Americas and "distance learners" back in Europe.[45] And it also demonstrates the penetration of Indigenous learning within the most elite intellectual circles of early modern Europe—filtered and perhaps distorted, to be sure, but inarguably present, a source of arcane and valuable knowledge that powerfully reinforced the authority of those who mediated it.

Early modern knowledge of the New World was, this chapter suggests, polyphonic and multivocal, even as it was powerfully shaped by incentives, practices, instruments, and techniques originating among Europeans. Colonial and imperial naturalists and administrators relied extensively on local knowledge and expertise to cultivate their own. Grasping American nature was a collaborative enterprise—performed within a highly unequal power dynamic, but one nonetheless reliant on sequences of exchange, adaptation, and appropriation that demanded sustained participation by marginalized people. Fully understanding the expansion and reorganization of natural knowledge in this period requires the recognition and restoration of those traditions and practices both distinct to and shared by European and other practitioners elsewhere, whether Nahua illustrators, Batavian naturalists, or enslaved African midwives. For together these groups catalyzed a new global intellectual culture.[46]

NOTES

My heartfelt thanks to Nick Popper for his heroic help and customary good humor, both much appreciated.

1. Léry 1990, lx–lxi.

2. Quoted in Swan 2015, 623.

3. Exceptions include Grafton with Shelford and Siraisi 1992; Barrera-Osorio 2006; Bleichmar 2017.

4. This scholarship is too vast to summarize in a footnote. Pathbreaking and influential contributions include N. Thomas 1991 and Subrahmanyam 1997. More recent examples include Gruzinski 2010, 2015; Bleichmar and Martin 2016; and Subrahmanyam 2017.

5. Examples include Parrish 2006; Schiebinger and Swan 2005; Schiebinger 2007; Gómez 2017; Safier 2008; Norton 2008; Bauer and Norton 2017; Delbourgo and Dew 2008; and Bleichmar 2012. This work is connected to developments in scholarship in other regions; see, for instance, Cook 2007; Raj 2007; and Breen 2013.

6. The classic studies are Crosby 1972 and 1986.

7. My account of Oviedo is based on Carrillo Castillo 2003, 2004; and Myers 2007. On the use of images in particular, see Carrillo Castillo 2004, 2008; and Myers 1993.

8. On *autopsia* and the New World, see Adorno 1992; Mason 1990, 179; and Pagden 1991, 91. On Peter Martyr d'Anghiera and his treatment of nature, see Gerbi 1985, 50–75.

9. Fernández de Oviedo (1526) 2010, 30 (ch. 9) and 11 (dedication); my translation.

10. On the education of the Nahua elite at Tlatelolco, see Duarte 2008 and SilverMoon 2007; on the college's library, see Mathes 1985. On Nahua intellectuals more generally, see McDonough 2014. On science and conversion, see Pardo Tomás 2013; more broadly, on postconquest natural history, see Pardo Tomás 2016.

11. For a modern facsimile and helpful essays, see Cruz and Badiano [1964] 1991; see also Emmart 1940. A transcription of the entire Latin text (from the Lincei copy) and its English translation are available in Clayton, Guerrini, and Ávila Blomberg 2009, 60–226. Scholarly analyses include Afanador Llach 2011; Gimmel 2008a, 2008b; and Pardo Tomás 2013.

12. Clayton, Guerrini, and Ávila Blomberg 2009, 60, 226.

13. Rey Bueno 2004, 256–57; Viesca Treviño 1995, 491–93; Zetina et al. 2011, 223–24.

14. For a comprehensive description and analysis of the works' manufacture, see Zetina et al. 2011.

15. It is not known whether Badiano penned the text himself or it was the work of a scribe. Stols (1964, 234–35), suggests that the handwriting is very close to the model provided by a popular calligraphy manual of the time.

16. Ávila Blomberg 2009, 49.

17. Clayton, Guerrini, and Ávila Blomberg 2009, 158.

18. Several modern scholars have to some extent shared this dismissive attitude; see Emmart 1940, 46; and Garibay 1964, 6.

19. Ávila Blomberg 2009, 96 and indirectly on 76 and 199; see also Emmart 1940, 22.

20. Ávila Blomberg 2009, 80, citing ch. 2, fol. 11r.

21. Clayton, Guerrini, and Ávila Blomberg 2009, 210, citing ch. 11, fol. 57v.

22. On its early Florentine reception, see Markey 2011.

23. The standard English edition is Sahagún 1950–82. From the extensive scholarship,

my account draws particularly on Magaloni Kerpel 2014; and Wolf and Connors with Waldman 2011. Connors gives a total of 2,686 images (2011, xi), while Magaloni Kerpel counts 2,486 (2014, 3). On the images as a third text, see Terraciano 2010. On the relationship between text and image, Magaloni Kerpel 2014, 9–12. On Sahagún, see León Portilla 2002.

24. Garone Gravier 2011, esp. 185–87; Magaloni Kerpel 2014, 2.

25. On Huitzilopochtli, see *Florentine Codex* (Biblioteca Medicean Laurenziana, Ms. Med. Laur. Palat. 118–20), vol. 1, book 1, fol. 1r; Sahagún 1950–82, 2:1; Aguilar-Moreno 2006, 148.

26. See Aguilar-Moreno 2006; on hummingbirds, Montero Sobrevilla 2015; on flowers, Alcántara Rojas 2011.

27. *Florentine Codex* (Biblioteca Medicean Laurenziana, Ms. Med. Laur. Palat. 118–20), vol. 1, book 2, ch. titled "Tlasuchimaco," fols. 58v–61r; Sahagún 1950–82, 3:101–3.

28. Sahagún 1950–82, 12:1.

29. Sahagún 1950–82, 12:12.

30. The 1577 printed questionnaire is translated in Mundy 1996, 227–30; questions quoted here on 229.

31. My account is based on Mundy 1996, esp. 20–27. The written responses are reproduced in Acuña 1981–88. See also Solano 1988; text of 1577 questionnaire on 80–86.

32. Mundy 1996, 30.

33. On the "communicentric view," see Kagan 2000, ch. 5, esp. 118–21; on "micropatriotism," see Lockhart 1992, 388; and 1993, 46.

34. On Mesoamerican pictorial conventions, see, among others, Boone 1998, 2000.

35. On the Spanish and Nahua conquest of Guatemala, see Restall and Asselbergs 2007. On the spatial and landscape implications of *congregación*, see Fernández Christlieb and Urquijo 2006.

36. On religious architecture in sixteenth-century Mexico, see Edgerton 2001.

37. On Hernández, see, especially, López Fadul 2015, ch. 3; Marcaida López 2014, ch. 2; Varey 2000, 200; Varey, Chabrán, and Weiner 2000.

38. Kusukawa 2012; on European botany at the time, see Ogilvie 2006.

39. Chabrán and Varey 2000, 5.

40. There is a third early modern version of Hernández's work, based on manuscripts found two centuries after his travels; see Hernández 1790.

41. On printing in Mexico, see Calvo 2003 and Chocano Mena 1997; on Peru, see Guibovich Pérez 2001.

42. Chabrán and Varey 2000, 9.

43. Quoted in Chabrán and Varey 2000, 16.

44. Guerrini 2009, 23–24.

45. This fascinating and complicated story is detailed in Varey 2000, esp. xvii–xix, 3–25.

46. An earlier version of this essay was published as the first chapter of Bleichmar 2017.

CHAPTER FOUR

Beyond East and West

ALEXANDER BEVILACQUA

What does it mean to write about "East" and "West" in the first quarter of the twenty-first century? Over the past two decades, scholars across the humanities have brought into view manifold interactions between late medieval and early modern European, Asian, and African peoples, nature, and culture. Yet a gap persists between the lively studies produced in cultural and intellectual history, art history, the history of science, literary studies, food studies, historical musicology, and elsewhere, and the creaky macroscopic terms that this area of inquiry has inherited. "East" and "West" are problematic concepts not only because, as many have established, they reify difference in ideologically charged ways. As scholars have also argued, it is artificial to consider the so-called "Old World" after 1500 apart from the Americas and the Atlantic economy, whose ramifications spanned the globe. Indeed, as major elements of research into different geographical and cultural areas and their interactions become more integrated, the study of early modern Europe increasingly forms just one part of a broader scholarly conversation about the early modern world.

As a result of these developments, the master narrative of globalization has become the most popular alternative to "East and West." Yet, as some have suggested, the concept of globalization deserves to be questioned and qualified by cultural and intellectual historians rather than treated as a template for the cultural transformations of the early modern world. In this essay, I reflect on the disjunction between this area of scholarship, which has enjoyed a renaissance, and the unsatisfactory concepts—both old and new—that seem to structure it. Limiting myself to cultural and intellectual history, I discuss three domains of research to identify shared trends and opportunities in the contemporary study of early modern Afro-Eurasia: microhistory and the history of go-betweens; the decorative arts and material culture; and intellectual history.

The study of interactions between Africa, Asia, and Europe will always be part of the genealogy of what now goes under the rubric of global. Thanks to Fernand Braudel and many others, the Mediterranean basin in particular has long been a site for thinking about cross-cultural interactions. If there remains a purpose today in considering the scholarship on early modern Europe's relations with Africa and Asia outside the category of global—and it is an open question whether there does—then it is to offer pathways to modifying and advancing the broader emergent historiography of global interactions. The pages that follow propose that the best current work in cultural and intellectual history has done so. This scholarship disrupts the master narrative of commercially driven "globalization" by acknowledging the central role of religion in cultural interactions, by challenging classic historiographic accounts of the European Renaissance as a profusion of global goods, and even by decentering Europe further than the globalization paradigm can do.

"East" and "West" are old and powerful historiographical categories. Their meaning has shifted many times since Herodotus, in the opening pages of his *Histories*, delineated an ancient conflict between Greeks and barbarians, or between "Europe" and "Asia."[1] Yet they remain common categories of analysis even today. In the 1990s, scholars such as Bernard Lewis and Samuel Huntington promoted the notion of an East-West clash, which became an even more widespread political narrative after the terrorist attacks of September 11, 2001. In the first two decades of the twenty-first century, much scholarly ink was spilled to undermine this thesis by revealing medieval and early modern connections and interactions that crossed supposed civilizational fault lines.

Nor did this set of engagements emerge ex nihilo. For an earlier generation, the study of "East and West" was ineluctably shaped by the controversies in the wake of Edward Said's *Orientalism*, published in 1978. Much has been written about these debates, which I will not revisit here.[2] Said's concepts, elaborated largely in reference to nineteenth-century British and French empires, aimed to capture the intersections of colonial power dynamics with cultural and intellectual production. Early modernists overall agree that Said's argument about the meshing of knowledge and colonial power does not apply readily to the period before modern European colonialism in the Middle East.[3] They continue to differ, however, over the relevance of Said's methodological insights to the early modern era, though the disagreement these days is at a simmer rather than a full boil. If some scholars, like Natalie Rothman, still seek to revise and extend Said's arguments, others, like Noel Malcolm, continue to reject their relevance to the early modern world.[4] Yet others, like Sanjay Subrahmanyam, aim to consider the "concrete and institutional conditions" of knowledge production, but understand their own work as fundamentally "post-Saïdian."[5] Indeed, the phrase "post-Saïdian"—intended in the rich, not merely chronological sense—seems an apt description of the state of scholarship as a whole: whether they see themselves as working in the spirit of Said, against it, or despite it, scholars of our time combine a sensitivity to the workings of power in intellectual production with an avoidance of reductive or untenably broad interpretations.

Altering the premises of discussions of "East and West" is not merely a political endeavor; it has also involved fundamental methodological rethinking. The generation that included (but was not limited to) Said and his immediate followers made much hay of the insight that, whether it was possible or not to establish what representations revealed about their referents, the same representations could easily be read as statements about those who produced them. This realization underpinned a large body of work that analyzed the semiotics of culture. For instance, an art historian might read a French depiction of a Chinese man for what it said about French self-conceptions.[6] The resulting output was part of a broader effort inspired by anthropology and structural linguistics to interpret "culture as a text."[7]

The present-day effort to think about the diffusion, circulation, transit, and transformation of culture reflects a shift away from the assumptions of this earlier scholarship. Most strikingly, those striving to see how ideas, objects, and practices moved from one part of Afro-Eurasia to another have exposed the limitations of the concept of representations and sometimes explicitly repudiated it.[8] The semiotic approach tended to direct attention away from contact points between languages, religions, and regions; it rendered linguistic, cultural, and national groups as solipsistic, forever dramatizing themselves rather than responding meaningfully to external empirical realities. The recent reorientation is best thought of as accretive: scholars today would agree that representations of the foreign—such as the "Turkish"-themed "Water-Fight" (naumachia) performed on the Thames in 1610 for the Jacobean court—referred to domestic affairs in one respect or several.[9] They would argue, however, that such representations also expressed a view of what constituted Ottoman culture, and that these depictions drew on, and disseminated, information about the Ottomans, however mediated or misunderstood.

As a result of this reorientation, the most frequent keywords of the new studies have been "connection" and "translation." As early as 1997, in an essay that proved to be hugely influential, Sanjay Subrahmanyam enjoined scholars of the early modern world to study connected histories, suggesting that scholars take "as [their] point of departure the notion of connectedness."[10] As for "translation," it encompasses representation and its politics, but also acknowledges and draws attention to an external referent.[11] Moving beyond the limitations of "representations," which left no room for mediation and contact, scholars have sought to emphasize an early modern world of connections and interactions. The "clash of civilizations" thesis has served this scholarship as a galvanizing provocation. As Francesca Trivellato has observed, studies of individuals moving across the world (to which we might add studies of objects doing the same) have most often been written to show that it was possible to shift between peoples, religions, and languages—in other words, to undermine the "clash of civilizations" thesis.[12] Indeed, the desire to oppose that argument still animates much scholarly work on the supposed East-West divide. Yet there are diminishing returns in repeating well-intentioned pieties about "cultural crossings," or in digging empirical holes in simplistic categories. (Nor does merely avoiding the terms "East" and "West," or replacing the word "civilization" with the word "culture," perform the necessary conceptual work, since concepts can and do still operate in the background of scholarly writing purged of banned words.)[13] Overcoming the oppositional terms of analysis so loosely deployed in popular parlance requires instead a long-term, collective project of rethinking. The best scholarship on the subject of "East and West" has sought to offer new vocabularies for cultural interaction and its macroscopic narratives, from revealing the ways that Africa and Asia went into the making of European culture to developing a phenomenology of interaction across cultural boundaries and investigating how early modern notions of cultural difference were created and sustained.

Scholars seeking to move beyond merely affirming the reality of cross-cultural interactions have tackled two related challenges. The first is to describe the variety of forms cultural interaction took: rejecting generalizations about "cultures," scholars have reconstructed in rich detail the mechanisms by which specific cultural formations were adopted and adapted. The second challenge is to explain the historical constitution of units such as civic, imperial, religious, or linguistic communities, as well as their limits. Cultural boundaries are not natural but are formed and maintained by human agency.[14] To support these lines of inquiry, many scholarly developments have been mobilized: the history of information, of networks, of humanistic knowledge, of the book, and of material culture.[15]

Whereas scholars of "representations" felt liberated from the need to assess how a European representation of a foreign culture rendered its referent, scholars of "connections" need to be able to understand the relationship of representation and referent. This is no straightforward task and often requires considerable technical expertise. To recognize medieval Chinese silks in Italian paintings, one has to be familiar with their patterns.[16] To evaluate whether a visual depiction of an Ottoman figure is distorting or not, one has to locate and identify its sources.[17] To know whether a seventeenth-century translation from Arabic succeeds or fails, it pays to read Arabic and to be familiar with the original text.[18] Inevitably, then, the new scholarship has examined sources from not one but several historical contexts in order to determine the relationship between them. Often this effort achieves remarkable results, as when things that were deemed merely peculiar or distorting turn out to be attempts, however maladroit, at rendering some aspect of the referent practice, person, object, or idea. Even the fantastical creatures in Sir John Mandeville's travels, for example, bear a relation to the gods of South Asia.[19] Clearly, the goal cannot be to award points for accuracy, which would be an anachronistic measure of success. The identification of the referent ideally allows a more serious appraisal of an interaction's motives and functions, of the interpretations it advances, and of a choice to reproduce something with (more or less) care, or else to transform it, whether in content or in contextual or figurative meaning. Thus the emphasis on translation incorporates the insights of the earlier cultural history even as the genealogy it reconstructs affords the object of study more complexity and depth.

The tensions surrounding how best to practice connected history are on view in a novel genre of historical writing: the "global microhistory." The phrase may be something of a misnomer—"global lives" comes closer to describing the main emphasis of the genre, at least in its current form. The lives described in these works are typically not those of Marco Polo or Robert Clive; instead, historians have more often used this method to recover "marginal" figures, such as women, converted Muslims, or Jews.[20] A number of the most prominent examples in this new genre focus on people who crossed boundaries in Afro-Eurasia.[21]

These global microhistories have served a heuristic or pedagogical function in showing "a World in a Grain of Sand," in William Blake's phrase. A figure such as Linda Colley's Elizabeth Marsh, who traveled to North Africa and South Asia, serves her historian by revealing the global reach of a (relatively) ordinary life.[22] The study's power derives precisely from the disjuncture between Marsh's lack of power and influence and her peripatetic progress. At the same time, such demonstrations cannot be repeated infinitely. Historians have therefore sought other ways of giving meaning to global lives, for instance, by scaling up through aggregation of individual biographies of both the eminent and the unsung.[23]

If connected "global" and "Eurasian" microhistories have had a macrohistorical agenda, it has been to overcome the "clash of civilizations" narrative. Yet microhistory can do more than puncture hoary grand narratives; it can, thanks precisely to its intrinsic empirical precision and to its carefully considered choice of focus, offer alternative accounts. "Micro-," after all, refers to the lens of analysis, not to the size or importance of the object of study.[24] Moreover, microhistory chooses its object advisedly: the peculiarity of the specific case should illuminate something broader—this is the valence of Edoardo Grendi's famous formulation "the exceptional 'normal.'"[25] Close investigation can be designed to offer, in Trivellato's words, "a richer basis for further comparison across time and space."[26] Once connection is no longer understood as an alternative to comparison, historians will be able to reject the false opposition between macrohistorical narratives, with their simplifications, and a purely local focus. Microhistory, including connective or global microhistory, can be designed instead to redefine and rewrite broader categories, periods, and narratives.[27]

The following example shows the power of local analysis to redefine the broad historical narratives that organize our understanding of Afro-Eurasia. Recent scholarship has drawn attention to a set of communities at the intersection of Europe and Asia: the Christians of the Levant and Egypt. These people—Maronite Arabs, Armenians, Copts, Melkites (Arabophone Orthodox), and others—have played an unheralded role in the interactions between Western Christians and Muslims. For instance, the *Corpus Toletanum*—the first major collection of Latin translations of Islamic texts, including the Qur'an—composed in twelfth-century Toledo, contained the *Risāla* (Letter) of al-Kindī, an Arabic Christian polemic against Islam. This work provided Western Christians with many of the talking points that would organize anti-Islamic polemic for centuries.[28] In the early modern era, Eastern Christians served as intermediaries between the Levant and both Catholic and Protestant Europe. As individual figures and their networks are recovered, they are revealed as key operatives in confessional and missionary struggles of their era, both in the Levant and in Europe.[29] Studying them draws attention to the ways Europe and the Levant were bound not only by long-standing enmity between Christians and Muslims or by modern commercial ventures, but also by the expansion of post-Reformation European interconfessional polemic to the Eastern Mediterranean. The reverse was also true. As John-Paul Ghobrial has pointed out, global interactions could be prompted by "local phenomena related to confessional change among Christian communities in the Ottoman Empire in the late seventeenth century."[30] The case of the Eastern Christians shows that Christian-Muslim relations were not a simple binary opposition but deeply entangled in the history of Christianity and its Levantine origins. Again and again, the Christians of the Near East, by traditional accounts marginal figures in early modern Europe (or entirely neglected), served as key mediators between Western Europe and the peoples of the Eastern Mediterranean. Some of them played this role again in the Napoleonic period.[31]

The history of the Eastern Christians undermines any facile generalizations about "Islam and the West," or any simplistic equivalence of Arabs and Levantines with Muslims, or Christians with Europeans. It shows the mechanisms by which the relationship between Western Christianity and Islam was mediated: for example, even at the "Year Zero" of the Western study of Islam, knowledge making in Toledo was mediated by al-Kindī's polemic. Out of measure to their small number, Eastern Christians contributed to the constitution of perceived boundaries between East and West, between Christianity and Islam. They also frequently exploited these boundaries professionally. Recovering the Christians of the Levant illuminates the local forces at work in places of mediation, and the sometimes parochial concerns and economies of the "trading zones" that helped to make the modern world.[32]

Another area of research only infrequently incorporated into the study of global cultural interactions is early modern Judaism. Not only was Judaism the most enduring form of religious difference in the predominantly Christian societies of early modern Europe, but Jewish communities dotted and connected medieval Africa and Eurasia.[33] After 1492, the Sephardic diaspora wove networks of Jewish émigrés that linked northern Europe, Italy, North Africa, and the Ottoman lands, and soon enough the Americas as well.[34] Early modern Jews undermine any simplistic distinction between Europeans and the rest of the world by revealing the alterity at the very heart of Western Christendom.[35] Whether focused on post-Tridentine Rome or on the German lands in the age of the Enlightenment, attention to Jewish history disrupts commonplaces about the Christian hegemony in European culture and its early modern transformations.[36] As local populations, European Jews have too often been studied apart from New World anthropologies or the Western Christian study of Asian languages that historians have recently reconstructed. Yet, as some of the more mobile and networked populations of the early modern world, these communities help to reveal many aspects of the history of early modernity as a whole. In sum, Eastern Christian and Jewish history (as well as the histories of other diasporic minorities, whether Armenians, Parsis, or others) promise to overturn conventional understandings of early modern cultural interactions if only their insights can be assimilated into a new, integrative view.[37]

The history of material culture has offered a different route around the dichotomy of "East and West."[38] Recent studies of eighteenth-century Ottoman architecture have emancipated it from earlier dismissals as merely derivative or Westernizing, showing them to be self-conscious creations of globally minded patrons and builders.[39] But decorative objects that traveled widely and gained status from their provenance have especially appealed to modern scholars as a way of studying Eurasian exchange. These objects have been privileged for several reasons. Some are true global products and seem emblematic of Eurasian connections in the period 1400–1800.[40] Porcelain, for example, developed by Chinese craftsmen, was admired worldwide for its durability and fineness. Ottoman and Western European artisans imitated it with varying degrees of approximation (see fig. 4.1).[41] Porcelain was, besides a technology, a Eurasian art form whose appreciation connected diverse elite publics that shared neither language nor religion. On account of their mobility and popularity, porcelain objects tell a history that transcends the geographic and especially temporal limitations of individual lives.

Decorative objects have also appealed to scholars of our time because they bring the foreign or the global right into the heart of urban and courtly elite culture. If recent studies of travelers and go-betweens such as Levantine and African Christians have stressed "contact zones" and borderlands as the sites of interaction, the history of the decorative arts offers a different lesson: global effects operated at a distance.[42] It suggests that the "contact zone" as a distinct space should be rethought: in an early modern world of networks, mobile agents, and mobile objects, centers and not just so-called peripheries were places of cultural interaction.

A further attraction of material culture lies in many historians and literary scholars' shared desire to escape the limitations of written and visual sources, which they associate with ideological representations of non-European others. By recovering a world of consumers reveling in the same material objects, scholars hope to reveal a reality occluded by period discourse. Whether an escape from the domain of ideas and discourse is possible or desirable seems, however, doubtful. Some of the best research on goods and objects has revealed how objects travel with ideas about their meaning and practices surrounding their use.[43] For example, the high heel came to seventeenth-century Europe from Persia, but, as an analysis of the differing techniques for producing the heel reveals, it was not the Persian object itself but merely the idea of the high heel that made the crossing, including the association of such footwear with horsemanship and manliness.[44]

Luxury goods have served as a central point of departure for understanding the European Renaissance as a global phenomenon. To make the Renaissance seem less parochial and make it speak to our time, scholars of early modern Europe have pushed hard on the conceptual pedal of Eurasian globalization.[45] They have placed commerce, and in particular consumer demand, at the heart of the process of the world coming together. Interpreting the Renaissance mainly as a demand for luxuries, however, makes protagonists of only the most elite actors, those who were able to commission the new goods or have them imported from afar. Likewise, the presentist attempt to link the "global" consumer of the eighteenth century to modern-day Western middle classes that define themselves by their forms of consumption denotes a failure in historical sensibility: in the delight of recognition lies a real risk of overstating the ways the past resembles the present. The scale of consumption and of the world of goods differs incomparably; its meaning likewise was then tied especially to the public display of status, which, while certainly not extinct, is nowadays commingled with the private pursuit of pleasure.[46] Moreover, the profusion of foreign goods was not necessarily a token of modern cosmopolitanism. As Alexander Nagel and Christopher Wood have remarked, geographically distant provenance was often misread as chronological distance. Thus, for instance, in Renaissance Italy, Byzantine icons were often treated as classical antiquities: distance in space was taken for distance in time.[47]

In part, these pitfalls stem from the survival bias of luxury goods, which have endured more readily than less valuable objects. Moreover, this set of narratives has been amplified by museum curators eager to use their collections to tell histories of globalization or, rather, to use this global macronarrative to assign meaning to their objects.[48] Yet noncuratorial scholars still have a lesson to learn from those who have long lived in the connected world of the decorative arts. Going back to ancient Greece, the history of ornament—of mutual inspiration and emulation—is Eurasian in scale.[49] Scholars of early modernity must consider more seriously this long history in order to pinpoint the peculiarity of cultural production in their period. In some respects, the global arts were continuous with what came before. For instance, courtly culture coexisted with new, commercially driven fashions and cultural forms. As Maxine Berg has pointed out, the porcelain produced by European royal manufactories for courtly patrons differed from the Chinese exportware that filled middle-class homes: new forms of courtly display were not necessarily viable products in a market economy, unlike "semi-luxuries" made for middle-class consumers.[50] Even as new commercial networks spanned the globe, courts continued to patronize the creation of prestigious (and emphatically noncommercial) objects and to prize those imported from afar. In this respect, then, porcelain was a child of courtly patronage, not just a product of commercial modernity.

Other cultural forms were instead manifestly new in the early modern era, and have therefore lent themselves to narratives about the origins of modern consumption. New consumer goods, such as sugar, tobacco, and the stimulant social drinks chocolate, coffee, and tea, became central to novel social rituals that inspired the design of a host of related implements, furnishings, and, most important, social spaces, from private parlors to coffeehouses.[51] Crucially, they all became products of the emergent European colonial economy in the Atlantic and elsewhere: if in the seventeenth century Europeans were importing coffee, in the early eighteenth they began to produce it in colonial settings by means of enslaved labor, on the model of sugarcane.[52] Although art museums have fastened on art patronage and consumption as ways to tell a celebratory history of early modern interactions, the study of consumption, then, ought never to be purely an aesthetic history; it is deeply tied into changes in the world economy from 1400 to 1800, many of which came at an enormous human cost.[53]

Yet the prominence of commodities and of consumption has encouraged scholars to emphasize a macronarrative emphasizing the commercial economy as a globally unifying force. In our present moment, such narratives suggest themselves readily. It is not a coincidence that so many current metaphors to describe the transit of culture draw on finance: flows (of capital), circulation (of currency), exchange (of specie).[54] The most significant, wide-ranging concept to have taken hold of the historiography of Eurasia in the past two decades is indeed of financial origin: globalization.[55] Global history, of course, should not be reduced to the history of a process of global integration, but the latter notion represents global history's most powerful macronarrative.[56] The new macronarrative of globalization has a particular appeal for those studying what was once called "East and West," for it seems to overcome that outdated model of evaluating cultural exchanges and a host of normatively loaded descriptions such as "exoticism" and "Westernization."[57] "Globalization" appears to put different parts of the world on equal footing, without forcing nineteenth-century outcomes and normative judgments onto the earlier period as earlier macronarratives such as the "rise of the West," "expansion of Europe," and "modernization" did.[58] It seems to afford, at least in principle, a multicentric history of contact and interaction.

Globalization, however, constitutes a problematic master concept if it implies the expansion of Europe or conceals it in disguise.[59] Indeed, the role of Europe in global integration, far from something to be taken for granted, is debated by experts in economic history.[60] Cultural and intellectual historians need to resist uncritically adopting terms and concepts about which the relevant specialists find no agreement. Moreover, globalization, then as now, has appeared to many to have ravaged and destabilized the world. In such a perspective, narratives about consumption bringing the early modern world together seem naïve at best, and at worst to deliberately ignore the massive violence and the enslaved labor that made possible the production of new goods.[61] As a maimed enslaved person reminds the protagonist in Voltaire's *Candide* (1759), "This is the price that has to be paid so you can eat sugar in Europe."[62] The objects themselves do not necessarily speak eloquently about their origins: for example, Madeline Dobie has shown that the Asian-themed decorations of eighteenth-century European marquetry often concealed the Atlantic origins of the tropical hardwoods of which these pieces of furniture were made.[63]

There is no need to give up on telling the story of cultural globalization. But cultural and intellectual historians should take advantage of their unique ability to reconstruct the past from the ground up.[64] They should pay attention not only to the obvious analogies between culture and capital or between early modern and contemporary processes, but also to global interactions that were unintended consequences of other efforts, to those motivated not by trade but by other ventures, and to those that failed, as well as to the many early modern specificities of which their sources speak.

A third area of research offers some lessons in this respect. In 1997, Subrahmanyam reminded his readers that not only bullion and technology but also "ideas and mental constructs" circulated across boundaries.[65] On the surface, intellectual history might appear to tell a history different from that revealed by material culture: one of wars of religion and ideological cleavages, rather than of global publics bound by kindred consumer choices. In fact, intellectual history benefits from closer connections to the history of material culture. Book history provides a valuable missing link. The study of Asian languages in early modern Europe, for instance, cannot be explained if one does not pay close attention to how Asian manuscripts were acquired abroad and collected, organized, and consulted in European repositories.[66] Significantly, the circulation of books proceeded along the same human networks that transmitted other items—both prestigious gifts such as spices and clocks and intangible things like rumor, gossip, and news. Reconstructing the circulation of the one helps to understand the circulation of the others.[67]

What would a unified Afro-Eurasian intellectual history of 1400 to 1800 look like? Several directions have been essayed. In one, scholars investigate genres of writing that sought to make sense of the world and its peoples: travel writing, geography, costume books, universal history, and others.[68] These genres are not exclusively or primarily European.[69] Further practices of scholarly inquiry, such as antiquarianism and collecting itself, likewise make promising starting points for comparison and even connection.[70] The risk of such comparative studies, however, is that they might overemphasize similarities or overstate the prominence of a particular genre in a given tradition. For instance, the chronologically organized history (as opposed to biography and prosopography) in Islamic historiography has been overrepresented in the West since the seventeenth and eighteenth centuries precisely because it seemed convergent with its European analogue.[71] As historians of the Ottoman, Mughal, and Safavid empires expand their studies of intellectual history, European historians will benefit from a comparison that will qualify any reflexive claims about European exceptionalism.

Similarly, the discipline of philology now enjoys renewed global salience among scholars. In Sheldon Pollock's words, "The problem of how to make sense of texts . . . has been at the heart of [the learned traditions of Eurasia] for centuries."[72] Comparison helps to distinguish between the peculiarities of a tradition and structures common to them all; it has also led to the hypothesis of a common early modern transformation of philology across Eurasia.[73] Textual practices essential to philology, such as collation, excerpting, glossing, canonization, and interpretation have likewise been scrutinized in a deliberately comparative setting so that multiple philological traditions can illuminate one another.[74] Information management, too, has been considered in Eurasian perspective. Ann Blair has shown that many features of early modern and modern reference works were present in scribal contexts, which invalidates any claim that printing inspired large-scale compilation, alphabetical indexing, or consultation reading.[75] Employed in this way, the Eurasian perspective can make the study of a given tradition less parochial and, in the European case, undermine assumptions about the uniqueness of the European path.

The revival of the history of scholarship, and its emphasis on knowledge rather than ideas, has contributed to and inflected the new cultural history of connections. In this respect, it has proceeded along parallel lines to the history of science. This is no coincidence; the new developments in the history of science have inspired historians of humanistic scholarship.[76] For both fields, pointing out the connections between Europe and the wider world has seemed to offer a way of avoiding the reification of European intellectual hegemony—and the developments in natural philosophy traditionally known as the Scientific Revolution, as well as the Enlightenment.

Historians of scholarship have revealed how Christian humanists' obsession with origins, whether Biblical or classical, and with internal debates about theology and liturgy—the very things that appear most local and arcane today—pushed them outward into ever new fields of scholarly inquiry. Over time, these questions made early modern scholars take an increasingly broader view of their remit. The study of Biblical Hebrew led them on the one hand to the study of rabbinic Judaism and on the other to the cultivation of all the Semitic languages.[77] Eventually, Christian scholars also tackled such challenging languages as Persian, Ottoman Turkish, Tibetan, Chinese, Japanese, and many South Asian and Southeast Asian tongues.[78] The past twenty years' research has revealed the extraordinary intellectual reach of Christian erudition.[79] As these wide horizons are recovered, the later achievements of the secular thinkers of the Enlightenment have come to seem less distinctive and more of a piece with the learned traditions that preceded them.[80] The study of connections, then, yields dividends even for European intellectual history and can offer an earnest, rather than opportunistic, way of recovering a globally influenced Enlightenment.

The new intellectual history faces some of the same challenges that other fields do. For example, scholars must acknowledge the significant intellectual achievement of seventeenth-century European scholars' joint study of Arabic, Persian, and Turkish, the three literary languages of the Islamic tradition.[81] Yet they must do so without turning the study of late humanism into a presentist celebration of early modern European interest in other peoples (along the lines of the celebration of early modern consumer culture as cosmopolitan). The motivations for all these studies may have been missionary ambitions or the need to engage in interconfessional polemic, or they may have been more esoteric—the search for an ancient theology.[82] They do not necessarily herald the beginning of modern *Wissenschaft*, a secular point of view on religion, or the emergence of modern disciplines. While some of their results did offer a legacy for nineteenth-century European scholars of Asia to build upon, they were reached within the presuppositions and limits of the intellectual and scholarly horizons of their time. Giving them their due means recognizing the capaciousness and audacity of early modern Christian erudition without assimilating it to modern, secular humanistic study.

Recent intellectual histories conceived as histories of knowledge offer some lessons that can fruitfully be applied to the European reception of Asian cultural and intellectual production. In particular, the new intellectual history stresses the irregular and nonlinear path of new knowledge and new ideas. Some thoughts had to be forged anew by each generation. Bad ideas—misrepresentations of foreign religions, for instance—continued to be repeated even long after they had been debunked by critical scholarship.[83] This textured account of intellectual life, which redirects attention away from radical breaks and toward continuities, afterlives, and iteration, can benefit the study of the multiple forms of understanding of Asian culture that coexisted at any given moment in European history. Instead of identifying a master discourse that exhausted all forms of thought about foreign culture and religion, historians have revealed the plural, often contradictory views that contemporaries could hold.

Historians of science, too, have expanded their Eurasian remit. In part, this has been enabled by an extension of chronologies, so that medieval Arabic astronomy, for instance, can seem part of a long-term reevaluation of astronomy, rather than a "Copernican revolution."[84] Likewise, relations between China and Europe are viewed not so much in terms of specific moments of encounter, as in the work of Joseph Needham, as in terms of the long-term development of

Eurasian science.[85] At the same time, the redefinition of science as knowledge of nature and recent attention to its plural forms—vernacular, tacit, embodied, artisanal, collaborative—have facilitated the transition to this new Eurasian frame of reference. Anti-Eurocentrism is often not merely the incidental outcome but the explicit aim of this new history of science. For instance, Harold Cook has shown the connections between Dutch commerce and overseas expansion and knowledge making.[86] Botanical, medical, astronomical, and geographical knowledge were also intertwined with philology: learned travelers who collected manuscripts abroad also gathered botanical samples and sought explanations in native texts of the medical properties of new goods such as coffee.[87] Recovering their efforts reveals a shared Eurasian structure of knowledge production, in this case defined by Galenic medical assumptions.

Intellectual history, finally, offers an approach for studying how religion changed from 1400 to 1800. This is a key way in which intellectual and cultural history can contribute to global history without merely strengthening a narrative of trade bringing about globalization. Traditionally in European historiography, the history of early modern religion was the history of the Reformations, Protestant and Catholic, and their long-term consequences. But the global effects of religion—the expansion of Christianity—run as a parallel narrative to that of religious reformation, as the two-part title of a recent overview suggests.[88] Recently, scholars have used the lens of cultural history to reexamine the effort to convert souls across the Asian continent.[89] More broadly, scholars of Christianity are rewriting the global history of early modern Christianity in a way that does not privilege the European context.[90] Other comparative and connective directions—for instance, to chart a Eurasian conjuncture of millenarian sentiments—have been investigated as well, both enlightening and decentering the history of European religion.[91]

Religion inspired knowledge production, whether by missionaries or by scholars who stayed at home. It stimulated trade, too; for example, the Russian Orthodox clergy made ecclesiastical garments out of imported Ottoman silks, which were used for the same purpose across Catholic Europe (see fig. 4.2).[92] Such textiles often came to Europe as diplomatic gifts and were then donated to churches, which fashioned them into ecclesiastical garments, but the Russian Orthodox patriarchate also had Ottoman textiles made to order.[93] Taking religion seriously as an engine of European expansion allows us to think of early modern Eurasia or the early modern "globe" without limiting them to mere anticipations of secular, commercial modernity.

What is next for the study of early modern Afro-Eurasia? An ongoing challenge to an integrative view remains geography. The view from Paris or London is often not the most revealing one. Privileging certain places as "centers" and treating others as "peripheries" reveals more about historians' biases than about the cultural geography of the early modern world. Until the conquest of Syria, Egypt, and the Arabian heartland in 1516–17, the Ottoman Empire was a predominantly European empire; Western Christendom and the Ottoman Empire should be considered much more frequently together. Treating Istanbul as a European capital or, indeed, taking a broader, Afro-Eurasian perspective, enables the recognition of connections and similarities that we might otherwise miss. Similarly, borderlands help to call into question misleading categories.[94] Borderlands were inhabited by entire populations of go-betweens, people whose everyday life involved multiple languages and interactions between adherents of different religions. In the Habsburg-Ottoman borderland, a vast area encompassing modern-day Hungary and the Balkans, intermittent warfare was only part of a history of constant relations between a variety of populations.[95] Likewise, the trading posts and factories of European chartered trading companies across Africa and Asia generated new societies that were highly diverse not just in religious but also in ethnic and racial terms. Manila, Batavia, and many other factories across Asia all witnessed significant cultural and racial pluralism and hybridity well before places in Western Europe did.[96]

I conclude with three possible areas for research into Afro-Eurasian history and beyond. First, what form did the common culture of early modern Afro-Eurasia take? Is it best described as a premodern courtly culture, with shared ideas about proper conduct, diet, and leisure activities such as riding, hunting, gardens, and flowers, or as a culture of urban elites, produced by new forms of commerce and sociability? Some comparative forays have been ventured, but there is much more to be said.[97] Similarly, is the peculiarity of cultural production after 1500 its global context, or have we been overstating the impact of global interactions? To what extent did the dynamics of cultural interaction produce global early modern culture? The answers to these questions cannot be reflexive assumptions—the possibility must be taken seriously that change was the outcome of the imposition of one group or another's practices and ideas or even of domestic dynamics already in motion.[98] Moreover, how much emphasis should be laid on distinctive early modern cultural forms that did not anticipate the global modernity to follow, such as astrological beliefs, Galenic medicine, emphasis on providential history or the millennium, and anticommercial prejudice?[99]

Second, from the European end, how should we distinguish and periodize European cultural responses to Africa and Asia from the fifteenth century to the end of the eighteenth? A world of difference separates the dignified turbaned men in Vittore Carpaccio's paintings at the outset of the sixteenth century from the comical, patronizing ones in the Comte d'Artois's *cabinet turc* panels, installed at Versailles in 1781.[100] How to make sense of the changes over time that give representations of the same referent completely different valences? And how, especially, to explain the simultaneous layering of different traditions of knowledge, of representation, and of creative appropriation—not just adversarial and admiring but also informed and imaginary—in a given place and time? A further demanding but crucial topic for European history is the change, in the eighteenth century, from the European sense of being a part of the Eurasian world, however distinguished by the good news of the true religion, to a sharp, secular sense of European exceptionalism and even superiority.[101] This transition has more often been described than explained. Cultural and intellectual historians can do better than make reference to ambient geopolitical change in order to interpret this diffuse and yet definite shift.

Finally, in the emphasis on the supralocal, are connections getting too much attention? What were the limits on connectedness? Intentional human efforts to hamper movement were widespread, systematic, and long-standing. Early modern Eurasia was a place of barriers and borders, sustained and created by human beings who, in different contexts, sought to prevent both imports and exports. From the Index of Prohibited Books to the "economy of secrets" to the deliberate seclusion from most of Europe of Tokugawa Japan, intentional efforts to hinder the movement of culture and ideas proliferated.[102] Nor were all forms of culture endowed with mobility—some things remained local. They, too, belong in the history of early modern Afro-Eurasia, and not just as bizarre exceptions. In addition, more attention could fruitfully be paid to the dynamics of culture in motion—what we might call friction, the process of transformation that something undergoes as it shifts contexts. While we have benefited from many individual studies of such friction, we still lack a generalized vocabulary for characterizing this process.[103]

Reading the period from 1400 to 1800 as an anticipation of the modern world, or even as its beginning, is inevitable. In some ways it is even desirable. At the same time, we should hold fast to the peculiarity of early modern Afro-Eurasian cultural interactions. This does not mean renouncing the power of the early modern period to help explain the age of nineteenth-century empire to which it yielded; the two goals of recognizing its distance and its teleology are not in mutual contradiction. If anything, as Suzanne L. Marchand's study of German Orientalism in the nineteenth century indicates, there is much to be said for reading nineteenth-century culture through an early modern lens.[104] "Making strange" has implications not just for the period 1400–1800 but for our own purported modernity as well. The cultural and intellectual history of early modern Afro-Eurasia is by no means an antiquarian project, but an undertaking our own time urgently needs.

NOTES

This essay was written in January 2017 and revised in June 2019. To my regret, I have been unable to do justice to scholarship published since then. For helpful feedback on earlier versions, I am grateful to Lauren Cannady, Anthony Grafton, Jan Loop, Valeria López Fadul, Helen Pfeifer, and my colleagues at the Harvard Society of Fellows, especially Abhishek Kaicker and Andrew Ollett, as well as the editors of this volume.

1. Strassler 2007, 3–5.

2. Said 1978. Bernard Lewis responded to *Orientalism* first in a review and in the exchange with Said that issued from it, and later in Lewis 1982. For an introduction to the *Orientalism* debates, see Macfie 2001.

3. Grafton 1997d. Among many responses by scholars of the early modern era, see Jardine and Brotton 2000, 60–61; G. MacLean 2005, esp. introduction, 1–28, here 7–8; Cowan 2005, 116; Meserve 2008.

4. E. Rothman forthcoming; Malcolm 2019, 415–17.

5. Subrahmanyam 2017, xi–xiii, here xiii.

6. Among many other works, see Rousseau and Porter 1990; Grosrichard (1979) 1998; Yeazell 2000.

7. Veeser 1989; Gallagher and Greenblatt 2000, introduction (1–19), with slippage between "culture as text" and "cultures as texts." These assumptions were common to cultural history as practiced by a wide variety of interpreters, from Said to Stephen Greenblatt and Simon Schama. See, for example, Schama 1987; Greenblatt 1990.

8. For disavowals, see E. Rothman 2006, 1–39; and Bevilacqua and Pfeifer 2013, 75–76.

9. Chew 1937, ch. 10, 460–61.

10. Subrahmanyam 1997, 760.

11. See, for example, Euben 2006; Burke and Hsia 2007; Flood 2009; and Subrahmanyam 2012. See also the discussion in Moyn and Sartori 2013, 10–13.

12. Trivellato 2011, section VI.

13. "Cultures" can easily be used as a shorthand for civilizational alignments. See also Sewell 2005, ch. 5, esp. 153–54.

14. For example, E. Natalie Rothman (2012) has investigated the population at the interface of the Venetian and Ottoman empires—commercial brokers, religious converts, and diplomatic interpreters—to reveal their role in shaping and sustaining the boundaries that they purported to mediate.

15. For example, Ghobrial (2013, 14) acknowledged his debt to Robert Darnton.

16. Mack 2002, 35–37; Monnas 2008, 70–73. See also Wunder's contribution to this volume.

17. See, for example, Avcioğlu and Flood 2010; and, in particular, Smentek 2010; and Avcioğlu 2010.

18. See, for example, Vrolijk 2010.

19. Mitter 1977.

20. Spence 1988; García-Arenal and Wiegers 2003; Davis 2006; Colley 2007; Valensi 2008; Subrahmanyam 2011.

21. Others look at the Americas; for example, Townsend 2006.

22. Colley 2007.

23. See, for example, Ogborn 2008; Rothschild 2011.

24. Ginzburg 2015, 462. See also Ginzburg (1993) 2012 and Revel 1995.

25. Grendi 1977, 512.

26. Trivellato 2011, section VIII; Ginzburg 2015, 471 ("An individual case may contribute to the reformulation of an infinitely larger question").

27. Examples from the history of trade include Goitein 1967–1993; Trivellato 2012; and Trivellato, Halevi, and Antunes 2014. Examples from intellectual history are discussed below.

28. Burman 2007.

29. Heyberger 1994, 1995, 2010a, 2010b; Hamilton 1994, 2006; Girard 2011; Ghobrial 2014. The collective research project "Stories of Survival: Recovering the Connected Histories of Eastern Christianity in the Early Modern World" promises to expand our understanding. See http://storiesofsurvival.history.ox.ac.uk (accessed June 16, 2019).

30. Ghobrial 2014, 59.

31. Coller 2011.

32. For the concept of the trading zone and its application to early modern Europe, see Galison 1997, 781–844; and P. Long 2015.

33. Goitein 1967–93.

34. Y. Kaplan 2000; García-Arenal and Wiegers 2003.

35. B. Kaplan 2010; Ruderman 2010; Jütte 2013.

36. Caffiero 2011; Feiner 2002; Sorkin 2008.

37. The literature on go-betweens is too great to be cited here, but see references in Ogborn 2013. See also Travers 2015.

38. See, for example, Brook 2008; Findlen 2013; and Gerritsen and Riello 2016. I leave aside the study of visual culture, although its contributions have been no less rich; see, for example, Fromont 2014.

39. Hamadeh 2007; Rüstem 2019. See also Dadlani 2019.

40. Finlay 2010; E. Jenkins 2013.

41. Similar histories can be told of other techniques, such as lacquering, known as "japanning." See Berg 2005, 81–82.

42. Pratt (1992) coined the term "contact zone."

43. See, for example, Norton 2006, 2008.

44. Semmelhack 2009, 80–90, and 2015, 9–33.

45. Freedman 2008; Jardine 1996; see also Brotton 2002.

46. Trentmann 2016, 32–33.

47. Nagel and Wood 2010, 97–107.

48. A prominent example among many is Peck 2013.

49. Boardman 1994.

50. Berg 2005, 80–81, 87–88, 126–27.

51. A precursor of these studies is Mintz 1985. On chocolate and tobacco, see Norton 2008; on coffee, Ellis 2004 and Cowan 2005. On new forms of sociability, see Bayly 2002, 54: "sociability revolutions."

52. Trentmann 2016, 24–27.

53. Scholars have reassessed the importance of eighteenth-century luxury industries, in Britain as well as in France, to the dynamism of these economies. See, among a vast litera-

ture, McKendrick, Brewer, and Plumb, 1982; Berg, 2005; de Vries 2008; Sewell 2010, references on 82.

54. Rockefeller 2011; on this language, see Rodgers 2014, 6, 8–11.

55. "Globalization," now an unavoidable word, entered common parlance only in the late 1980s and early 1990s. See Stearns 2016, 2; Hopkins 2005; Bentley, Subrahmanyam, and Wiesner-Hanks 2015. See also Osterhammel 2011.

56. Hunt 2014, 44–77; Conrad 2016, 92–98.

57. Hamadeh 2004.

58. See, for example, McNeill 1963; and P. O'Brien 2006, 33–39. By contrast, recent histories of early modern global interactions do not make Europe into the central force bringing them about; see, for example, Wills 2015.

59. F. Cooper 2001.

60. Pomeranz 2000. See also Rosenthal and Wong 2011.

61. See Conrad 2016, 210–214. Not all scholars have been so naïve, of course. For example, Mintz (1985) told the history of sugar production and consumption with a keen eye for their human cost.

62. Voltaire 2000, 43.

63. Dobie 2010; see also Anderson 2012.

64. See Hunt 2014, 63–71, citing Marcy Norton and Francesca Trivellato as examples of writers of global history "from the bottom up."

65. Subrahmanyam 1997, 748.

66. Ghobrial 2016.

67. See, for example, Ghobrial 2013; P. Miller 2015; S. Roberts 2013.

68. Among many such works, see Osterhammel (1998) 2018; Popper 2012; and Ramachandran 2015. On universal history, see Robinson 2003, 74–79, 118–23, 134–38; Subrahmanyam 2005, 2015.

69. Alam and Subrahmanyam 2007.

70. P. Miller and Louis 2012; Schnapp 2013; Bleichmar and Mancall 2011.

71. Robinson 2003, 74–75.

72. Pollock 2015, 1. See also Bod et al. 2016; in their introduction, the volume editors stress a global and comparative perspective (211–13).

73. Pollock, Elman, and Chang 2015.

74. Grafton and Most 2016a.

75. Blair 2010, ch. 1, and specifically p. 46.

76. Grafton and Most 2016b, 4–5.

77. For an overview, see Dunkelgrün 2017.

78. Hamilton 1985; Toomer 1996; Hamilton and Richard 2004; Hamilton 2005; Burman 2007; Sheehan 2005; Loop 2013; Tommasino 2013; García-Arenal and Mediano (2010) 2013; Bulman 2015; Levitin 2015; S. Kelly 2015.

79. App 2010; McManus 2016.

80. For a more extensive discussion of this theme, see Bulman's contribution to this volume.

81. See, for example, Bevilacqua 2016, 216–18, and 2018, 116–18.

82. App 2010.

83. See, for example, Bevilacqua 2018, 89–92 and esp. 91.

84. Smith 2009 (esp. 368–72), 2013, 2014; Saliba 2007; Hasse 2016; Feldhay and Ragep 2017.

85. Needham 1954–; Elman 2005, 2006.

86. Cook 2007.

87. For example, Bevilacqua and Pfeifer 2013, 91–92.

88. Hsia 2007.

89. See, for example, Brockey 2007.

90. See, for example, Ditchfield 2010.

91. Subrahmanyam 2003.

92. Sardar 2013, 74; Atasoy et al. 2001, esp. 180.

93. Sardar 2013, 74.

94. On the concept of the borderland, see Hämäläinen and Truett 2011.

95. Radway 2017.

96. See, for example, J. Taylor 2009. These communities were connected to Europe by both mercantile and scholarly networks. For the English case, see Games 2008.

97. Andrews and Kalpakli 2005; Subrahmanyam 2012; Duindam 2016.

98. On the latter point, see Blair 2010, 11–12.

99. For other examples, see Bayly 2002, 71; Smith 2014, 123–24.

100. New York, Metropolitan Museum of Art, accession numbers 07.225.458a and b.

101. Osterhammel 2018; Thomson 1987.

102. Jütte (2011) 2015.

103. See, for example, Cook 2014. For the concept of friction, see Tsing, 2005; and Rodgers 2014.

104. Marchand 2009.

PART THREE

DISCIPLINARY AND
GENERIC HORIZONS

CHAPTER FIVE

Reconfiguring the Boundary between Humanism and Philosophy

JILL KRAYE

In *The Culture of Correction in Renaissance Europe*, based on his 2009 Panizzi Lectures at the British Library, Anthony Grafton describes the role of the Italian humanist and bishop of Aleria Giovanni Andrea Bussi as a corrector for the Roman press of Sweynheim and Pannartz.[1] Focusing on Bussi's preparation of the 1470 edition of Pliny the Elder's *Natural History* (*Historia naturalis*),[2] Grafton recounts a friendly and erudite disagreement between Bussi and the Byzantine scholar Theodore Gaza "about what Aristotle had said about the spines of fish that sparrows used to make their nests."[3] Thanks to the notes that Bussi and Gaza left in the margin of the Vatican Library manuscript that served as the base text for the 1470 edition of Pliny, Grafton enables us to eavesdrop on their conversation. Next to Pliny's statement on the subject (*Natural History*, 10.47.91), Bussi alluded to Aristotle's view; but Gaza was able to correct his account by citing the relevant passage, in Greek, from the *History of Animals* (616ᵃ32). Gaza may have been working on this treatise at the time, for six years later he published a Latin translation of it which was so successful that it held a "virtual monopoly" until the nineteenth century.[4] For Grafton, this anecdote is telling because it shows that "in the new world of print," just as in the era of manuscript transmission, "correction remained a social, collaborative activity";[5]

it is therefore a small piece of evidence in support of his belief that collaboration was a central feature of Renaissance scholarship.[6]

There is more to be learned, however, from this glimpse of two fifteenth-century humanists going about the business of correcting an ancient text. That this text was a treatise on natural philosophy—one that just over two decades later would attract the critical attention of another humanist, Ermolao Barbaro[7]—and that both Bussi and Gaza turned immediately to Aristotle in order to gloss Pliny's meaning will come as no surprise to those who have been working at the coalface of humanist scholarship. Yet science and natural philosophy have no place in the formulation, endlessly repeated in the English-language secondary literature on Renaissance humanism, that the *studia humanitatis* comprised a cycle of five disciplines: grammar, rhetoric, poetry, history, and moral philosophy.

This well-worn notion originally derives from a celebrated article by Paul Oskar Kristeller, "Humanism and Scholasticism in the Italian Renaissance," published at the end of World War II and reprinted several times with updated notes,[8] which has been described as "the single most influential article on Renaissance humanism" of the twentieth century.[9] In this article Kristeller, attempting to counter the then current interpretation of "humanism as the new philosophy of the Renaissance, which arose in opposition to scholasticism, the old philosophy of the Middle Ages," maintained that humanists were "professional rhetoricians" and were "neither good nor bad philosophers, but no philosophers at all." The same applied to other disciplines such as science, law, medicine, and theology: although humanists eventually had an impact on all these fields, they did so not as professionals but as "amateurs"—in other words, they were moving into new domains outside their acknowledged sphere of competence. Kristeller was without doubt right to insist that humanists were not professional philosophers, just as they were not professional scientists, jurists, physicians, or theologians. But he overstepped the mark when he maintained that they had "nothing to do with philosophy even in the vaguest possible sense of the term."[10] Humanism, in fact, had a good deal to do with philosophy, in all its branches and not merely moral philosophy, the one philosophical discipline that, according to Kristeller, belonged properly to humanists.[11]

As John Monfasani has rightly insisted, Kristeller always "stressed the transformative and salutary effect of Italian humanism not simply on philosophy, but on virtually every aspect of educated European culture and learning."[12] Yet he never abandoned his programmatic definition of humanism, which continues to be regularly invoked in Anglophone scholarship.[13] In the past few decades, however, specialist evidence has gradually built up making it increasingly clear that this narrow and restrictive definition belies the complexity of humanism's relationship to a whole range of disciplines and that the rigid disciplinary boundaries separating humanism and philosophy, in particular, need to be reconfigured in order to allow space for a zone of shared intellectual territory.

That territory was ancient philosophy, in which both Renaissance philosophers and Renaissance humanists took an active interest, though their aims and methods were very different. Philosophers saw their primary role as seeking timeless answers to universal questions; they went about doing this by analyzing the logical coherence and rational demonstrability of the arguments presented in the writings of Aristotle, still regarded as "the Philosopher" in the Renaissance—though interest in Plato, which began to gain ground in the fifteenth century largely through the efforts of Marsilio Ficino,[14] expanded the range of ancient thought on which they were able to draw. In their search for speculative truths, they often turned for help to medieval philosophers and commentators who had grappled with the same problems in much the same way. Humanists, on the other hand, were firmly anchored in the historical context of antiquity, regarding philosophy as one part of the lost ancient world that they wanted to recover and reconstruct in all its manifold aspects. It was by fully understanding this larger framework that philosophy's interpretative problems could be solved. Yet even though Renaissance philosophers and Renaissance humanists were coming from different directions and headed for different destinations, their paths crossed on the common ground of ancient philosophy.

The Philosopher as Humanist and the Humanist as Philosopher

In 1994, an exhibition was held in the Laurenziana library marking the fifth centenary of the deaths of both Giovanni Pico della Mirandola and Angelo Poliziano and emphasizing Florentine humanism as the intellectual background to the scholarly activity of the two friends.[15] In the same year, a colloquium devoted solely to Pico was also held in Florence, and Sebastiano Gentile, one of the contributors to the exhibition catalog,[16] was asked to deliver a paper on the philology of texts written by Pico. He decided, however, that it would be more interesting to turn the topic on its head and to see instead if it was possible to treat Pico himself as a philologist.[17] The exhibition had drawn attention to the connections between Pico and Poliziano,[18] and Gentile, too, framed his question in terms of their relationship, asking whether "Poliziano became a bit of a philosopher, and Pico, for his part, become a bit of a philologist."[19]

To highlight Pico's neglected philological side,[20] Gentile pointed out certain traits he shared with contemporary humanists. First, there was his lifelong interest in libraries, not just his own and that of Lorenzo de' Medici, which he helped to enlarge, but those of his friends whom he frequently importuned to lend him books, along with any other collection that held out the prospect of intellectual riches. After being released from prison in Paris, for example, he made a detour to see the famous library of Nicholas of Cusa in Germany before returning to the safe haven of Florence. Then, there was his enthusiasm for learning ancient languages: Hebrew, Arabic and Aramaic, as well as Greek. But it was above all in Pico's final work, the *Disputations against Divinatory Astrology* (*Disputationes adversus astrologiam divinatricem*), that Gentile found evidence of his mature—if this term can be applied to someone who died at age thirty-one—philological approach. Here, Gentile found Pico consulting Greek astrological treatises and commentaries in the original, sometimes as soon as the manuscripts arrived in the Medici library from Byzantium; indulging in the favored humanist sport of attacking the incompetent Latin translations made in the Middle Ages—in this case of Ptolemy—by comparing them to the Greek text (he claimed to have identified hundreds of mistakes in the medieval Latin version of the *Tetrabiblos* that was still in common use in the fifteenth century); and casting a critical eye on unfounded attributions of works to illustrious ancient authors.[21] Gentile admitted that this was perhaps not enough to make Pico a philologist *stricto sensu*, but he argued that it revealed a spirit which could be called philological or which was, at any rate, inspired by the philological methods of the time.[22] The most important representative of those methods was Pico's great friend Angelo Poliziano, and Gentile underlined how much of Pico's philological activity was done in collaboration with the most learned of Quattrocento humanists.[23]

This idea was clearly in the air, for just a few years earlier, Grafton, in one of his 1992 Jerome Lectures, had presented his own take on how Pico and Poliziano became "colleagues in philology."[24] Since the theme of the lectures was the history of reading in the Renaissance, Grafton laid even more stress than Gentile on Pico's obsession with books and libraries; and he, too, focused on the *Disputations*. Grafton demonstrated that Poliziano had introduced Pico to his own "fanatically precise philology,"[25] which helped him to untangle abstruse Greek astrological treatises, and how he had taught Pico to treat the context and chronology of ancient texts with scholarly rigor, enabling him to write a genuinely critical history of astrology.

Grafton, like Gentile, regarded the influence between the philosopher and the humanist as reciprocal: Poliziano's tutelage helped Pico hone his philological skills, while Pico whetted Poliziano's appetite for philosophy.[26] Just as some scholars were belatedly recognizing Pico's philology, others were investigating Poliziano's engagement with ancient philosophy, starting in 1986 with Ari Wesseling's edition of Poliziano's inaugural lecture for his 1492 course on Aristotle's *Prior Analytics* at the Florentine *studio*. The lecture was entitled *Lamia*, after a mythical bloodthirsty vampire, a monstrous figure Poliziano used to caricature those who were attacking him for daring to teach philosophy.[27] In his commentary, Wesseling identified all of Poliziano's sources, including not only the Greek and Roman literary works, both well known and recondite, which were his stock in trade, but also an impressive array of Greek philosophical writings, with the *Life of Pythagoras* and *Exhortation to Philosophy*, by the Neoplatonist Iamblichus, prominent among them. The 2010 translation and collection of essays edited by Christopher Celenza did even more to bring out the philosophical erudition by covering the ancient Peripatetic and Neoplatonic commentary traditions on display in the *Lamia*.[28]

All this knowledge counted for nothing, however, in the eyes of Poliziano's critics who complained that he was not a philosopher but a "trifler" (*nugator*) because he lacked the training and credentials that had, since the late Middle Ages, been demanded of university teachers of philosophy. In other words, he was not, strictly speaking, a philosopher but rather, to use Kristeller's word, an "amateur." In Poliziano's view, the fact that he was not a paid-up member of the philosophers' guild and did not philosophize in the time-honored manner by no means disqualified him from lecturing on Aristotelian philosophy; however, he insisted, he did so not as a philosopher but as an interpreter of Aristotle.[29] Describing himself as a *grammaticus*—what we might call a philologist— he believed that he was entitled to lecture not only on Aristotle's *Organon* but on any text, whatever the subject matter—poetry, history, rhetoric, philosophy, medicine, or law—written in antiquity.[30] Although he had previously taught disciplines that belonged to Kristeller's five-part scheme,[31] he was now declaring that his humanist expertise on the ancient world knew no disciplinary boundaries.

Some thirteen years earlier, Poliziano had given a practical demonstration of how the *grammaticus* could bring his philological abilities, encompassing both textual criticism and the interpretation of ancient authors, to bear on a philosophical work. In 1479, he produced a Latin translation of the *Enchiridion* of Epictetus, which was frequently printed in the sixteenth century,[32] and which has occasioned three studies since 2001 alone.[33] In the dedicatory letter to his patron, Lorenzo the Magnificent, Poliziano explained that he had repaired missing or corrupt passages in the two Greek manuscripts available to him in the Medici library by consulting the lemmata from the commentary on the *Enchiridion* by the late ancient Platonist Simplicius. But Poliziano's intervention went further than restoring the bricks and mortar of the text; his practice of reconstruction included pointed philosophical interpretation. Although Poliziano does not mention it, he also used material from Simplicius's preface to his commentary to give a Platonic slant to Epictetus's Stoic handbook, making the work more appealing to Lorenzo and his circle, who, under the philosophical leadership of Ficino, were strongly inclined toward Platonism.

Both Poliziano and Pico were close friends of Ermolao Barbaro, another humanist whose interests in antiquity—encompassing a philological commentary on Pliny's *Natural History*, as we have seen,[34] and a textbook on Aristotelian natural philosophy[35]—do not fit into Kristeller's pigeonholes. In 1485 Pico and Barbaro corresponded about the relationship between philosophy and rhetoric, a problem dating back to Plato's *Gorgias*. Their letters, full of learned wit and playful punning, have defied straightforward interpretation, generating what Luca Bianchi has aptly described as an "interminable" amount of literature.[36] The main point at issue is how to resolve the paradox of Pico's eloquent humanist attack on humanist eloquence and his powerful defense of scholastic philosophers despite their barbarous Latin. It is unlikely that a consensus will be reached any time soon.[37] Rather than carrying on with the vain attempt to determine which side Pico was on—philosophy or humanism—it would be more fruitful to recast the question and ask instead why he sought to make a compelling case for both sides at the same time.[38]

Ficino, unlike Pico—his former protégé, sometime friend, and occasional opponent—belonged firmly to the philosophical camp. Yet his lifelong interest in classical literature, his elegant Latin style, and his high-level philological skills are hallmarks of the influence exerted on him by the humanist movement. Ficino's groundbreaking Latin translations of the works of Plato and the Neoplatonists, accompanied by learned commentaries, are testimonies to the advances and achievements of Renaissance humanism as well as of Renaissance philosophy.[39]

Research over the past forty years or so on Pico, Poliziano, Barbaro, and Ficino can serve as a model for intellectual and cultural historians seeking to gain a deeper understanding of what Renaissance humanism has to do with Renaissance philosophy. Paying heightened attention to the collaboration and disciplinary cross-fertilization between humanists and philosophers and also to the impact that humanist methods had on the practice of philosophers has the potential to create new perspectives on the problem that reflect the complicated situation better than Kristeller's formulaic and discipline-bound scheme.[40]

Expanding the Corpus of Ancient Philosophy

Humanists not only applied their philological practices to the interpretation of philosophical texts but also played a key role in the expansion of the corpus of ancient philosophy available in the early modern era. The study of this process is currently a lively area of research, and there is no shortage of evidence of humanist activity. What is so far lacking, however, is an attempt to bring together what we know and use it as a basis for rethinking or recalibrating our view of how humanism fits into the history of Renaissance and early modern philosophy.

As a test of how this might be done, I look at some current scholarship on the growing interest in Hellenistic philosophy in the Renaissance and early modern era.[41] The revival of Epicureanism, Stoicism, and skepticism could not have taken place without the philological abilities of humanists, who used their specialized skill set to recover ancient Greek and Latin texts belonging to each of the Hellenistic schools and prepare them for consumption by philosophers and nonphilosophers alike. The Presocratics were among the other ancient thinkers put on the map of ancient philosophy through the efforts of humanists, and I glance briefly at this episode as well.

Our knowledge of Epicureanism in the Renaissance has been considerably augmented in the past few years by a spate of publications on the *fortuna* of Lucretius's Epicurean poem, *On the Nature of Things* (*De rerum natura*), of which Stephen Greenblatt's best-selling monograph, *The Swerve*,[42] is merely the tip of the iceberg.[43] Fueled by the current fashion for classical reception studies,[44] this research has underscored the central place of humanists in the saga of the rediscovery, manuscript circulation, printing, editing, and interpretation of Lucretius. In carrying out these typically humanist activities, however, they also assisted Renaissance readers in comprehending and assimilating Lucretius's Epicurean philosophy.

To cite just one well-known instance: a copy of the French humanist Denis Lambin's 1563–64 Lucretius edition, with handwritten annotations by Michel de Montaigne, came to light in 1989 and was published in 1998.[45] We know from the dates Montaigne entered in his copy that he purchased the book soon after publication and read it immediately, pen in hand; and his annotations show him responding both to Lucretius's language and to the challenging Epicurean doctrines he espoused. Lambin's much improved text of the poem, interspersed with his extensive notes, which, though heavily weighted toward textual criticism, also explained philosophical doctrines where necessary,[46] enabled Montaigne to get to grips not only with Lucretius's poetry but also with his philosophy and physics. This material Montaigne of course adapted to his own purposes in the *Essays*; but we can now see how his understanding of Epicureanism began to take shape under the guidance of Lambin's edition.[47]

The historiographical fate of Pierre Gassendi, the pivotal figure in the seventeenth-century revival of Epicureanism, further suggests that just as philosophy has not been seen as within the remit of humanists, so humanism has not been considered an essential component of Renaissance and, especially, early modern philosophy.[48] Although a major figure in the scientific community of his day, Gassendi's reputation has not aged well, especially in comparison to that of his contemporary Descartes, who has been accorded vastly more airtime by historians of philosophy; Gassendi studies are a niche market, while scholarship on Descartes has mass appeal. By the mid-seventeenth century, philosophers increasingly opted to write in the vernacular; but Gassendi, a humanist to his fingertips,[49] composed and published all his Epicurean works in Latin.

He also edited and extensively annotated the Greek text of the "Life of Epicurus" in Diogenes Laertius's *Lives of the Philosophers*—a work that had been circulating in humanist Latin translations since the 1430s;[50] and, in the course of his voluminous Epicurean writings, he quoted almost all seven thousand–plus lines of Lucretius.[51] Although this humanist style of presentation offered distinct advantages—widespread accessibility, since Latin was still the international lingua franca of scholarship, and the provision of in-depth examinations of the ancient sources on which early modern Epicureanism was based—it was, nevertheless, a major factor in Gassendi's decline from an "insider" to an "outsider" in the history of philosophy.[52] We should not, however, allow the skewed perspective of later generations to diminish our appreciation for what Gassendi and his fellow humanists achieved in their own day. Epicureanism could not have been revived without their efforts, which, it is worth reiterating, involved practices and channels of transmission that went far beyond Kristeller's circumscribed conception of humanism.

The Flemish humanist Justus Lipsius, who was as important for the Renaissance revival of Stoicism as Gassendi was for Epicureanism,[53] has also not fared well at the hands of those historians of philosophy who, like Poliziano's *lamia*, regard philosophy as the preserve of philosophers. For A. A. Long, Lipsius was "a brilliant classical philologist" but "not a philosopher in any deep sense of the term,"[54] while, for John Cooper, he was "not a philosopher's philosopher."[55] These anachronistic judgments fail to do justice to the impact Lipsius had, *qua* humanist, on Renaissance and early modern philosophy. In connection with his magisterial 1605 edition of Seneca's philosophical writings, the product of decades of hard philological labor, Lipsius published two treatises the previous year, *A Guide to Stoic Philosophy* (*Manuductio ad Stoicam philosophiam*) and *The Natural Philosophy of the Stoics* (*Physiologia Stoicorum*), in which he provided readers with the background needed to get the most out of Seneca's doctrines. Both treatises are in the form of a *cento* of quotations and testimonies, which he compiled by plundering his immense storehouse of knowledge, comprising the whole of Greek, Roman, and patristic literature known in his day. Although Lipsius, in line with other Renaissance thinkers, was primarily interested in the late Roman Stoics Seneca and Epictetus, he also sought out the fragmentary remains of early Greek Stoicism in order to present a complete picture of the sect's history and doctrines. The availability of a readily consultable and well-organized anthology of fragments made it much easier to study the initial stages of Stoic philosophy; and this was one of the catalysts for a gradual shift of interest, starting in the second half of the seventeenth century, from late Roman to early Greek Stoicism, which saw philosophers as well as historians of philosophy increasingly devoting their energies to what came to be regarded as the creative early period of Stoic philosophy at the expense of its derivative later phase.[56]

Natural philosophers also took a keen interest in the Greek sources of Stoicism opened up by Lipsius. Rudolf De Smet and Karin Verelst, following up a hypothesis put forward almost forty years ago by Betty Dobbs,[57] found that passages from Philo, Clement of Alexandria, and other ancient Greek authors quoted by Lipsius in support of his dematerialized and Christianized version of the Stoic *pneuma* can be systematically linked to statements in Newton's *Scholium generale*, perhaps through the intermediary of the Cambridge Platonists. It is therefore possible that, as they argue, Newton's "immaterial explanation of the functioning and cause of gravity and gravitation" may have been indebted to Lipsius's humanist erudition and cut-and-paste methods of composition.[58]

The ancient Stoics were not the only group given new life by early modern humanists. Grafton drew attention to another humanist collection of philosophical fragments in an appendix to the 1988 *Cambridge History of Renaissance Philosophy*. He asked: "How did you study the Presocratics in the Renaissance?" To which he answered: "by collecting and analysing fragments quoted by Clement of Alexandria, Simplicius and Sextus Empiricus," as the French humanist Henri II Estienne, with the assistance of his friend Joseph Scaliger, did in his 1573 anthology, *Philosophical Poetry, or at any rate, the Remains of Philosophical Poetry (Poesis philosophica, vel saltem, Reliquiae poesis philosophicae)*.[59] We now know more about the volume's precedents, aims, methods, and sources, thanks to a detailed study published in 2011 by Oliver Primavesi.[60] Placing Estienne's anthology of fragments of Greek philosophical poets in its intellectual context, Primavesi showed that it was related to one of the many sixteenth-century literary controversies sparked by the Renaissance recovery of Aristotle's *Poetics*.[61] According to Aristotle, a philosophical theory expressed in metrical form did not count as poetry, the essence of which was imitation; he had therefore asserted that while Homer was a poet, Empedocles was instead an investigator of nature (*physiologos*).[62] Estienne's fragment collection was intended to provide the evidence needed to decide the question, though his preface makes it clear that he disagrees with Aristotle, as does as his decision to entitle the volume "philosophical poetry." In spite of the title, however, the collection also included prose fragments of Heraclitus, who was believed to have plagiarized earlier Orphic poetry, and of Democritus, because of his habitual association with Heraclitus, as in the "laughing Democritus" and the "weeping Heraclitus." Only a scholar of Estienne's caliber had the familiarity with the entire body of surviving Greek texts[63]—which he had demonstrated the previous year in his five-volume Greek dictionary, the *Thesaurus linguae Graecae*[64]—and the philological proficiency necessary to compile such a collection. Thoroughly humanist in conception and execution, the volume was also a contribution to the study of ancient philosophy. It remained the standard edition of the earliest Greek philosophers until the late eighteenth century[65]—Ralph Cudworth, for instance, drew on it in his *True Intellectual System*,[66] a work we know was carefully studied by Newton.[67]

Among the most important sources mined by Estienne for his collection of fragments were the writings of Sextus Empiricus, whose *Outlines of Pyrrhonism* and *Against the Professors* were—in addition to serving as manifestos for the Pyrrhonian brand of skepticism to which he subscribed—a treasure trove of information about ancient philosophy and other disciplines, from grammar and rhetoric to music and astrology. Starting with the pioneering work of Charles Schmitt,[68] and carried forward by Luciano Floridi, scholarship on the Renaissance recovery of Sextus Empiricus has shown that the fifteenth-century Italian humanists who initially gained access to his treatises and made use of them were "driven by antiquarian, historical, and philological concerns" rather than any philosophical interest in skepticism.[69] Poliziano, for instance, ignoring Sextus's skeptical arguments, extracted historical data from his writings in order to compile his own personal "encyclopedia of the arts and sciences";[70] and among the material he supplied to Pico for his *Disputations* was a passage from Sextus about the Chaldean names for astrology and astrologers.[71] More than a century later, Lipsius, though treating Sextus as an authority specifically on ancient skepticism, still adopted the humanist technique of excerpting information about the sect's history from the intricate philosophical argumentation.[72] It was not until the mid-seventeenth century, with Thomas Stanley's *History of Philosophy*, that the philosophical doctrines of Greek skepticism were given serious attention in a work of humanist erudition.[73] Yet it was his predecessors' painstaking and meticulous philological mining of ancient texts for nuggets of information about the sect that laid the essential groundwork for his achievement.[74]

Estienne, in addition to quarrying Sextus for his Presocratic fragment collection, produced the first Latin translation of the *Outlines of Pyrrhonism*, published in 1562, making the complete text of the treatise available to the learned public of his day, supplemented by two additional ancient sources about skepticism: Diogenes Laertius's "Life of Pyrrho" and Galen's *On the Best Method of Teaching* (*De optimo docendi genere*) in a Latin version by Erasmus.[75] Furthermore, in the preface to his translation, Estienne presented a sympathetic account of Pyrrhonism that, while stopping far short of a full endorsement of systematic doubt, prescribed a humble suspension of judgment as an antidote to the proud dogmatism of contemporary philosophers, which he believed was the road to impiety.[76]

In much the same way that Montaigne's Epicureanism was informed by his study of Lambin's Lucretius, his understanding of Pyrrhonian skepticism rested on his reading of the *Outlines* in Estienne's translation.[77] The degree to which Montaigne absorbed and adopted what he read in Sextus Empiricus has, however, been called into question,[78] as part of a more general backlash against the influential thesis of Richard Popkin that the rediscovery of Pyrrhonism in the sixteenth century triggered a "Pyrrhonian crisis," the chronological boundaries of which Popkin progressively extended in the three editions of *History of Scepticism*.[79] Most prominently, Dominik Perler has challenged Popkin's scenario by arguing that there were other sources of skepticism at the time, including anti-Aristotelianism and late medieval scholastic thought, and that thinkers such as Montaigne and Descartes exhibit forms of skepticism that differ in profound ways from the Pyrrhonist variety: Montaigne because he never reaches a state of tranquility (*ataraxia*) in which he recognizes that there is an equally powerful counterargument to every argument,[80] and Descartes because he doubts the existence of material things rather than, as the Pyrrhonian skeptic, their appearance.[81]

Yet even those modern scholars who want to scale back the overstated claims of Popkin and his adherents do not deny that the recovery of the ancient sources of Pyrrhonism was one of the determining factors in the rise of early modern skepticism. True, the Huguenot Estienne and the Counter-Reformation controversialist Gentian Hervet, whose Latin translation of Sextus's *Against the Professors* was published together with Estienne's version of the *Outlines* in 1569,[82] were pursuing their own agendas when they let the Pyrrhonist genie out of the bottle. Likewise, when other humanists edited and commented on Cicero's *Academica*,[83] they did not realize that the information about the Academic school of skepticism they were disseminating would become one of the preconditions for the multiple forms that early modern skeptical thought eventually assumed.

As each of these cases testify, to the extent that Renaissance and early modern philosophers worked within the parameters of ancient philosophy, they were, knowingly or not, reliant on the services of humanists. Whatever use philosophers chose to make of skepticism, Epicureanism, Stoicism, and Presocratic philosophy, this newly available material was encountered, directly or indirectly, in a form that had been shaped—through commentaries, translations, or collections of fragments—by humanists. To understand Renaissance and early modern philosophy, we need to consider not just its internal logic and formal reasoning but also the textual and material conditions in which these cognitive processes took place.

The ancient philosophical schools that humanists helped to recover were marginal to mainstream Renaissance philosophy, which stayed more or less solidly Aristotelian throughout the period. A strong tradition of scholastic translations and commentaries, many of them inherited from the Middle Ages, continued to provide access to Aristotle's writings and thought;[84] but even here, humanists made their mark. Gaza's Latin version of the *History of Animals* was a notably successful humanist translation, but there were many others.[85] In the preface to his collection *Philosophical Poetry*, Estienne explains that he had decided to compile the volume after coming across a long fragment of Empedocles in Aristotle's *On Respiration*, which he was reading in preparation for editing all of Aristotle's works in Greek and Latin. These plans fell by the wayside, but he did bring out a small edition of selected treatises by Aristotle and Theophrastus in 1557.[86] It was left to another French humanist, Isaac Casaubon, to produce a complete Greek-Latin *Opera omnia* of Aristotle. Casaubon was well qualified for the task, having studied philosophy for three years with the noted Aristotelian Giulio Pace. His first publication, as a twenty-four year-old, moreover, had been a volume of notes on Diogenes Laertius's *Lives of the Philosophers*.[87] Casaubon's 1590 Aristotle, as John Glucker has written, "became a standard text for more than a generation and was reprinted . . . in many later editions,"[88] some of which were emended by his former teacher Pace.[89]

Charles Schmitt has taught us that various different "Aristotelianisms" co-existed in the Renaissance.[90] Yet among these strands, scholastic Aristotelianism has garnered the lion's share of attention, both from early modern opponents of Aristotle and from historians of early modern philosophy;[91] and this emphasis has tended to eclipse humanist Aristotelianism, which, thanks to the efforts of scholars such as Gaza and Casaubon, helped to sustain the viability of Aristotle's philosophy over many centuries by ensuring that his works could be read, studied, taught, and interpreted in ever more accurate and reliable editions and Latin translations. That Aristotelianism managed not merely to survive but even in some quarters to thrive until well into the seventeenth century, despite its increasing vulnerability to philosophical, scientific, and religious attacks,[92] was due in no small measure to the philological bulwark constructed and assiduously maintained by humanists.

The weight of contemporary scholarship does not support the view that humanism can be associated with a particular set of disciplines or a particular branch of philosophy. It suggests instead that a more workable definition of humanists would describe them as experts on antiquity with transferable skills that, as Poliziano insisted, could be applied to any ancient text no matter the discipline. Humanists regarded it as their remit to recover, restore, and, as far as possible, reconstruct whatever their world had inherited from antiquity. The philosophy of ancient Greece and Rome was an important part of that legacy, but one that humanists shared with philosophers; and although there was a rough division of labor between them, we are becoming increasingly aware that it was fluid, fluctuating, and punctuated by instances of interchange and interaction. Separated by a disciplinary boundary that recent and ongoing research is reconfiguring as shifting and permeable, humanism and philosophy each made essential contributions to a fuller understanding of the philosophical traditions that had flourished in antiquity.

Although this picture has become much clearer over the past few decades, it is far from complete. There is doubtless more evidence of collaboration and cross-fertilization between humanists and philosophers to be uncovered if we are alert to the clues lurking in the margins of manuscripts or hidden away in learned treatises, orations, philosophical commentaries, lecture notes, correspondence, and the like. Moreover, present-day historians of humanism and of philosophy need to abandon hidebound distinctions between the two fields and, venturing outside their disciplinary silos, engage in the kind of collaborative study that, as we have seen, enabled their Renaissance forerunners to gain a new understanding of ancient philosophy.

NOTES

1. Grafton 2011b, 46–49.

2. For another account of Bussi's labors on this edition, see M. Davies 1995, 242–47.

3. Grafton 2011b, 47.

4. Monfasani 1999, 205. For Gaza's Latin translation, see Aristotle 1476.

5. Grafton 2011b, 47 and n121. The notes are next to the passage in the *Natural History* where Pliny speculates about the nests of *halcyones* (kingfishers, not sparrows).

6. For other examples, see the collaborative notes by Isaac Casaubon and Jacob Barnet in Grafton and Weinberg with Hamilton 2011, figures 5.10–5.12; for the "shared, or collaborative, activity" of Casaubon and Joseph Scaliger, see ibid., 289. For the "collaborative system of art production" devised by Alberti, see Grafton 2000, 138–39.

7. Barbaro (1492–93) 1973–79.

8. Kristeller 1944–45. I refer below to the reprint, with updated notes, Kristeller 1961.

9. Black 1998, 246.

10. Kristeller 1961, 99, 101.

11. For a fuller version of this argument, see Kraye 2016.

12. See John Monfasani's review of Rocco Rubini, Monfasani 2014, 133.

13. In Italy, however, the interpretation of Eugenio Garin, who emphasized the social engagement and civic values of humanism, has been more influential; see Garin 2009. In Germany, the main focus tends to be on the philosophical content of humanist thought, as can be seen in the massive study of Thomas Leinkauf (2017).

14. Garfagnini 1986.

15. Viti 1994b.

16. Viti 1994b, 127–47 ("Pico e Ficino").

17. Gentile 1997, 465.

18. Viti 1994a.

19. "Il Poliziano fosse diventato un po' filosofo e il Pico, per sua parte, un po' filologo." Gentile 1997, 465.

20. Of the seven hundred or so studies published on Pico since the mid-nineteenth century, the majority are concerned with his philosophy and, to a lesser extent, his subsidiary interest in Cabala and rejection of astrology; see Quaquarelli and Zanardi 2005.

21. On this aspect of humanist scholarship, see Grafton 1990a, esp. ch. 3.

22. Gentile 1997, 490.

23. Mattioli (1965) had hinted in this direction by examining a letter in which Poliziano asked Pico to resolve a philological dispute between him and Ermolao Barbaro over the spelling of a Greek philosophical term.

24. Grafton 1997b, 128.

25. Grafton 1997b, 134. For these techniques and methods, see Grafton (1977) 1991.

26. Grafton 1997b, 128–29, and Gentile 1997, 469–70.

27. Poliziano 1986.

28. Poliziano 2010.

29. "Ego me Aristotelis profiteor interpretem. Quam idoneum non attinet dicere, sed certe interpretem profiteor, philosophum non profiteor." Poliziano 2010, 240.

30. "Grammaticorum . . . sunt hae partes, ut omne scriptorum genus, poetas, historicos, oratores, philosophos, medicos, iureconsultos excutiant atque enarrent." Poliziano 2010, 244.

31. Poliziano had lectured on poetry (e.g., Homer, Virgil, Persius, and Ovid), history (Suetonius), and moral philosophy (*Nicomachean Ethics*); his philosophical opponents, however, thought that the *Ethics* belonged to their territory. See Kraye 2016, 13–14.

32. For a list of early modern printed editions of Poliziano's translation, see Oldfather 1927 and 1952, 160.

33. Kraye (2001) 2009; Mercier 2006; Viti 2015.

34. Barbaro (1492–93) 1973–79. Although Pliny's encyclopedic work covered a very wide range of topics, Renaissance scholars focused primarily on its natural history content, and humanists were particularity interested in it as a repository of ancient scientific terminology; see Nauert 1979.

35. Barbaro 1545; see also Bianchi 2004, 351–62.

36. "Sulla polemica fra il Barbaro e Pico la letteratura è sterminata." Bianchi 2004, 357n45. For the most important items, see Kraye 2008, 13–14n2.

37. For two recent contributions to the debate, see Candido 2010; and Giglioni 2015, 252–53.

38. Elsewhere (Kraye 2008) I have suggested that his motives are best understood in terms of the eclecticism he championed in the *Oration*: "What scholastic philosophy had to offer was the vigor and attention to detail of Duns Scotus, the solidity and sense of proportion of Thomas Aquinas, the terseness and precision of Giles of Rome. . . . What humanist rhetoric had to offer was exceptional powers of persuasion. Pico brought these seemingly discordant movements together . . . using scholastic philosophy to discover the truth and humanist rhetoric to make others see the light" (36).

39. See Robichaud 2018, whose bibliography contains references to the fundamental contributions of Michael J. B. Allen, James Hankins, Henri Dominique Saffrey, and, of course, Kristeller, among many others.

40. For another case of collaboration between a philosopher and a humanist, see Pomponazzi 1970, where he refers to a passage in Philoponus's commentary on the *Sophistici Elenchi*: "quem cum essem Ferrariae feci mihi traducere a nostro excellentissimo domino

Marco" (2:11). "Marco" was most likely, according to Poppi, Marcus Musurus, a Byzantine scholar who emigrated to Venice, where he edited books for the Aldine Press. See Ferreri 2014.

41. See, for example, Osler 1991; J. Miller and Inwood, 2003; Boros 2005; and Kraye 2007b.

42. Greenblatt 2011. See Anthony Grafton's review (2011c).

43. The most important bibliography is provided by Kraye 2014a, 619–25, and 2014b, 1038–40. See also Gentile 2013; Hankins 2013; and Palmer 2014.

44. See, for example, Martindale and Thomas 2006; and Hardwick and Stray 2008. Norbrook, Harrison, and Hardie's edited volume (2016) recently appeared, in the most prolific monograph series in this field, Classical Presences.

45. Screech 1998; see also Ford 2008, 33–36.

46. See Butterfield 2019, and the literature cited in 98n6.

47. MacPhail 2000; Hoffman 2005; Krazek 2011; Passannante 2011; Williams 2016.

48. Some early modern philosophers such as Descartes and Hobbes deliberately distanced themselves from their Renaissance predecessors by announcing that they would no longer build their systems on Aristotle or other ancient philosophers. It is now recognized that these claims were exaggerated; see Muratori and Paganini 2016.

49. On Gassendi's humanism, see Joy 1987, chs. 2–4.

50. Gigante 1988.

51. Wolff 1999.

52. Osler 2010.

53. See Morford 1991 and the recent survey by Lagrée 2016.

54. A. Long 2003, 16.

55. J. Cooper 2004, 28n11. For Anthony Grafton ([1987] 2001), Lipsius was also not a humanist's humanist, given his "willingness to abandon the core of the humanistic enterprise, the effort to understand the past on its own terms, in his desire to make ancient experience accessible. Here . . . we see a humanist at his most inhumane" (240–41).

56. Kraye 2007a, 22–23, and 2017.

57. Dobbs 1988, 68, and 1991, 203.

58. De Smet and Verelst 2001, 12.

59. Grafton 1988, 767. Elsewhere Grafton (1983–93) specifies that Scaliger "supplied Estienne with advice and brief critical appendices for his collection of the fragments of the pre-Socratics" (1:127).

60. Primavesi 2011. See also Céard et al. 2003, 316–22 (no. 85).

61. See, in general, Weinberg 1961.

62. Aristotle, *Poetics*, 1447^b16–19.

63. Scaliger's knowledge was even more extensive. His working copy of Estienne's 1573 edition, now in the Bodleian Library, Oxford, includes "many additional philosophical fragments in Scaliger's hand" (Grafton 1983–93, 1:289n159); and an unpublished manuscript now in the Leiden University Library contains Scaliger's own compilation of fragments, which overlaps with Estienne's but is enriched by numerous new quotations, especially from Simplicius's commentary on the *Physics*, an invaluable source for Parmenides and Empedocles not exploited by Estienne. Primavesi 2011, 170–72.

64. Céard et al. 2003, 286–310 (no. 83).

65. Primavesi 2011, 159. A new wave of interest in the early Greek philosophers began

at the end of the eighteenth century, at which time the term "Presocratics" was coined by Eberhard (1788, 47: "Vorsokratischer Philosophie").

66. Cudworth (1678) points out that a fragment of Parmenides in Simplicius "was not taken notice of by Stephanus in his Poesis Philosophica" (388).

67. McGuire and Rattansi 1966, 134.

68. Schmitt 1983b.

69. Floridi 2010, 278, and 2002. Cao (2001) notes that "the fifteenth century witnessed a revival not of sceptical philosophy but of sceptical texts" (263).

70. Martinelli 1980.

71. See Cao 2001, 260, 277–79, correcting Gentile 1997, 479n47.

72. Lipsius 1604, 69–76.

73. Stanley 1659. See also Malusa 1993.

74. Stanley was by no means an isolated case. As Levitin (2015) has shown, humanist scholarship on the ancient Greek sects, together with the work of Scaliger and other late humanists on Zoroastrian, Chaldean, Egyptian, and Hebrew traditions, was keenly pursued and extensively drawn on by late seventeenth-century philosophers, scientists, and theologians.

75. Céard et al. 2003, 89–94 (no. 28); see also Maclean 2006.

76. Naya 2001.

77. Villey 1933, 1:242–43.

78. Rosaleny 2009.

79. Popkin 1960, 1979, 2003.

80. Berry (2004) claims, however, that "the argumentative style and tone of the 'Apology' [of Raymond Sebond] is completely indebted to the skeptical philosophy of Sextus Empiricus" (507).

81. Perler 2004. For another critical perspective, see Ayers 2004.

82. Maillard et al. 1999, 239–40. The Greek text of Sextus Empiricus's writings was not published until 1621 (ibid., 270–72).

83. Schmitt 1972.

84. Garrod 2014.

85. Schmitt 1983a, ch. 3, "Translations."

86. Primavesi 2011, 185, 187; Céard et al. 2003, 29–32.

87. Grafton and Weinberg with Hamilton 2011, 11, 13.

88. Glucker 1964, 275.

89. Schmitt 1983a, 83–85.

90. Schmitt 1983a, ch. 1.

91. Des Chene 1996; Ariew and Gabbey 1998; Gaukroger 2006; Leijenhorst 2002; Ariew 2011.

92. C. Martin 2014.

CHAPTER SIX

The Varieties of Historia in Early Modern Europe

FREDERIC CLARK

Many grand narratives have held together the study of early modern Europe, and one of the most resilient has posited its discovery of history.[1] According to this narrative, the history that fifteenth-, sixteenth-, and seventeenth-century scholars discovered was not just res gestae, or the preservation of past events in written form. That had existed since antiquity. Rather, early modern scholars discovered methods that anticipated the *historicism* that supposedly blossomed in the nineteenth century and accompanied the professionalization of history as a discipline of the modern academy. In other words, early moderns discovered the *pastness* of the past. They waged war against anachronism. They apprehended differences when they examined cultures, societies, and polities that had flourished long before their own moment. To use a favored aphorism of twentieth-century origin, they discovered that the past was a foreign country. And at the same time, they stopped ascribing historical causality to divine providence or other supernatural forces. These discoveries made them modern, or at least earned their awkwardly named period the enviable status of protomodernity.

Although most scholars today would not deny that the practice of history underwent profound changes between 1400 and 1800, they have also significantly revised and complicated the above narrative. They have challenged the teleologies inherent in reading early modernity as the prehistory of historicism.

And in doing so, they have enriched our understanding of what it meant to write and read history in early modern Europe. They have reminded us that early moderns did not necessarily invent the historical techniques and genres they used. And they have made clear that not everyone in this period who practiced what we today might term history self-identified as producers or consumers of the Latin term *historia* and its vernacular equivalents. In early modern usage, *historia* could sometimes translate as "history" in our sense of the term. But it could also signify a story, an account, or a narrative in a more general sense—true to its etymological roots, it could connote anything that was the product of an inquiry. The wide semantic range of *historia* can indeed cause confusion when we encounter the term in early modern sources (or in ancient and medieval writings, for that matter). However, our confusion should not forestall our inquiries. As argued here, many early moderns were aware of the multivalence of *historia*, and they creatively exploited this quality in a fashion worth recovering. *Historia* was simultaneously an object of inquiry and a mode of inquiry, a specific branch of learning and a more general way of knowing. But this multivalence did not necessarily lead to contradiction or incoherence; rather, for many early moderns, it made *historia* good to think with.

In this essay I explore some of the ways twentieth- and twenty-first-century scholars of early modernity and early moderns themselves understood the capaciousness of *historia*. I examine what closer attention to actors' categories might tell us about early modern approaches to the past. Many early moderns who did *not* claim to write *historia* nonetheless researched and wrote about past cultures and contexts. They identified not as historians, but rather as critics, grammarians, commentators, jurists, and antiquarians, to cite but a few of the available alternative labels. I consider how scholars defined the relationship between *historia* and these parallel enterprises and sought sometimes to combine them. I also investigate how early moderns used categories of *historia* developed across antiquity and the Middle Ages—such as universal history (*historia universalis*), ecclesiastical history (*historia ecclesiastica*), or natural history (*historia naturalis*)—and stretched them to accommodate new circumstances and new epistemologies. Future research along these lines promises not only to shed new light on how early modern scholars interpreted their pasts but also to offer guidance to our own practitioners of historical scholarship in the present.

This essay contains two parts. The first examines how modern scholars have assessed the place of early modernity in the history of historical thought, from the middle decades of the twentieth century to today. Specifically, it reconstructs several strands of the grand narrative sketched above, and then examines attempts—both in the past century and the current one—to revise it. Moving backward, the second part examines some examples of historical theory from early modernity itself. Building upon recent studies of early modern guides to the *ars historica*, or "art of history," it considers how some well-known early modern scholars—from Juan Luis Vives to Francis Bacon—discussed the nature and boundaries of *historia*.[2] It closes with a case study, considering how a far less famous scholar extended the reach of *historia* in this innovative fashion—across time, space, and various other branches of learning. Here I examine the *Systema*, or bibliographic *System*, of the neglected seventeenth-century Parisian Jesuit and librarian Jean Garnier, and the ways this library catalog defined various types of *historia*. Although Garnier's approach to history was certainly idiosyncratic—even by the standards of the methodological diversity that defined the late seventeenth-century Republic of Letters—it is nonetheless illustrative of how history could organize wider systems of knowledge.

Garnier and his more famous predecessors, I argue, suggest a specific "new horizon" for early modern studies. Rather than favoring one mode over the other, vast amounts of early modern historical scholarship were simultaneously "historicist" and "antihistoricist" in equal measure. But what seems contradictory to us was not necessarily so in early modern Europe. More important, it was the very expansiveness and eclecticism of early modern *historia* that allowed it to embrace so many different areas of inquiry and different epistemic styles. Many early modern scholars and readers believed they could simultaneously appreciate the past on its own terms—mining it for antiquarian details—and render it exemplary or useful for their present, with an avowed instrumentalism that might shock our own sensibilities. We might be tempted to celebrate the demise of these seemingly presentist approaches to the past as a prerequisite to the birth of "proper" history. Yet, if we reverse our perspective, the story of the rise of historicism is also the story of the narrowing, and the compartmentalization, of *historia*—both cause and consequence of its transformation into the professional discipline of history. Whether we consider ourselves partisans or critics of historicism (and the various guises under which it has been made "new"), we would do well to recover its early modern alternatives, including those dependent upon historicism's simultaneous practice and denial.

An Early Modern Historical Revolution? Some Modern Assessments

To many scholars in the middle decades of the twentieth century, it seemed a given that a sense of protohistoricism could be located in early modern Europe.[3] Perhaps no scholar expressed this view more programmatically than the art historian Erwin Panofsky, beginning with his 1933 collaboration with Fritz Saxl and his 1944 article published in the *Kenyon Review*. While Panofsky acknowledged that the Middle Ages had witnessed innumerable classical revivals, he christened them mere transitory "renascences." The capital-R Renaissance of the Italian Quattrocento differed from these assorted medieval "renascences" in its newfound sense of perspective. This Renaissance was first to apprehend the gulf separating the present from the Greco-Roman world it sought to emulate.[4] "The classical past," he wrote, "was looked upon, for the first time, as a totality cut off from the present; and, therefore, as an ideal to be longed for instead of a reality to be both utilized and feared." Panofsky finished his argument with an anthropomorphic analogy: "The Middle Ages had left antiquity unburied and alternately galvanized and exorcised its corpse. The Renaissance stood weeping at its grave and tried to resurrect its soul."[5]

According to Panofsky's logic, what was new about the Renaissance was not that it imitated classical antiquity but rather that it invented a new vision of history in order to do so. Consider, for instance, how one of Panofsky's contemporaries characterized Petrarch, the early Italian humanist par excellence. In a 1942 essay, Theodore Mommsen (a descendant of the nineteenth-century German classical scholar of the same name) explained how Petrarch had inaugurated a historiographical revolution.[6] Mommsen quoted the Italian philosopher Benedetto Croce's contention that medieval historiography had been universal in scope and character, and as a result, ahistorical. "The medieval historians almost without exception wrote universal history," according to Croce, "'a history of the universal . . . a history in labor with God and toward God.'"[7] By this logic, medieval and modern historiography differed fundamentally from each other; the latter was real history whereas the former was not.

According to Mommsen, medieval chroniclers, following the example of early Christians like Eusebius and Jerome, had incorporated even the most local of histories into a sweeping, unbroken narrative that extended from the Book of Genesis to their own times; they united past, present, and future in the drama of Creation, Christ, and Apocalypse. But Petrarch purportedly discarded this frame and set his sights on the history of one *particular* culture that time had swept away: namely, Roman antiquity.[8] By eschewing medieval universal history, Petrarch liberated history from theology. To adopt Panofsky's formulation, Petrarch—much like the visual artists of the Renaissance—conceived of classical Rome as a totality cut off from the present. Universal history, on the contrary, had by its very nature extended into the present; it could leave no antiquity, however old, unburied. It is not clear whether Petrarch actually intended to challenge this vision—and most medievalists today would not accept this simplification of medieval historical writing—but for Mommsen and Panofsky, its Renaissance overthrow seemed a prerequisite for the advent of "modern" historical thought.[9]

But if early modern historical thought emerged from this rejection of the Middle Ages, mid-twentieth-century scholarship suggested that the specifics of this transformation had occurred in some unlikely ways. Neither Panofsky's study of visual artists nor Mommsen's study of Petrarch concerned writers of histories in the most conventional sense of that term. Even if Petrarch, Giotto, and the like manifested a newfound historical consciousness, they did not self-identify as *historians*. For some mid-twentieth-century scholars, this raised a crucial question: Did this purported change in historical consciousness actually change the way early moderns wrote and read histories? Did historians go on doing what they had always been doing, even as their nonhistorian colleagues changed their methods? And if so, what might explain this apparent paradox? A number of important scholars—including Arnaldo Momigliano, J. G. A. Pocock, and George Nadel—took up this problem directly.

Perhaps no mid-twentieth-century scholar more thoroughly weighed the difference between history and other forms of investigating the past than the classical scholar Arnaldo Momigliano. In his seminal 1950 article, "Ancient History and the Antiquarian," Momigliano explored how, in early modern Europe, a new type of scholar pioneered the use of material, nontextual sources—coins, inscriptions, archeological remains, and the like—as historical evidence.[10] He named these scholars antiquarians. Momigliano distinguished between historians and his antiquarians not only in terms of what kind of evidence they used but also in terms of how they presented such evidence. Historians employed diachronic narratives, whereas antiquarians preferred synchronic surveys, catalogs, and compilations. For Momigliano, the legacy of early modern antiquarianism lived on not in history departments but rather in the modern social and cultural sciences. Antiquaries were, if anything, more "historically" minded than their historian colleagues, who continued to write circumscribed narratives of high politics—involving kings, emperors, battles, diplomacy, and the like—based solely on prior written sources.

Momigliano also drew important distinctions *within* early modern definitions of history. As the title of his article suggests, he distinguished between early modern views of ancient history and its nonancient (i.e., medieval and modern) counterparts. In his view, early modern scholars believed that *historians* could write of the latter, but not the former. Scholars of Greco-Roman antiquity were instead *antiquarians*, who "commented on historians and supplemented historians, but usually did not claim to be historians."[11] Why did these students of "ancientness" not claim the historian's mantle? According to Momigliano, it was because ancient history was a closed corpus, sealed by its very canonicity: "Roman history had been written by Livy, Tacitus, Florus, Suetonius, the *Historia Augusta*. There was no reason why it should be written again, because in the main it could be written only as Livy, Tacitus, Florus and Suetonius had written it."[12] Whether or not early modern scholars apprehended the ancient world with the newfound distance and precision that Panofsky and Mommsen ascribed to them, Momigliano made clear that they would not have dreamed of writing new *histories* of antiquity.

A few years later, in his now classic 1957 study, *The Ancient Constitution and the Feudal Law*, J. G. A. Pocock explored when and among whom early modern Europe first experienced a historical revolution. Although Pocock focused the bulk of his book on seventeenth-century English constitutionalism, he devoted his first chapter to sixteenth-century French legal scholars, including François Baudouin, Jean Bodin, and François Hotman. Titling this introduction "The French Prelude to Modern Historiography," he explained why this milieu was key to the history of historical method. And in its opening pages, he looked backward still further, to what he termed the "paradox of humanism." Here he drew not on Panofsky but on another influential recent study, R. R. Bolgar's *The Classical Heritage and Its Beneficiaries*. Following Bolgar, Pocock highlighted one of the key contradictions in the humanist project, whose very desire to imitate antiquity ended up revealing the ultimate impossibility of said imitation. As he put it, the humanists "showed that Greco-Roman civilization formed an independent world, a world of the past, but they did not, indeed could not, rob the European mind of its sense of being deeply and vitally affected by the fact that the past, in some way, still survived."[13] It was against the background of this paradoxical relationship between past and present, loss and survival, that he defined the importance of sixteenth-century legal scholarship:

> This book has been written in an attempt to throw light upon one aspect of the rise of modern historiography, a movement whose beginnings in general may with some assurance be dated from the sixteenth century. For it was then that the historian's art took on the characteristic, which has ever since distinguished it, of reconstructing the institutions of society in the past and using them as a context in which, and by means of which, to interpret the actions, words and thoughts of the men who lived at that time. That this is the kernel of what we know as historical method needs no demonstration; that it distinguishes modern from ancient historiography may be seen by means of a comparison with the historical methods of the Greeks and Romans.[14]

In Pocock's view, the revolution in historical thought that Petrarch and Mommsen had located in the early Renaissance actually occurred several centuries later. In the sixteenth century, history ceased simply to narrate the past; now it also used various pasts as *contexts* for interpreting texts, persons, or events in said pasts. But this change did not originate among those who called themselves historians. Rather, Pocock argued, it was the work of jurists, who felt the paradoxes of humanism most acutely. Jurists worked at the intersection of past and present. They sought to contextualize the development of Roman law, its accumulation of precedents, and the production of compilations like the Justinianic Code, and thereby assess whether such traditions should still carry normative, prescriptive force in their own sixteenth-century world. Meanwhile, "history as a literary form went serenely on its way, neither taking account of the critical techniques evolved by the scholars nor evolving similar techniques of its own," at least until the skeptical crisis of the late seventeenth century.[15] Hence, modern historians neglected the story of how these "critical techniques" had arisen in realms beyond history narrowly understood: "The history of historiography has been studied as if it could be identified with the history of those literary works which bear the title of histories, and in consequence a one-sided view has arisen which ascribes not nearly enough importance to the work of scholars who did not write narrative histories."[16] Jurists and other erudite scholars developed innovative historico-critical techniques, but their contributions remained underappreciated because they had not related them in a recognizable form of history as genre.

Precisely what would have been recognizable as history in this world? In "Philosophy of History before Historicism," published several years after Pocock's *Ancient Constitution*, George Nadel examined what he took to be the most recognizable form of history in the sixteenth and seventeenth centuries. Nadel's view of the matter diverged from the vision of historical revolution posited by Mommsen and Panofsky. He argued that some profoundly anti-historicist assumptions still flourished centuries after Petrarch. These centuries instead marked the heyday of the so-called exemplar theory of history. In Nadel's view, the exemplary history championed by humanists was an alternative to scholastic philosophy. Its practitioners trumpeted the superior efficacy of *exempla* over precept in teaching both virtue and statecraft. As they argued, an arid universal was less compelling than a vivid lesson plucked from the Rome of Cicero, Livy, or Tacitus. Yet exemplary history was also universalizing in its own way: it was not concerned so much with the causal links between events as with the broader utility and profit to be derived from each event itself. Historical particulars were to be studied not for their particularity but for their generalizable lessons, and no particular was so alien that it could not be molded for present ends.[17] Contra Panofsky, Nadel's early moderns had not yet buried antiquity in the grave.

If we accept Nadel's categories, history as example—the notion that the past could furnish a normative guide to the present—might seem the antithesis of history as criticism—the notion that the past could be reconstructed as a context to adjudicate the truth-value of its sources. How did exemplar history survive for so long, even as early modern scholars began to highlight the radical distance separating the original contexts of such *exempla* from the present? As Momigliano, Pocock, and Nadel all realized, the coexistence of these seemingly antithetical early modern approaches demanded further study.

In the 1960s and 1970s, scholars continued to explore the history of early modern historical thought. Donald Kelley's work proved particularly pioneering. Expanding upon Pocock and others, he argued that it was sixteenth-century

French jurists—influenced in part by the example of Italians like Lorenzo Valla—who laid the "foundations of modern historical scholarship."[18] Figures like François Baudouin, who programmatically called for a union of jurisprudence and history, loomed large in Kelley's account. Early modern legal scholarship, as Kelley made clear, required that jurists think seriously about the historical contexts of the laws they studied. He, along with Julian Franklin, George Huppert, and others, also investigated the early modern *ars historica*, which proved a popular genre in the sixteenth and seventeenth centuries. Through the late 1990s and early 2000s, both Kelley and Pocock continued to work on topics in the history of historical thought, providing an important strand of continuity between midcentury debates about historicism and the current direction of the field.[19]

The study of the history of early modern historical thought has expanded at an astounding rate, aided by both the renewed strength of intellectual history and the integration of methods from fields such as the history of science and book history into history's broader ambit. Although space does not permit us to consider all of them in the detail they deserve, the 1990s and early 2000s saw a proliferation of landmark studies devoted to early modern historical thought and method.[20] Even if the grand narrative outlined by the likes of Panofsky and Mommsen was no longer taken as a given, the topics they opened up continued to flourish.

Scholarship has also turned now to actors' categories: it has examined just what *historia* meant in early modern Europe and how its pursuit fit within the ecosystems of the early modern university and humanist Republic of Letters.[21] The volume *Historia: Empiricism and Erudition in Early Modern Europe* (2005), edited by Gianna Pomata and Nancy Siraisi, exemplifies this approach. If some of the work surveyed above focused on "historical" enterprises that early moderns did not label history, Pomata and Siraisi's volume attended to forms of inquiry that, even if we might not label them as such, early moderns *did* consider history. "The starting point of our inquiry," Pomata and Siraisi wrote, "is the ubiquity of *historia* in early modern learning—the fact that it featured prominently in a wide array of disciplines ranging from antiquarian studies and civil history to medicine and natural philosophy."[22] The presence of these final two categories—which many today would group under the label "science" and perhaps thus even imagine as antithetical to a humanistic discipline like history—demonstrates not only the capaciousness of early modern *historia* but also the corresponding parochialism of our own category of history. Yet starting in the 1990s—particularly with Paula Findlen's pioneering *Possessing Nature*— and continuing until today, a new group of scholars have demonstrated in rich detail how early modern natural history was just as much a part of *historia* as fields that we now would unambiguously label "historical."[23]

Scholarship of the past several decades has also trained its sights on the nature of early modern historical *method*. This turn has been inspired above all by the work of Anthony Grafton, who—as a student of Arnaldo Momigliano—has expanded upon Momigliano's own inquiries in the history of scholarship. Grafton's earliest work, which began to appear in articles in the 1970s, concerned early modern technical chronology. Although it might seem alien to modern notions of history, chronology (especially its attempts to reconcile Biblical time with pagan sources) constituted a key element of *historia* in early modern Europe.[24] In his influential *Forgers and Critics: Creativity and Duplicity in Western Scholarship* (1990), Grafton also challenged the notion that early modern humanism marked the unambiguous triumph of historical criticism, arguing instead that criticism and falsification had always existed in a dialectical relationship with each other.[25] Even such astute Renaissance scholars as Erasmus and Carlo Sigonio had forged pseudo-ancient texts, and many of their compatriots in the supposedly "critical" world of early modernity fell for ancient and modern forgeries alike. Grafton, Walter Stephens, Christopher Ligota, and others have reconstructed how one of the Renaissance's most notorious forgers—the Dominican friar Annius of Viterbo, author of a collection of forged ancient texts, some of which he passed off as *historiae*—furnished his champions and adversaries alike with valuable rules of historical criticism.[26] Hence, even unambiguous examples of critical methods—such as the rules for assessing the *fides* of a source that Jean Bodin actually took from Annius—turned out to have surprising, and disquieting, origins.[27]

Taken together, such work has questioned the provenance of early modern historical methods that prior interpretations celebrated as protohistoricist. Sometimes, as in Annius's case, history's rules of the game derived from outright forgeries. As Grafton has shown, early moderns did not invent a new critical science *ex nihilo*. Rather, they adapted tools incubated by premodern traditions, from the grammatical exegesis of Hellenistic Alexandria to the scholastic theology of medieval Paris.[28] And even some of the most modern-seeming of methodological triumphs (such as the humanist scholar Isaac Casaubon's debunking of the Hermetic Corpus) were informed by underlying paradigms—such as the primacy of Biblical revelation—that were anything but. Although Casaubon certainly used historicizing methods—such as identifying anachronistic language—to indict Hermes, theology as much as historical criticism led him to reject the authenticity of the supposed Egyptian sage. As a Protestant attacking his Catholic antagonist Cesare Baronio's notion of "pagan revelation," he could not accept the proposition that Hermes had preceded Moses or that any such pagan *prisca sapientia* antedated Scripture.[29]

In addition to questioning these genealogies of modern historical method, Grafton and others influenced by him have also devoted new attention to forms of early modern *historia* whose flourishing belies tidy modernization narratives. They have stimulated new study of early modern genres—namely *historia ecclesiastica* and *historia universalis*—that had their origins in the early Christian world.[30] As discussed earlier in this chapter, Mommsen had identified the advent of the Renaissance with the demise of universal and ecclesiastical history. Momigliano, by contrast, recognized the methodological significance of "confessional historiography," which stimulated an efflorescence of ecclesiastical scholarship in early modern Europe. Building upon Momigliano's suggestions, Grafton and others have shown just how eclectic and methodologically inventive early modern church history was.[31] Whether practiced by Protestant and Catholic partisans or by irenically minded moderates, ecclesiastical history often combined history with other enterprises, such as jurisprudence or philology, and promoted the development of newer disciplines, such as paleography and diplomatics. It stimulated large-scale studies of the Church Fathers, whose texts humanists edited and printed just as they did works of pagan antiquity.[32] And sometimes the zeal for recovering Christian antiquity was met by forgery, as the case of the lead plates devised to prove that Granada was the first Iberian Christian community attests.[33]

But even genuine works of Christian history were predicated upon assumptions—about everything from divine providence to the exemplarity of the early church for the present—that might not sit easily with modern notions of historicism. And indeed, if recent scholarship has taught us anything, it is that early modern *historia* embraced a startling diversity of domains and methods. Whether it traced the fortunes of Christian churches, cataloged new species of plants, debunked forged ancient texts, or expounded moral lessons, *historia* both adumbrated historicism and affirmed old tropes of exemplarity. The following section considers how early modern scholars attempted to order and define this reach.

What Was *Historia*—And How Far Might It Stretch?

Early modern scholars were ever fond of categorizing and compiling forms of knowledge in encyclopedias and reference works. In doing so, they frequently defined *historia* and thence divided it into multiple parts.[34] These definitions elucidated everything from chronological and geographical differences to epistemological distinctions. And while such taxonomies often appeared in formal *artes historicae*, they were not limited to that genre. For instance, Petrarch did not write an *ars historica*, but he left us a fundamental distinction—crucial to early modern definitions of antiquity—that would easily have fit into the many *artes historicae* of the sixteenth and seventeenth centuries. In a letter to his friend Giovanni Colonna, he distinguished between "new [i.e., modern] history" (*historia nova*) and its counterpart "ancient history" (*historia antiqua*). The former occupied the time "before the name of Christ was venerated at Rome and celebrated by Roman emperors," whereas the latter reached "from that point up to this age."[35]

Many early modern definitions of *historia* combined a desire for historical exemplarity with recognition of the reality of historical difference—that paradox of humanism that Pocock highlighted in his *Ancient Constitution*. One of the clearest attempts to reconcile the two is found in the 1531 *De tradendis disciplinis* or *On Teaching the Disciplines* of the Spanish humanist Juan Luis Vives. In Book V, in a section devoted to history, Vives defended *historia* against those whom we might term radical historicists—those who argued that the present differed from the past so profoundly that the latter possessed no usefulness whatsoever for the former. As he explained:

> Nevertheless there are those who may persuade themselves that knowledge of ancient history (*veteris memoriae cognitionem*) is useless, because the method of living all over the world is changed, as e.g. in the erection of elegant dwellings, the manner of waging war, of governing people and states. . . . To be sure, no one can deny that everything has changed, and continues to change, every day, because these changes spring from our volition and our industry. But similar changes do not ever take place in the essential nature of human beings, that is, in the foundations of the affections of the human mind, and the results which they produce on actions and volitions. . . . Indeed, there is nothing of the ancients so worn out by age and so decayed, that it may not in some measure be accommodated to our modes of life.[36]

After making this defense (what we might term a defense of exemplarity with an acknowledgment of partial or mitigated historicism—i.e., the argument that the outward manifestations of human nature change but human nature itself does not), Vives pointed out just how much other disciplines needed history. Medicine and moral philosophy made use of history; and law was but history under another name—in Vives's judgment, "law, whether the Roman or any other law, is nothing else than that part of history which investigates the customs of any people." And although he did not wish to "seem impudent," Vives maintained that even a great part of theology was history. *Historia* was essential to diverse other branches of learning.[37]

Baudouin, the subject of Pocock's and Kelley's studies and also a significant protagonist of Grafton's *What Was History?* (2007), made similar arguments about the eclectic nature of *historia*. In 1561 he published his *De institutione historiae universae et eius cum iurisprudentia coniunctione* (*On the Institution of Universal History and Its Conjunction with Jurisprudence*), in which he called for combining history with law. But this "interdisciplinary" synthesis depended on other acts of union within *historia*. Baudouin also argued that only by combining civil and ecclesiastical history could one produce a truly complete (*integra*) or universal history.[38] Moreover, he put his theorizing into practice, consulting for the so-called Magdeburg Centuriators in their plan to produce a systematic Protestant church history.[39]

In 1566, the French scholar and jurist Jean Bodin published his *Methodus ad facilem historiarum cognitionem* (*Method for the Easy Comprehension of Histories*).[40] He began by dividing all of history according to a conventional threefold schema: "There are three types of history—that is true narration—namely, human, natural, and divine. The first pertains to man, the second to nature, and the third to the parent of nature."[41] Bodin defined the difference between human history and its natural counterpart as a question of constancy versus contingency: "Natural history presents an inevitable and steadfast sequence of causes unless it is checked by divine power or for a brief moment abandoned by it."[42] Like Vives, he affirmed the inherent mutability of *human* custom: "But human history mostly flows from the will of men, which ever vacillates and has no objective; rather, each day new laws, new customs, new institutions, new manners arise."[43] For Bodin, *historia* encompassed the regular and the irregular, the immutable and the mutable, the natural and the human. In other words, it encompassed some domains in which historicist methods might seem germane and others in which they might not.

In his *De augmentis scientiarum* (1623), an expanded Latin version of his *Advancement of Learning* (1605), the English philosopher and polymath Francis Bacon also defined the number and types of history—in a manner that, at first glance, might not seem to differ much from how Bodin had. He offered rather traditional distinctions between such well-known fields as natural history, civil history, and ecclesiastical history. But he also added a fourth category, one that had been spottily treated in previous decades and rarely had received the attention Bacon now gave it: *historia literaria*, or what we might best translate rather clumsily as the "history of letters and learning." According to Bacon, whereas enterprises such as civil and ecclesiastical history were well established, *historia literaria* remained at present a desideratum. But fulfilling its promise was necessary for the completion of *historia* itself, as he made clear through a memorable analogy to the Cyclops Polyphemus. Bacon lamented that "the history of the world (*historia mundi*), if devoid of this part [i.e. *historia literaria*], can be judged not unlike the image of Polyphemus, with its eye plucked out." Restoring *historia* to its ideal two-eyed state meant pursuing *historia literaria*, and that involved reconstructing "which forms of learning and which arts flourished in which regions and ages of the world."[44] Building upon the sixteenth-century *artes historicae*, Bacon sought to define a new and largely unnamed branch of learning by expanding the reach of *historia* itself. As it came to be practiced, mostly by late humanists who wrote dense Neo-Latin tracts, *historia*

literaria eventually embraced everything from bibliography and literary studies to nascent discussions of something like intellectual culture, all within the
overarching rubric of history. Not for the last time, *historia* gave coherence to a
motley assortment of fields and enterprises that had not hitherto been joined
together.

I close with an example of how one early modern scholar, today far less famous than Bodin or Bacon, not only combined disciplines that we might think
distinct under the overarching rubric of *historia*, but also explicitly joined
methodological styles—both "historicist" and exemplary—and even types of
knowledge—that is, fact and fiction—when doing so. Jean Garnier (1612–88)
was a Jesuit theologian and patristic scholar who taught at the Parisian Collège
de Clermont, later renamed the Collège Louis-le-Grand.[45] Garnier composed
his *Systema Bibliothecae Collegii Parisiensis Societatis Jesu* (*System of the Library
of the Parisian College of the Society of Jesus*; 1678) while he served as the library's
curator.[46] His *System* was at once a guide to the contents of the Collège library
and, more abstractly, a map of the proper organization of knowledge. *Historia*
occupied an outsized importance in Garnier's scheme. While Garnier divided
the *Systema* into four main subject areas, including theology, philosophy, history, and jurisprudence, history contained more subdivisions and occupied more
pages than any of its counterparts. Although volume alone is not indicative of
hierarchical importance, Garnier's typology of *historia* also absorbed other disciplines that modern classificatory schemes have kept outside its reach. What
we might label political science or sociology, and what other humanists might
have understood as moral philosophy, Garnier folded under the rubric of *historia artificialis*. And patently fictional texts, instead of earning their own category like "literature" or "belles lettres," Garnier assigned to *historia fabulosa*.

Granted, Garnier's approach was not exactly conventional, and not everyone
agreed with his classification. A half century later, when Johann Köhler reprinted the *System* at Frankfurt, he explicitly took Garnier to task for including
historia in his fourfold division of knowledge, and remarked that theology, jurisprudence, medicine, and philosophy was the more common fourfold scheme.[47]
Similarly, Köhler objected to Garnier's "peculiar" category of *historia artificialis*.[48] Yet Garnier lived in a world of profound intellectual tumult, and debates
over *historia*—however unconventional—proved central to it. As Catherine
Northeast noted in her study of Parisian Jesuits, Garnier played a role in the
development of so-called positive theology—a movement within Catholicism
that sought (in contrast to Thomist philosophy) to accord new emphasis to the
historical exegesis of doctrine.[49] And he also associated with figures who made
his own departures from historical orthodoxy seem rather dull by comparison.
Garnier's colleague at the Collège and successor as librarian was none other
than the notorious Jesuit classical scholar Jean Hardouin, who in a wild flight

of historical Pyrrhonism would go on to maintain that nearly every ancient text was a forgery.[50]

Garnier recognized the limits of more traditional categories of *historia*. His entry for *historia universalis* made clear that not even a universal history could contain everything. In a definition that possessed antecedents in Baudouin and others, Garnier noted that three qualities made a history universal.[51] It could be so by virtue of time, if it contained "matters of each testament" (*res utriusque Testamenti*), that is, the Old and New. It could be so by virtue of place, if it contained "matters of the whole world" (*res totius Orbis terrarum*). Finally, history could be universal by virtue of its very contents, if it featured "matters at once ecclesiastical and profane" (*res Ecclesiasticas simul et profanas*).[52] In this last aspect was that foundational distinction, repeated by Baudouin, Bodin, Bacon, and so many others before him, between sacred history and its civil or secular counterpart. Garnier then suggested dividing writers of universal history into three classification schemes: first, writers of matters both sacred and profane in the whole world; second, those who combined the two, but only in Europe; and third, those who combined them, but in Asia. Of course, this was hardly the whole *Orbis terrarum*. His definition of *historia universalis*, as he freely admitted, did not include everything. "Matters which were conducted in Africa and the New World (*Orbe novo*) pertain to *historia peregrina*, whether sacred or profane."[53] Garnier thus conjured an altogether different category—*historia peregrina*, or foreign, exotic history, the stuff of travel and travelers—for histories of Africa and the Americas. Instead of simply incorporating such "new" worlds into the broader domain of *historia universalis*, he chose to endow *historia* itself with yet another category—a recognition, perhaps, that old genres could not always absorb new circumstances.

Likewise striking are Garnier's discussions of three fields he deemed the "appendices" of history: *historia gentilitia*, or genealogy; *historia literaria*, which embraced both bibliography and Bacon's notion of a record of learning; and *icones historicae*, pictorial representations of the past, including collections of coins, inscriptions, emblems, portraits, and the like. Garnier explained that he labeled these three pursuits history's appendices "because they complete the parts of historical philology, and philology is indeed the appendix of the disciplines."[54] He understood them a bit like the way we might understand the so-called auxiliary disciplines. Noting that "each of these three rightly bear the name of history," Garnier went on to explain their exemplary uses.[55] Although genealogical tables, encyclopedic reference books, and collections of coins and inscriptions did not constitute narrative expositions of the past, they nonetheless belonged under Garnier's broad tent of *historia*.

Garnier concluded his catalog of *historia* with a final trio: *historia naturalis*, *historia artificialis*, and *historia fabulosa*. As described above, the first was a rather conventional category, but Garnier defended it with gusto against some new challenges: encroachments from natural philosophers and physicians. The "works of nature" (*opera naturae*) were to be contemplated with two eyes (*duobus quasi oculis*): the one was historical, and the other philosophical and medical. The physician used natural phenomena, the philosopher expounded nature's wondrous achievements, but the historian simply narrated them.[56] This last office was the broadest and most prominent, and hence Garnier, citing Pliny and Aristotle as precedents, consigned natural history to *historia*. He then complained that many physicians tried to usurp anatomy, but that it most properly belonged to *historia*, since the "history of the human body" (*historia corporis humani*) was but a component of *historia naturalis*.[57]

Garnier's next category was more idiosyncratic. *Historia artificialis* covered everything "made by human art."[58] Moving from an individual as a single unit to a state or society as a whole, Garnier divided *historia artificialis* into four groupings. The first concerned "man himself" and featured texts on birth, life and its ages, food, clothing, the home, death, and funeral rites. The second touched upon "man in the family" and examined marriage, husbands, wives, children, servants, and other unions. The third considered "incomplete societies of men" (*hominum societates imperfectas*) such as guilds, confraternities, associations and commercial unions. The fourth pertained to the "complete society of men" (*hominum societatem perfectam*) and embraced kingdoms, states, sacred rites, and law.[59] Garnier's categories were akin to the ones Vives had outlined for the malleable world of human custom. In this fashion, Garnier united high politics, moral philosophy, and the study of culture under the overarching rubric of *historia*. The Jesuit scholar even defined this branch of history by its *lack* of exemplarity: "*Historia artificialis* is distinguished from the *artes liberales*, which labor for the utility of men (*hominum utilitati*), because whereas the latter teaches what must be done and how, the former chiefly teaches what was once done by the ancients."[60] Whereas the liberal arts were prescriptive, *historia artificialis* was above all descriptive. Much like the Baroque *Wunderkammer*, it illuminated foreign and exotic worlds—including that of the "ancients." And unlike those liberal arts that sought to extract instructive *exempla* from the distant past, it was blithely unconcerned with furnishing any normative guide to the present.

Garnier closed with *historia fabulosa*, an unlikely union of the frequently opposed domains of *historia* and *fabula*.[61] Encompassing epic poetry, tragedy, and comedy, *historia fabulosa* consisted of patently fictional texts that, though literally false, nonetheless contained some approximation of truth. Here Garnier returned to the notion of exemplarity he had just eschewed. Unlike *historia artificialis*, Garnier considered *historia fabulosa* to be historical insofar as it supplied *exempla*, implanted prudence, and informed mores. He explained that this type of fictive writing rightly bore the name of "fabulous history" insofar as it "pertains to the same faculty of man as true history (*vera Historia*), it follows the same mode of eloquence, it tends to the same end, and it answers to true history just like a shade to the body."[62] Hence, *historia* could cross into the domain of fiction, just as it extended everywhere from nature to society, the universal to the particular. It generated new fields of inquiry and combined the most traditional of fields in unexpected ways.

New Horizons for the History of History

This seeming difference between history as a discipline or a narrative one wrote, on one hand, and *historia* as a tool used by nonhistorians of various stripes, on the other, was hardly unique to early modern Europe. Although it may seem obvious or trivial when stated explicitly, this remains the case today. History has become one of the standard disciplines in modern academic institutions, even as the adjective "historical" applies to a wide variety of methods used in everything from literary studies to the qualitative social sciences. And there are also disciplines—sometimes described as "auxiliary" not unlike Garnier's "appendices," and sometimes housed in separate departmental structures—that are nevertheless integral to historical practice: these range from philology and paleography to lexicography and numismatics. If there is anything that this survey of early modern uses shows us, it is that *historia* has always been more encompassing, and more eclectic, than the discipline it signifies. While this is surely the case with all disciplines, it is one exacerbated by history's double pose as both a method and an object of analysis.

As argued here, the primary significance of early modern *historia* is not to be found in whether or not it adumbrated historicism—or any other modern "ism" for that matter—but rather in its eclectic nature, which allowed it to simultaneously illuminate the past and instruct the present, even when these two aims were at odds. Perhaps this might strike us as paradoxical, but some paradoxes—as the definitions of Garnier's *Systema* demonstrate—were both practically and theoretically useful. They could still order books on a library shelf and schematize the concepts within each book's pages. *Historia* was ipso facto capable of diversification—of methods, aims, and even philosophical assumptions—that make it a category error to credit it with the formation of any one "ism," worldview, or historical consciousness at the expense of all others. It was always inquiry and the product of inquiry, whether such inquiries were "natural," "human," or "divine." *Historia* was both endlessly divisible and hence endlessly multipliable. Early modern scholars who defined it could not resist breaking it apart and putting it back together again. They divided the discipline into subcategories and subdisciplines, and sometimes reassembled these parts into new and different wholes. Thinking about *historia* in explicit terms could create new categories of thought and inquiry. Some of these might seem prophetic, and others might seem outmoded. But whatever our judgments of the specific categories it generated, *historia* was itself fundamentally generative, ever seeking to square the circle of experiencing the past in the present. At the risk of advancing a sentiment that may itself seem ahistorical—and one that, unlike Panofsky's Renaissance, attempts to galvanize the past's corpse—we as twenty-first-century scholars may do well to make early modern historical thought a little less foreign. As we debate the best way to organize and constitute modes of humanistic inquiry—especially when the humanities themselves face an uncertain future—the capacious and eclectic nature of early modern *historia* offers us many *exempla* to heed.

NOTES

I wish to thank Ann Blair and Nick Popper for their expert guidance of this project and their insightful readings of this essay, which have immeasurably improved the final product. The anonymous reviewers for Johns Hopkins University Press also generously suggested many important improvements. I am grateful to Tony Grafton for both his erudite comments on earlier versions and the example of his own work on the multifaceted nature of early modern *historia*. Finally, I thank David Bell, in whose 2010 French history seminar I first explored the figure of Jean Garnier.

1. The discovery of history is hardly the only organizing grand narrative that scholars have told about the early modern period: others, to name but a few, have included the Scientific Revolution and the triumph of the New Philosophy, the advent of secularization, the military revolution, and the formation of the modern state.

2. On the early modern *artes historicae*, see Grafton 2007b (discussed later in this chapter).

3. For variations on this narrative, stretching back to the Renaissance itself, see, especially, Ferguson 1948.

4. See, first, Panofsky and Saxl 1933 and Panofsky 1944. Panofsky then delivered a set of lectures on the theme in 1952. They were later published as Panofsky 1960.

5. Panofsky 1960, 113.

6. Mommsen 1942.

7. For this distinction, see Mommsen 1942, 237–38, where he cites Croce 1921, 206.

8. This dichotomy ignored the fact that much of universal history was itself Roman-centric, reflecting the importance accorded Roman *imperium* in narratives of Christian salvation history.

9. More recent scholarship has pushed back against the notion that Petrarch marked a definitive break from medieval thought and methods. See, for example, Black 1995 and Quillen 1998.

10. Momigliano 1950. See also the more recent perspectives collected in P. Miller 2007.

11. Momigliano 1950, 291.

12. Momigliano 1950, 291. I discuss the implications of Momigliano's point in Clark 2018, esp. 192–93.

13. Pocock 1957, 5, citing Bolgar 1954.

14. Pocock 1957, 1.

15. Pocock 1957, 6.

16. Pocock 1957, 7.

17. Nadel 1964. See also Koselleck 1984.

18. Kelley 1970.

19. See, for instance, Kelley 1998 and Pocock 1999–2015.

20. For a by no means exhaustive selection, see Ginzburg 1999; Levine 1999; Fasolt 2004; Soll 2005; and Schiffman 2011.

21. Two important edited collections on these themes are Kelley 1997 and Pomata and Siraisi 2005. The latter volume drew inspiration from Seifert 1976.

22. Pomata and Siraisi 2005, 1.

23. For a selection of works in this vast field, see Findlen 1994; Ogilvie 2006; Egmond 2010, and Kusukawa 2012.

24. Grafton 1983–93, vol. 2.

25. Grafton 1990a, esp. 101–4.

26. See Stephens 1979 and Ligota 1987. For an exploration of falsifications in the realm of art history—and a challenge to Panofsky's vision of Renaissance historical "lucidity"—see Nagel and Wood 2010.

27. For Bodin's co-opting of Annius's rules, see Grafton 1990a, 113–15.

28. Grafton 1990a. On historical methods in the Middle Ages, see, for example, Guenée 1980.

29. Grafton 1990a, 62–63, 87–93. See also Grafton 1983 and 1985. Some later works of Grafton that touch upon these themes include Grafton 2014a and 2016. See here Soll 2016.

30. For studies of early modern ecclesiastical scholarship, see Ditchfield 1995; Lyon 2003; Quantin 2009; Haugen 2011; Van Liere, Ditchfield, and Louthan 2012; Olds 2015, and Hardy and Levitin 2020. On early modern universal history, see, for instance, Griggs 2007; Popper 2012; Nahrendorf 2019; and Lotito 2019. On early modern Biblical scholarship, see Hardy 2017.

31. Grafton and Weinberg with Hamilton 2011, esp. 290. For the example of how François Baudouin combined philology and jurisprudence with ecclesiastical history in similarly eclectic fashion, see Kelley 1964 and 1970, 116–36.

32. For the example of Erasmus's critical work on Jerome, see Vessey 1994 and Pabel 2008. On the early modern reception of Augustine, see Visser 2011.

33. On Granada's forged history, see Harris 2007.

34. On early modern encyclopedism, see Zedelmaier 1992 and Blair 2010. On Theodor Zwinger's attempts to define *historia* and distinguish it from theory and precept, see Blair 2005.

35. "Quid ergo? multus de historiis sermo erat, quas ita, partiti videbamur, ut in novis tu, in antiquis ego viderer expertior, et dicantur antique quecunque ante celebratum Rome et veneratum romanis principibus Cristi nomen, nove autem ex illo usque ad hanc etatem." Petrarch 2017, 6.2.16, 70–71.

36. "Sunt tamen qui veteris memoriae cognitionem inutilem esse sibi persuadeant, quod mutata sit uniuersa ratio victus, cultus, habitandi, gerendi bella, administrandi populos, et ciuitates . . . nimirum negare nemo potest omnia illa esse mutata, et mutari quottidie, nempe quae sunt voluntatis nostrae atque industriae. Sed illa tamen nunquam mutantur,

quae natura continentur nempe caussae affectuum animi, eorumque actiones et effecta . . . nihil est enim veterum adeo desuetum et abolitum, quod nostris viuendi moribus accommodari quadamtenus non queat." Vives 1531, fol.126r. For the translation used here, with slight modification, see Vives 1913, 232–33.

37. "Ut ius vel Romanum vel cuiusuis alterius gentis nihil sit aliud, quam ea historiae pars, quae mores alicuius populi persequitur." Vives 1531, fol. 126v. For an English translation, see Vives 1913, 234.

38. See Grafton 2007b, 106–7; and Kelley 1970, 134–35.

39. See Grafton (2001) 2009; Lyon 2003; and Bollbuck 2014.

40. For various treatments of the *Methodus*, see, for instance, Franklin 1963; Kelley 1970, esp. 136–38; Couzinet 1996; and Schiffman 2011, 188–98.

41. "Historiae, id est verae narrationis, tria sunt genera: humanum, naturale, diuinum. Primum ad hominem pertinet, alterum ad naturam, tertium ad naturae parentem." Bodin 1566, 9.

42. "Naturales enim necessariam habent et stabilem causarum consecutionem, nisi diuina potestate prohibeantur, uel paulo momento ab ea deserantur." Bodin 1566, 12. For the translation used here, with slight modifications, see Bodin 1945, 17.

43. "At humana historia quod magna sui parte fluit ab hominum voluntate, quae semper sui dissimilis est, nullum exitum habet: sed quotidie nouae leges, noui mores, noua instituta, noui ritus oboriuntur." Bodin 1566, 12. For the translation used here, with slight modifications, see Bodin 1945, 17.

44. "Atque certe historia mundi, si hac parte fuerit destituta, non absimilis censeri possit statuae Polyphemi, eruto oculo. . . . *Argumentum* non aliud est, quam ut ex omni memoria repetatur, quae doctrinae et artes quibus mundi aetatibus et regionibus floruerint." Bacon (1623) 2011, 2.4, 502–3. For a selection of the extensive literature on *historia literaria*, see Schmidt-Biggemann 1983; Gierl 1997; Nelles 2000; and Carhart 2007. An important sixteenth-century precedent for the genre was Christophe Milieu's *De scribenda universitatis rerum historia*, discussed in Kelley 1999. On early modern disciplinary historics, see Popper 2006.

45. On Garnier, see Kane 1940. Garnier also appears in Carlos Sommervogel's (1890–1932) massive bibliography of Jesuit writings. On this world of French erudition, see Barret-Kriegel 1988.

46. Garnier 1678. On Gabriel Naudé, whose own library system influenced Garnier and itself drew upon Baconian categories, see Nelles 1997.

47. Köhler 1728, sig.)(2v.

48. Köhler 1728, sig.)(3r.

49. Northeast 1991, 56.

50. Northeast 1991, 16–17, 82. On Hardouin's theory of forgery, see Grafton 1999.

51. See Baudouin 1561, 41, and discussion of this passage in Kelley 1970, 135.

52. Garnier 1678, 54. On the interplay of *historia sacra* and *historia profana* in early modern scholarship, see Mulsow 2005.

53. "Res, quae in Africa, et in Orbe novo actae sunt, pertinent ad Historiam Peregrinam, sive Sacram sive Profanam." Garnier 1678, 54.

54. "Dicuntur tamen appendices tantum, quoniam implent Philologiae Historicae partes; Philologia vero est doctrinarum appendix." Garnier 1678, 79.

55. "Unaquaeque trium fert merito Historiae nomen. Siquidem et memoriam perficit

ut Historia, et quaedam est temporum doctrina: nam eruditum est scire, ex quibus orti sint, qui nobiles existunt; et quo tempore scripta monumenta Literarum, in quibus immortalem vitam heroes vivunt; et videre oculis subjectas res multis retro saeculis peractas, defunctorumque heroum vultus adhuc in metallis et marmoribus spirantes, etc." Garnier 1678, 79.

56. "Medicus enim rebus Naturalibus utitur; Philosophus earumdem admirabilem effectionem exponit; Historicus, quid factum sit, narrat." Garnier 1678, 86.

57. "Ad Historiam naturalem reduci debuit Anatomia corporis humani, quam sibi Medici usurpare solent, sed jure non admodum merito; cum ipsimet Anatomiam, tum universalem, tum particularem, nomine *Historiae corporis humani* appellare soleant." Garnier 1678, 87. In this connection, see also Pomata 2005, esp. 105 on anatomies.

58. On early modern approaches to the concept of *ars*, see Bredekamp 1995.

59. Garnier 1678, 88–89.

60. "Artificialis praeterea discernitur ab Artibus liberalibus hominum utilitati adlaborantibus, quod hae doceant quid sit agendum et quomodo; illa potissimum, quid olim factum a veteribus." Garnier 1678, 89.

61. On the opposition between these categories of *historia* and *fabula*, see Bietenholz 1994.

62. "Primum, Fabulosum hoc genus scribendi merito ferre Historiae nomen: nam et ad eamdem hominis facultatem pertinet, ad quam vera Historia, et eumdem modum eloquentiae sequitur, et ad eumdem finem conducit, et verae Historiae respondet velut umbra corpori." Garnier 1678, 90.

CHAPTER SEVEN

The Knowledge of Early Modernity
New Histories of Sciences and the Humanities

NICHOLAS POPPER

This chapter seeks to consolidate recent work in early modern intellectual history and advance a method emphasizing a fine-grained notion of practice as the motor of transformation in the production of knowledge. While intellectual histories of early modern Europe have long been divided into discrete strands such as the history of science, history of humanism, and history of political thought, scholarship in the past two decades, particularly in the history of science, has exposed the artificiality of these demarcations and illuminated the process by which they were erected, entrenched, and naturalized over time. Such work has intimated that the modern disciplines that emerged in the nineteenth century have served as a distorting lens into the past, imposing anachronistic boundaries between forms of knowledge while obscuring vital patterns of transmissions between individuals, communities, and texts.[1] Accordingly I argue that the analytic of "practice" is particularly suited for excavating such transmissions and, accordingly, for highlighting the generative, undisciplined fluidity of early modern intellectual culture.

Fully appreciating the role of exchange, border crossing, entanglement, contingency, and even the stochastic in knowledge production has momentous consequences for early modern intellectual history. While it enables a fuller apprehension of the complex work of historical subjects on their own terms, it also undercuts master narratives that have long lent significance to the period. These developments—for example, the Scientific Revolution, the rise of modern liberalism, and the emergence of historicism—have often been portrayed as propelled by autonomous internal logics whose gradual realization served as the engines of intellectual change. For the history of science in particular, denaturalizing the modern disciplines that emerged from this process also demythologizes internalist narratives of their formations.

To reject the claim that the Scientific Revolution marked the historical articulation of an ideal form of knowledge production, however, should not inhibit examination of its traditional components, nor prevent scholars from trying to build better macronarratives of historical change. Indeed, early modern European intellectual culture unquestionably witnessed a suite of momentous transformations. But these changes were not governed by forces of necessity, and the task ahead involves rethematizing the epochal shifts of this period. In what follows, I propose a history of discrete knowledge practices as an alternative to intellectual histories that presuppose demarcation and discipline as both driver and outcome of early modern intellectual work, and I offer the sliver of a case study applying this model. The framework described here aims to provide a program for interrogating intellectual change that prioritizes contingency and entanglement; while it is here directed toward enhancing understanding of early modern Europe (and predominantly its examinations of the natural world), it aspires to broad applicability as a way of discerning and interrogating the unique forms of intellectually productive complexity and messiness across places and periods.

I

When the discipline of the history of science coalesced in the early twentieth century, its practitioners honored early modern Europe with an exalted place.[2] Beginning with Copernicus and accelerating from the early seventeenth century, they claimed, luminaries like Galileo Galilei and Isaac Newton thrust forward a scientific revolution that, along with its ancillary consequences the Enlightenment and Industrial Revolution, dragged Europe from medieval darkness into the light of modernity. Their presiding concern, accordingly, was to ascertain when specific fields of study crossed into the realm of the scientific. To these twentieth-century admirers, moreover, the improved methods that early modern figures devised for knowing nature spurred the ability to predict and control it, and this mastery underlay Europe's assertion of power over the globe. The significance these early historians of science assigned to their discipline was thus predicated on a story of modernization, for they posed emergent science as the engine of European dominance.[3]

Most scholars anatomizing the development of modern science paid little attention to early modern intellectual enterprises such as rhetoric, history, philology, or law that lay outside the remit of modern scientific disciplines.[4] Contemporaneous scholarship examining these former pursuits in early modernity tended to emphasize the impact of Renaissance humanism, which was distinguished from scholasticism by its attention to rhetoric and philology and devotion to subjects outside the university curriculum, while allegedly differing from "the New Science" in its textual basis and veneration of the ancients. Through much of the twentieth century, histories of Renaissance humanism competed with the history of science to claim the dominant intellectual explanation for modernity. Hans Baron, most prominently, characterized the global conflagrations of the twentieth century as articulating a tension in political philosophy between authoritarianism and the liberal humanism elicited by communion with ancient authors.[5]

Studies of humanism and early modern science were far from parallel. For historians of science, science was an autonomous agent of modernity, its arrival marking the appearance of a wholly novel mode of producing knowledge.[6] Scientific disciplines were assumed to possess an underlying unity that, for admirers, guaranteed their practitioners' objectivity and rigor. Historians of humanistic endeavors—such as historians of political thought, of philosophy, and of historiography—operated with very different questions and goals.[7] In particular, their works infrequently addressed the formal epistemological terrain claimed by the history of science. Similarly, they were not obsessed with disciplinary origins, only occasionally debating whether the modern humanities arose from Renaissance predecessors.[8] Instead, they often portrayed their subjects as perennial, typically suggesting that humanistic modernity reconfigured long-standing conflicts—though often depicted as neglected between antiquity and the Renaissance—rather than producing new techniques of investigation for forging superior knowledge. And, by and large, such scholars were utterly unconcerned with demarcating the boundaries of humanities as a whole, for those studies in which twentieth-century scholars directed their attention to the early modern development of fields like anthropology and comparative religion did not consider whether the disciplines of the humanities or even the social sciences held shared foundations.[9]

The contrasting approaches to histories of science and the humanities reinforced the depiction of science as a species of learning separate from and sovereign over other forms of knowledge. And most efforts to correlate these competing narratives enhanced science's significance, for while scholars of humanism typically depicted their subjects as successfully revolting against medieval scholastics, historians of science portrayed their subjects as supplanting humanism.

This hierarchy only solidified in the wake of World War II, whose techno-logically driven outcome was magnified by works like C. P. Snow's *The Two Cultures*.[10]

Historians of science have continued to demarcate the sciences from other modes of knowledge even as they have reconsidered many of the discipline's other foundational assumptions.[11] For example, even the full-throated chal-lenge of the sociology of science beginning in the 1970s tacitly cordoned off the disciplines of sciences from other modes of inquiry, seeking social and political forces structuring the production of knowledge and adjudication of conflict, but rarely turning its gaze toward modes of knowing whose importance did not derive from their participation in a traditional Scientific Revolution narrative.[12] While this approach mandated the reconsideration of fundamental aspects of science—undermining the presumed unity of its disciplines, the necessity of conventional notions of progress, or the surety that any local episode conforms to a grand narrative—it also implicitly suggested that the contours of early modern science could be recognized from direct descendants. At the same time, a persistent stream of surveys covering alterations in early modern natural knowledge have hardly reconfigured the traditional boundaries of the disci-plines, practices, and communities of science.[13] Global histories of early mod-ern science situating European developments within broader context have often likewise suffered from internalizing European-derived characterizations of sci-entific knowledge.[14] Within the history of science, that is, even the most pow-erful challenges to the traditional account of the Scientific Revolution have continued to define science on the basis of modern categorizations, boundaries, and disciplines.

From the middle of the twentieth century, however, a diverse body of work has emphasized the entanglement of endeavors marked as progenitors of science with putatively extrinsic fields of knowledge. For example, Marie Boas Hall posed the Scientific Revolution as dependent on the Renaissance; Frances Yates and Betty Dobbs argued that early modern sciences were shaped by fields such as alchemy and magic seemingly incompatible with modern scientific ideals. Lisa Jardine demonstrated the significant impact of Francis Bacon's rhetoric training on his empirical philosophy; and Anthony Grafton has shown, among other examples, that Johannes Kepler's understanding of nature relied on re-constructing systems from isolated astronomical fragments in ancient texts, a practice of reading widespread among contemporary philologists, and that Fran-cis Bacon's model research institute, Solomon's House, formalized techniques of collaboration in ecclesiastical history exemplified by the Magdeburg Cen-turiators.[15] Svetlana Alpers's examinations of the visual culture of early modern science, Barbara Shapiro's analyses of the historical and legal origins of the modern "fact," and Mordechai Feingold and Jed Buchwald's investigations of

Newton's Biblical chronology similarly fall within this domain. This literature has had some spectacular successes—most notably, historians of science now consider part of their ambit the study of operational enterprises like magic and alchemy that were long dismissed as pseudoscientific.[16]

This corpus does not reflect a shared theoretical framework. But in recent years, several studies highlighting cross-disciplinary unities rather than demarcation have emerged that might provide ways of weaving the studies together. Some scholars have sought to identify the specific modes of inquiry that anchored transformations across fields; for example, Peter N. Miller has argued in a series of works that antiquarianism provided a shared foundation for early modern scholarship, and James Turner has identified philology as this common basis.[17] Rens Bod, by contrast, has called for a discipline of "the history of the humanities," which now has a bibliography of foundational works, a regular conference, and a journal.[18] Scholars in this incipient field have assumed various stances toward the relationship between the humanities and the sciences. Some have continued to observe a strict partition between the two and trace trajectories of the modern humanities without reference to the sciences; others wish to capture the prestige of the sciences for the humanities by identifying when humanistic disciplines achieved paradigmatic stability putatively characteristic of scientific knowledge; still other works have delineated unexpected continuities and exchanges. Finally—and most in line with the arguments developed below—an emerging "history of knowledge" has examined the practices of creating and converting raw information into cooked knowledge, at the same time adopting an inclusive approach to household, craft, popular, and other forms of cognitive labor previously excluded from intellectual history.[19]

Both the earlier studies and this broader-reaching recent scholarship cumulatively demand resistance to the notion that categories derived from modern disciplines can be used to map the inexorably revolving kaleidoscope of early modern European intellectual culture. And more positively, this body of work has created the imperative to reappraise transformations of knowledge in early modern Europe released from the assumptions that underlie now-enfeebled master narratives. For it has confirmed that the intellectual culture pulsating through and beyond early modern Europe was characterized by vitality, precisely because of the fluidity and permeability between different types of inquiry in an ecosystem that promoted cross-pollination. Those alterations traditionally isolated as the Scientific Revolution should thus be seen less as the organic realization of an abstract scientific method than as signs of a chaotic, dynamic intellectual culture.

The power of such works suggests that early modernists now need to work toward a new synthesis, one that maps and interrogates a transformation in the history of knowledge that was not exclusively the "scientific revolution" as classically conceived or one that also swept up the nascent humanities, but a holistic reconfiguration of intellectual life under conditions of intense turbulence. This project would ideally (a) give a symmetrical foundation to future histories of science and the humanities, showing their transformations as resulting from common forces underlying the early modern production of knowledge, (b) properly reflect the processes of globalization at work in early modernity (which my case study below admittedly does not do),[20] and (c) not just present a case study in intellectual revolutions but develop a new foundation for the significance of early modern Europe.

II

Several efforts have sought to establish the structural foundations for an intellectual history focused on the production of knowledge. For example, Lorraine Daston has produced a splendid set of articles demonstrating how "epistemic virtues" have united intellectual authority and persuasiveness across disciplines, in particular how objectivity came to assume unprecedented significance throughout all domains of learning in the nineteenth century. Bod has argued that the new history of the humanities should trace the migration of formalisms, regularities, and laws across disciplines.[21] A spate of recent works have identified paper technologies as driving forces for intellectual change.[22] Daston and Bod have also suggested an array of other approaches, including a transdisciplinary history of practice that I develop further here.

Historians of science predominantly trace their conception of practice to Pierre Bourdieu, for whom it consisted of the repertoire of social, material, and political techniques that human actors implement and improvise with in order to reproduce or challenge social and cultural fields.[23] Subsequent scholars gravitated toward his conceptualization as a way to bundle objects, actions, and values into a form that allowed them to skirt the reductionist shoals of pure structural determinism and unconstrained human agency. But the term has also often been used to contrast a "reality" unfolding in time with unfulfillable ideals or subsequent rationalizations.[24]

A sharp division exists in early modern studies, though it is effectively unacknowledged, between practice-as-technique and practice-as-reality. Most prominently, some historians of alchemy use "practice" to capture their own method of reconstructing past laboratory procedures, prioritizing the value of reenactments of material transformation to argue that the experience of their historical subjects was conditioned by material reality and therefore not explicable by allegorical interpretation alone.[25] Book historians, on the other hand, have anatomized modes of reading that more closely hewed to Bourdieu's sense of practices as constituted by discrete techniques supplied by their *habitus* to show how readers produced contingent interpretations of texts distinct from authorial intention or modern interpretation.[26] The question of whether scholars of practice should see themselves as reenacting or reconstructing practices remains open. Both groups, however, investigate the epistemic and material techniques used to highlight, erase, cull, excise, sieve, isolate, summarize, consolidate, reorganize, and transport assembled collections of material that those producing knowledge momentarily suspended from an ineluctably Protean world, then manipulated over time to form new assemblies, themselves subject to continued iterations of practice.

While historians of science have spent considerable energy focusing on science in practice, Daston and Glenn Most have observed that "we lack . . . a history of practices unfiltered either implicitly or explicitly by anachronistic criteria as to what counts as scientific."[27] But again, existing literature offers models. In addition to the above-mentioned practices adopted by traditional paragons of the Scientific Revolution, recent scholarship has shown that projects of natural history reform hinged on antiquarian collecting; that luminaries of the Scientific Revolution depended heavily on contemporaneous methods of creating manuscript notebooks; and that even arguments for the supremacy of empirical knowledge rested on textually mediated ancient precedents.[28]

A short catalog of essential early modern practices of knowledge production would include reading, eyewitnessing, collaboration, collection, visualization (or graphic rendering), and quantification. These may be crucial to intellectual history of any place or period, but they played an outsized role in the transformations in early modern Europe's world of learning.[29] They are intentionally broad and, even so, may be too narrow. For example, several could be grouped under the category of inscription, which proliferated in early modern Europe to an extent never properly articulated (indeed one might propose a narrative in which the massive production and dissemination of paper, amplified by the invention of the printing press, enabled a symbiotic growth in collaboration and inscription that distinguishes the medieval from the early modern). Yet the practices are broad enough to encompass a spectrum of subsidiary techniques that require disaggregation. For one example, collectors chose between prioritizing accumulation or curation, juxtaposition or collation, description or classification.[30]

Similarly, historians have shown that practices of reading were capacious and could be shaped both by the normative catalogs of advice elaborated in manuals and by the more adversarial modes.[31] For example, the 1566 *Bibliotheca Sancta* of the Dominican Sixtus Senensis constituted an encyclopedic overview of proper communion with the Bible according to apostolic tradition in the wake of the Council of Trent. Its third book, a comprehensive examination of the art of expounding Scripture, can be read as a catalog of the repertoire of techniques early moderns mobilized as they transformed their sources.[32]

After examining the four traditional types of exegesis (literal, allegorical, tropological, anagogical) and methods belonging to *inventio* (discerning hidden elements of Scripture, such as its rhetorical foundations, chronology, or physical laws), Sixtus described at length what he called "dispositive" or "methodical" reading, which fashioned the crude insights of these lesser forms into a more powerful grasp of the sacred. He identified twenty-four types of intervention that materialized the ineffable divinity of Scripture. Some established a basic level of order for a source that lacked intrinsic material form, such as by partitioning Scripture into chapter and verse or arranging it in columns. Many involved exercising principles of selection to emphasize specific aspects, including producing reference aids such as indexes, or excerpting *loci communes* of the most instructive passages. Still other interventions rearranged the text, as by collating passages or collecting the opinions of previous commentators, potentially interspersed with Biblical passages, either serialized verbatim or paraphrased. Similarly, other techniques augmented brief extracts with expansive explanation to produce glosses, lexicographies, and commentaries. Sixtus also detailed numerous techniques for rephrasing the holy books, such as translating, epitomizing (which reduces a passage to the fewest possible words), paraphrasing, and producing homilies in colloquial style. Other practices worked the text into radically distinct forms, such as "sciography" or "pictorial exposition," which entailed rerendering textual descriptions graphically, or the poetic method that expounded Scripture in lyric form. These were joined by the kinds of intellective analyses one might expect, including contemplative exposition intended to transform one's soul, and the inquisitory method, which dissolved ambiguities or identified heresies through disputation. For each method, Sixtus elaborated the distinctive knowledge to be gained by its implementation and supplied examples—for instance, he enthused that the images enabled by sciography could generate architectural plans of Noah's ark or maps of the Holy Land that would allow the reader to more fully grasp Scripture's literal sense.[33]

Sixtus, in short, saw explicating Scripture as an interventionist practice requiring the accumulation of sources, establishment of priorities for attention, isolation of evidence, recasting and joining of linked material, and reembodiment and circulation of rearticulated knowledge. Several points here are worth emphasizing. First, his catalog helps to highlight how discrete practices process source material into a new form. Second, for Sixtus this transformation was achieved by using reading and inscription as methods to isolate, magnify, and reembed. Finally, the striking overlap between some of his methods and those of collectors of natural historical matter noted above should suggest that both applied an elastic set of techniques that could further be directed at cosmography, history, moral philosophy, and other pursuits. Although Sixtus's catalog is by no means comprehensive, it can illuminate the attentiveness that early moderns had toward their own repertoire of techniques for redistributing and disposing the raw material of texts, experiences, and interactions into evidence or argument. Identifying overlaps with other interventions in the period suggests the possibility of a history of collation, for example, as a transdisciplinary technique of knowledge making.[34]

Early modern Europe witnessed a proliferation in both the pace and range of techniques used to dissolve sources into fragments of various shapes and sizes and then convert these shards into resplendent mosaics of knowledge. And the application of specific practices did not mechanically re-form the sources—practices are not deterministic, but instead open possibilities. Historians should aspire to trace how individuals and communities improvised with their techniques— how exchanges, transmissions, adaptations, and slippages among practices of dissolution and reintegration constituted the dynamic of early modern intellectual culture. The emphasis on practice should magnify the kinesis rather than the *telê* of early modern knowledge.[35]

III

The transdisciplinary study of mathematization—of rerendering texts, natural phenomena, or any other aspect of the encountered world according to the language of number—can serve as an exemplary case. The mathematization of natural philosophy has long been recognized as essential to the Scientific Revolution, as historians have identified an emerging consensus in early modern Europe around the notion that the world was explicable by quantitative analysis and that the actions of earthly phenomena—including, by the end of the seventeenth century, human economies and populations—could be analytically reduced to purely mathematical relations.[36]

Mathematization, however, effected meaningful change in many of Europe's learned activities well before the seventeenth century. For example, double-entry bookkeeping, perspective painting, Copernican cosmography, and astrology all situated experience and observation within mathematical reckoning.[37] But there was another site of inquiry where scholars sought to find a common grammar for explaining the mechanics of change through mathematical analysis: the study of the past. Examining the study of history shows how the vitality of practices of quantification in the sixteenth century flowed into the transformations of the seventeenth.

The practice of quantification included a number of subsidiary techniques. The following analysis revolves on three distinct forms of practice used to plot phenomena as mathematical entities: first, an arithmetic technique assigning basic numerical identities to phenomena; second, a geometrical one of plotting them graphically as linear or nonlinear motions; and, third, a harmonic technique of abstracting arithmetical identities into proportional relationships.

Throughout the sixteenth century, the effort to mathematize history applied all these methods. In fact, the Christian study of the past had long relied on basic arithmetic in the form of elemental chronology.[38] The Reformation provoked a scholarly arms race in which figures throughout Europe argued for the superiority of their beliefs by shaping a thicket of sources into elegant world chronologies.[39] Their primary mode was correlating events across time, using the mathematical identifiers of dates. The sixteenth century saw feverish increase in this activity, spurred in part by the cosmographer Peter Apian's suggestion that obscure events could be dated by coordinating astronomical events across cultures. Copernicus himself, along with the Dominican Giovanni Maria Tolosani and the Lutheran Johann Funck, began to do so from the 1530s and he, among others, used historical accounts of eclipses to conflate (inaccurately) the Babylonian king Nabonassar with the Assyrian king Shalmaneser, an identification that forged an invaluable hinge between sacred and secular history. The cascade of chronological studies led scholars from the 1560s to complain that entering into the field was like descending into a labyrinth. The desire for an Ariadne's thread provoked Joseph Scaliger to devise the Julian period, a single dating system onto which chronologers could pin events designed to absorb all calendars first published in 1583; this tool provided a shared arithmetical framework with which to order and communicate all chronologies.[40]

To force Scripture, ancient historians, astronomical observation, and the burgeoning production of contemporary chronologers into one uniform time frame often required purely arithmetical speculation. Walter Ralegh, for example, behaved like other chronologers when, in trying to figure out when the ancient Egyptian king Osiris died, he made his solution work by simply subtracting twenty-three years from Osiris's reign. This decision was not supported by any specific sources about Osiris; Ralegh's logic derived solely from the conviction that establishing mathematical continuity across sources was the greatest imperative for chronology. Arithmetical concurrence outweighed the authority of any testimony.[41]

Astronomy and arithmetic thus flowed into technical chronology; indeed, many of the chronologers noted above treated astronomy as predominantly useful for reconstructing world history. And chronology was only one way in which mathematical reasoning was directed toward historical problems; others were exemplified by Jean Borrel's *Opera Geometrica* (1554), which taught geometry by using it to reconstruct Noah's Ark and the Ponte Sublicio with which Caesar's engineers bridged the Rhine.[42]

Some viewed the mathematical relations between events as not merely descriptive. In his widely read *Methodus ad facilem historiarum cognitonem* (1566), the French historian and political theorist Jean Bodin instructed readers to scrutinize the rise and fall of states to yield causal patterns and form advantageous counsel.[43] Like his contemporaries, Bodin believed that since political states had been created and were thus transitory and finite, they could be analyzed using the stages drawn from Aristotle's *De generatione*—birth, growth, maturity, division into realms, decline, and death. Bodin saw states' transitions along this continuum as a species of movement and, accordingly, took his task to be revealing their underlying mechanics. He was convinced that these rises and falls adhered to numerological rules and that, as he explained in a section entitled "changes in states correlated with numbers," the alterations visited by divine providence were governed by mathematical regularities that would be revealed by processing the data of world chronology into the language of number.

Bodin began his discussion by criticizing what he saw as similar projects. The Wittenberg lawyer Valentin Forster's *De historia iuris civilis* (1565), he recounted, insisted that states' declines could be understood as the gradual distortion of Plato's harmonic ratios in their judicial systems.[44] Bodin rejected Forster's interpretation for, as he argued, some discordant ratios united in harmony.[45] He also criticized the Copernican claim—articulated in Rheticus's *Narratio* and related by Forster—that "the changes of empire were related to the center of a small eccentric circle [belonging to the earth] . . . and to its motion."[46] Bodin rejected this proposition as presuming a "force exerted from the centers of celestial circles," which "no one was ever so lacking in knowledge as to think." Instead of these harmonic and geometrical replottings, he turned instead to arithmetical reasoning, claiming that pure numerology would decode the logic of change in the world. "I have noted," he explained, "not without wonder in various cases, that changes happen in states in multiples of either seven or nine, or in the squares of seven or nine multiplied together, or in the perfect numbers, or spherical numbers."[47] This was equally true for individual biographies, as many examples involving shrewd manipulation of these numbers demonstrated. But he then sought to show that "the square and the cube of twelve, which is called the great number of Plato by the Academicians" was even more essential to understanding the lives of states, as a vertiginous sequence of examples from Assyrian, Chaldean, Persian, Egyptian, Greek, Hebrew, Roman, and French history revealed.

Bodin expanded further on this numerological criticism in his *Six Livres de la République* (1576). In the section "whether there be any meane to know the changes and ruins, which are to chaunce unto Commonweales," Bodin again blasted Copernicus for tying the vicissitudes of political entities to the eccentric motion of the earth, in a passage that led Bodin's translator Richard Knolles to characterize Copernicus as "the great Astrologer of his time."[48] Bodin then declared:

> Of all things which we have yet brought to judge of the future changes and ruins of Commonweales, we see no rule (whether it be Astrologie or musike) certaine

and sure: howbeit that we have by them some probable conjectures, whereof yet none seemeth unto mee more certain or easie, than that which may be drawen from numbers. For why I thinke almightie God who with wonderfull wisdom hath so couched together the nature of all things, and with certain their numbers, meanes, measures, and consent, bound together all things to come: to have also within their certaine numbers so shut up and enclosed Commonweales, as that after a certaine period of yeares once past, yet must they needs then perish and taken end, although they use never so good lawes and customes."[49]

Bodin thus reinforced his certainty that the patterns of world history could be best understood by filtering them through numerology, which he contrasted to astrological geometry or musical harmony.

Elsewhere in *La République*, though, Bodin reestablished a footing for the study of political harmonics. In its final book, he insisted that the judicial and political logics of democracies, oligarchies, and monarchies were respectively arithmetical, geometrical, and harmonic. He found in ancient histories evidence that a state's adherence to strict geometric or arithmetical logic led to inevitable destruction, while absolute monarchy's harmonic combination of the arithmetical and geometrical presented an ideal type balancing the rigidity of the arithmetic with the pliability of the geometric. "Both in making of laws," he argued, "and in deciding of causes, and in the whole government of the Commonweale, we must still so much as possible is, observe and keepe that Harmonicall proportion, if we will at all maintain equitie & iustice; whereas otherwise it will be right hard for us in the administration of iustice, not to doe great wrong."[50] In Bodin's system, stability or discord resulted from the harmonic reverberations that governed the lives and motions of states.

Bodin's approach inspired a range of ways of reducing human history to mathematical analysis. His numerological chronology provoked studies like that of the Scottish reformer Robert Pont, whose *A Newe Treatise on the Right Reckoning of Yeares* (1599) showed that all human events were ordered by the fundamental logic of jubilee periods of forty-nine years. Sacred numerology also underlay Petrus Lindebergius's eccentric *De praecipuorum tam in sacris quam in Ethnicis scriptis numerorum Nobilitate, mysterio, et eminentia* (1591), which the author produced by consulting Scripture and other sources, recording any extract that mentioned a select range of important numbers, and then compiling them into a commonplace book. Lindebergius thus sought to decode the patterns of world history by directing the techniques of collection, extraction, and collation toward numerological explanation.[51]

Other scholars absorbed Bodin's plea for a mathematical basis for understanding the mechanics of states' transitions between stages, but instead geometrized them according to Aristotelian dynamics, which contrasted natural, circular motion with violent, rectilinear movement. Michael Eytzinger's *Pentaplus regnorum mundi* (1579) analyzed the transformations of empires as species of motion: "Progress [*progressio*] is a motion from the beginning proceeding through the middle, until it comes to rest in its end. Which motion is perceived to be found not only in natural things, but in civil things, such as realms or kings, and in eras too."[52] Eytzinger explained that the histories of empires proceeded by Aristotelian dynamics: "There are two kinds of progress. One is rectilinear progress, by which God the founder of all things founds and constitutes new realms, according to this: 'Behold, I make all things new' [Revelations 21:5]. The other is circular motion, by which God omnipotent renovates old realms and empire, according to this: 'The thing that hath been, it is that which shall be; and that which is done is that which shall be done' [Ecclesiastes 1:9]."[53] History was full of rectilinear kingdoms, which arose and disappeared as violent spasms in time, but the Four Monarchies and the Empire of Christ alone provided examples of circular motion that would persist eternally.

Gabriel Harvey developed a similar analysis in notes written in his copy of the Ramist Johann Thomas Freigius's *Mosaicus* (1583). Freigius had expounded a moral explanation for the mutations of empires: "Sacred Scripture alone indicates the superior and principal cause of conversions: that is to say, idolatry, injustice and tyranny of governors, sins of all kinds and shameful acts by princes, priests, and people."[54] To further refine this explanation, Harvey turned to Ramus's textbook *Scholae Mathematicae* (1569),[55] in which Ramus had famously pleaded for a practical mathematics. Harvey, like Bodin, took up this call to mathematize political science. Below Freigius's passage, he wrote: "Still, in Physics and Geometry, balance and equality is the cause of the rest and position of all things in all of nature, and imbalance the cause of all motion and ruin, as Ramus most beautifully philosophizes in book 2 of the *Scholae Mathematicae*: 'all things seek to rest at right, not at oblique, angles.'"[56] Harvey understood terrestrial events as legible within a sacred system of mathematical dynamics; the lives of kingdoms could be recast as divinely determined periods of violent motion and blissful rest.

Others translated Bodin's conception of harmonies in human societies to new spheres. At the end of book 3 of his *Harmonice Mundi* (1619), Johannes Kepler examined Bodin's employment of harmonies in *La République*. Surprisingly, Bodin was the most cited modern authority in the work, for Kepler singled him out for using harmonic principles to plot empirical relations between corporeal bodies. As Aviva Rothman shows, Kepler saw Bodin's project as uniquely in league with his own, writing, "Eventually Bodin compares the kingdom which he has described with the actual world, showing how God the creator has embellished this work of his by joining the ratios of equal and of similar in one concerted harmony. I agree with his purpose, as much as anyone, and what he or the preceding philosophers have not even touched on, which concerns the most accurate harmonic tempering of certain motions, I supply in the books which follow, and bring to light by the clearest demonstrations."[57] Kepler admired Bodin for seeking to impose order on a chaotic mass of particulars by rearticulating it as harmony, and accordingly he translated Bodin's attempt at a mathematically exact harmonic system to the heavens.

But while Kepler approved of the frame of Bodin's project, he argued that a tenuous grasp of technical details led Bodin to map them onto states confusedly. And he did not insist that more precise mathematical knowledge would address this weakness. As Rothman shows, even as Kepler sought a faultless score of the music of the spheres, he saw Bodin's efforts to mathematize prescriptive melodies of politics as forcing an elegant austerity on the irreducible complexity of human affairs. Thus even as Kepler revealed the harmonies governing the sky, he limited the mathematical ambitions articulated by Bodin.

The irony that Bodin was more optimistic about the universal applicability of mathematics than Kepler should not be lost as we appraise the long trajectory of the history of science. Indeed, following this arcane strain of early modern numerology reveals how smoothing out the texture of early modern intellectual culture distorts understanding of intellectual change. At the most basic level, the figures associated with the Scientific Revolution appear here in profoundly different light: Copernicus was a chronologer and an astrologer as much as an astronomer, and his revelation of deep history resonated as powerfully as his reordering of deep space; Ramus's applied mathematics could be directed toward the realm of the political as readily as to the natural; Kepler saw his astronomical harmonies as emanating from the same foundation as Bodin's arcane judicial theory, even as he was less sanguine than Bodin about the universal applicability of mathematization. Tracing shifting understandings of the stars requires looking beyond works and scholars solely concerned with heavens, lest we lose the ability to see the interplay of multiple fields involved in intellectual change.

The preceding analysis also shows the value of distinguishing between discrete practices of plotting events or motions mathematically, for what Kepler drew from Bodin was not merely the desire to quantify earthly phenomena, or an extension of the Copernican geometrical cosmology, but particular techniques of mathematization.[58] Early modern Europe generated many specific practices of quantification, and these were exchanged, borrowed, and modified as they crossed studies of law, history, reading, political theory, and astronomy. The exercise introduced here constitutes only one strand in a larger synthesis of the motility of practices that needs to be written.

Finally, tracing the use of quantification in early modern historical analysis illuminates a perhaps unexpected chronology. Stirrings of the mathematization of the human occurred at least contemporaneously with the nascent mathematization of nature, and those who examined the natural world adapted techniques initially developed to analyze social and political phenomena.[59] Much as the philosophy of Francis Bacon might be understood as an adaptation of the sixteenth-century culture of collection, some species of mathematization borrowed from the whirlwind of projects with humanistic ambitions. Crucial aspects of modern science emerged during the early seventeenth century, this point suggests, not through the forging of wholly novel modes of investigating nature but through innovative redirections—if not pathologies—of methods pioneered in the study of human societies toward the natural world. And the eventual divestment of forms of quantification from history points to the possibility of practice-oriented studies of the rise of disciplines, focused on how specific practices came to dominate study of particular phenomena and on the process by which disciplinary regimes formalized rules prioritizing canonical techniques, banishing others, and rechanneling the distinctively protean early modern circulation of practices.

IV

The intellectual fertilizations associated with modernization could not have happened without exchange and collaboration across what now appear to be disciplinary lines. It is certain, of course, that narrower content-related threads knit together specific traditions—early modern astronomy, to take one example, is in part a story of transit from Copernicus to Galileo to Newton. It should be no less certain, however, that for each of these figures, improving on their predecessors' inheritance entailed incorporating new practices of producing evidence and devising meaning adapted from other environments. Their work should be seen not as the unfolding of a natural lineage but rather as constituting creative syntheses in a condition of volatility.

Recognizing such transdisciplinary ferment in early modern Europe has significant consequences. Above all, it reveals that the conventional accounts of the Scientific Revolution have been made possible only by a distorting model of historical inquiry that effaces and denies such borrowings, a distortion parallel—though not morally equivalent—to how it has often elided the labor of "invisible technicians" whose gender, race, class, ethnicity, or social status inhibited their ability to claim credit for their work.[60] These exclusions reflect a broader intellectual tradition that has prioritized border drawing and demarcation over exchange, contingency, and hybridization, and whose assumption that ideas tend toward purifying self-realization has directed attention away from complex, disorderly constellations of people, labor, practice, text, and nature (often united by horrifically violent or viciously insidious asymmetries of power).[61] There remains always the need to tell macronarratives and to explain why specific individuals appear prominent within them. And there remains always the need to make distinctions—to discern the separation, for example, between the universes of Sacrobosco and of Newton. But we can improve our understanding by rejecting models that enshrine anachronistic hierarchies and demarcations, and instead strive to devise a model for anatomizing intellectual change that captures the dynamism and fluidity of human affairs. This ambition is especially critical when examining the early modern moment of global collision. Emphasizing practice constitutes an important dimension of this model, as it foregrounds the discrete techniques underlying moments of transformation and perpetuation rather than assuming that world-altering intellectual change arises from disembodied reason or structural necessity. It allows us to see the world as constantly subject to reconstitution.

Developing a firmer apprehension of intellectual change demands that we confront this distinctive weakness in our broader intellectual tradition. We need instead to configure a method that recovers the hidden practices and muted voices that catalyzed intellectual transformation, while honoring mutuality, exchange, and symbiosis as the foundations of human creativity. In this way, we can restore recognition of both the vitality of early modernity and the interrelationship of domains now cleaved into the humanities and the sciences, but we can also practice intellectual history in a way that can serve as a model for the production of innovative and experimental knowledge in the present.

NOTES

I would like to thank the other contributors to this volume, as well as Lauren Kassell, Christian Flow, and audiences at Cornell University's STS Seminar Series and King's College London for feedback and criticism that improved this chapter immeasurably.

1. Harman and Galison 2008; Daston and Most 2015; Marchand 2013.

2. Exceptions include Otto Neugebauer's (1951) vision of science as originating in the ancient Near East and imperial Greece and Pierre Duhem's (1913–59) prioritization of medieval Europe.

3. For the early history of science and the Scientific Revolution, see H. Cohen 1994. Note that even twentieth-century Marxist historians of science such as Edgar Zilsel, while emphasizing the role of labor and material factors in the Scientific Revolution, contested its causes but not its contours. The same holds for those such as Robert Merton who saw Protestantism as essential to the rise of early modern science.

4. On disciplinary thinking in intellectual history, see Latour 1993; Marchand 2013; and Rheinberger 2016. For early modern divisions between fields of knowledge, see Blair 2008.

5. The most prominent examples include Baron 1955; Pocock 1975; and Skinner 1978.

6. These claims have recently been revived in H. Cohen 2010 and Wootton 2015.

7. On the shifting lines of demarcation, see Rheinberger 2010 and Bouterse and Karstens 2015.

8. Most prominently, as Clark shows in this volume, some historians of early modern historical thought did seek to establish modern historicism in the early modern period, sometimes drawing on Thomas Kuhn to map shifts in paradigms. See, for example, Fussner 1962; Levy 1967; Kelley 1970; Huppert 1970; and Franklin 1977.

9. See, for example, Hodgen 1964 and Weinberg 1961.

10. Snow 1959; Ortolano 2009; Rheinberger 2016.

11. Smith 2009.

12. Most representative of this canon is Shapin and Schaffer 1985. This scholarship, furthermore, tended to frame the multiplicity of early modern knowledge as conflictual and as revealing social fractures. See, for example, Biagioli 1993.

13. See, for example, Shapin 1996, whose famous beginning—"There was no such thing as the scientific revolution and this is a book about it"—has distracted antagonistic readers from noticing that the book defends the notion of the Scientific Revolution. See also Smith 2009. Similarly, recent challenges from philosophers of science to the traditional narrative have contested the framework of interpretation more than the significance of the subject; see, for example, Garber 2016.

14. Cf. Huff 2011; Burns 2016; and Gómez 2017. More effective have been efforts to write the history of early modern science from a Mediterranean perspective; see Saliba 2007 and Zaken 2010, though these two have been perhaps excessively conditioned by modern demarcations.

15. See Grafton 1991b and Grafton (2001) 2009; other relevant works include Grafton 1990b, 1996, and 1999 and the coedited volumes Grafton and Siraisi 1999 and Grafton and Newman 2006.

16. Hall 1962; Yates 1964; Dobbs 1991; Alpers 1993; Shapiro 2000; Buchwald and Feingold 2012; Jardine 1974. For alchemy's place in the Scientific Revolution, see works by William Newman (e.g., 2004, 2006) and Lawrence Principe. For other examples of this sort of work, see P. Miller 2015; Levine 1977; Haugen 2011; and Serjeantson 2006.

17. See P. Miller 2007 and Turner 2014.

18. Indeed, it is striking that while the history of science has long flourished, there was until recently no such thing as a formal history of the humanities, and neither grand narratives nor disciplinary frameworks have emerged for formal disciplines examining the history of the humanities or social sciences. It is as yet impossible to say whether the new field marks a moment of revitalization or an epitaph. See Bod 2013; Bod, Maat, and Weststeijn 2010–14; Turner 2014; and the *Isis* Focus section Bod and Kursell 2015. See, as a forerunner, Grafton and Jardine 1986.

19. Burke 2000, 2015b; Mulsow 2019 (with comment by Daston and response by Mulsow); Jacob 2007–11, vol. 1; Smith and Schmidt 2008; Lässig 2016; Marchand 2019. I would like to thank Marchand for agreeing to share the manuscript with me prior to publication.

20. See, for example, Grafton with Shelford and Siraisi 1992; Bleichmar 2012, 2017; Sweet 2013; Ghobrial 2014a, 2014b; Gómez 2017; Strang 2018; and Bevilacqua's contribution to this volume.

21. See Bod 2015; Bod and Kursell 2015; and Bod, Maat, and Weststeijn 2010–14; Daston and Galison 2007; Daston 2014; Daston and Most 2015.

22. See Blair 2010; Mendelsohn with Hess 2010; Yeo 2014; Bittel, Leong, and Oertzen 2019; and Yale 2016.

23. Above all, see Bourdieu 1977, as well as Soler et al. 2014.

24. Perhaps most robustly, Andrew Pickering (1995) used the term to characterize science as a form of action rather than a body of knowledge and stressed empirical descriptions of the contextual, material, and tacit processes by which science intervenes in the world rather than actors' fulfillment of idealized cognitive norms.

25. See, for example, Rampling 2014; Principe, Fors, and Sibum 2016; and Newman 2006. Another prominent example is the Making and Knowing Project at Columbia University, led by Pamela Smith; see Bilak et al. 2016 and www.makingandknowing.org.

26. See Jardine and Grafton 1990; Blair 2010; and Raphael 2017. For an example within the realm of alchemy, see Kassell 2005.

27. Daston and Most 2015, 385.

28. See Findlen 1996 and 1997; Yeo 2014; and Siraisi 2007.

29. My suggestions here also draw on Pickstone 2000; Smith and Schmidt 2007; Biagioli and Riskin 2012; and Smith, Meyers, and Cook 2014.

30. See Findlen 1996 and 1997.

31. See Hankins 1990, 1:3–26; Blair 2010; Daston and Lunbeck 2011, esp. chs. by Park, Pomata, and Daston, 15–114; and Krämer 2014.

32. In a similar vein, see Flow 2015, esp. 69–73.

33. Sixtus Senensis 1566, 214–62.

34. For examples, see Grafton 2016 and Grafton and Most 2016a.

35. Smith 2014.

36. See the works of Edmund Husserl and Alexander Koyré, as well as Dear 1995.

37. See Soll 2014; Kemp 1990; Alberti 2011; Smoller 1994; and Azzolini 2013.

38. See Grafton and Williams 2006 and Nothaft 2012.

39. Van Liere, Ditchfield, and Louthan 2012.

40. Grafton 1983–93, vol. 2; Rosenberg and Grafton 2010.

41. Ralegh 1614, 241. See also Popper 2012 and Haugen 2012.

42. Buteo [Borrel] 1554.

43. On Bodin and history, see Grafton 2007b. On Bodin, see Blair 1997. For politics through the veil of practice, see Soll 2005; Popper 2012; and Keller 2012, 2015.

44. Forsterus 1565.

45. Bodin 1945, 223–24.

46. On Rheticus's *Narratio*, see Swerdlow and Neugebauer 1984.

47. Swerdlow and Neugebauer 1984, 234–35.

48. Bodin 1606, 454.

49. Bodin 1606, 457.

50. Bodin 1606, 784.

51. Pont 1599; Lindebergius 1591.

52. "Est autem progressio, motus a principio per media eousque procedens, donec in suo fine conquiescat. Qui quidem motus non in naturalibus dumtaxat rebus, verum & in civilibus, utpote Regnis, Regibus, ac temporibus quoque inveniri animadvertitur." Eytzinger 1579, 14.

53. "Progressionum autem duo sunt genera. Est enim progressio alia Recta, qua conditor omnium rerum Deus optimus maximus (ne sempere idem facere videatur) nova Regna condit atque constituit, iuxta illud. Ecce nova facio omnia. Alia Circularis, qua idem omnipotens Deus (ne quando rerum quas fecit intercidat memoria) vetera regna atque Imperia

renovat, iuxta illud, Quid est quod fuit? Ipsum quod futurum est. Quid est quod factum est? ipsum quod est faciendum." Eytzinger 1579, 14–15.

54. For Harvey's historical reading, see Jardine and Grafton 1990. Johannes Thomas Freigius wrote in 1583: "De causis vero mutationum & conversionum in monarchiis, regnis & Rebuspubl. multa sapienter disputat Aristot[eles] lib. 5 Polit[icorum libri] . . . tamen superiorem & principalem causam conversionum solae sacrae literae indicant: videlicet idolatriam, iniustitiam ac tyrannidem gubernatorum & omnis generis peccata & flagitia in principibus, sacerdotibus & populo." Freigius 1583, 115; Harvey's copy is now in the British Library, shelfmark c.60.f.4 (hereafter referred to as Harvey's Freigius).

55. In this context, see Goulding 2010.

56. "Etiam in Physiciis, et Geometricis, omniumque in universa rerum naturae; Aequalitas, Quietus, et Status causa est: contra, Inaequalitas, motus, et ruinae. Ut pulcherrime philosophat Ram[us] libro 2. Scholarum mathematicarum: Ad rectos angulas quiescere, consistereque omnia, non ad obliquos." Harvey's Freigius, 115.

57. Kepler 1997, 278; A. Rothman 2017.

58. For the later injection of probabilistic mathematical reasoning into history, see Stigler 1986 and Craig 1964.

59. William Petty's pioneering mathematization of the knowledge of human societies later in the seventeenth century—one focused on geometric rather than arithmetic or harmonic reasoning—thus appears as a subsequent moment of fertilization rather than an origin point. See McCormick 2009.

60. Shapin 1989; see also Blair's contribution to this volume.

61. Recent years have seen an emphasis in early modern history of science on "connected histories," following predominantly on the work of Sanjay Subrahmanyam (especially 1997). Exemplary works that reveal the significance of, for example, African and Indigenous knowledge and labor to early modern intellectual transformation include Parrish 2006; Cook 2007; K. Murphy 2011; Bleichmar 2012; Breen 2013; and Gómez 2017.

PART FOUR

EVIDENTIARY HORIZONS

CHAPTER EIGHT

Material Histories

Museum Objects and the Material Culture of Early Modern Europe

AMANDA WUNDER

The historical study of material objects and craftsmanship has surged in popularity in these first decades of the twenty-first century, at a historical moment when teaching, scholarship, and everyday life are increasingly tied to digital technologies. The field of early modern European history has seen a boom in material culture studies as scholars trained in various subspecialties have engaged in thinking about the relationships between early modern people and the things that they made, bought, traded, and collected in Europe and around the globe.[1] The material turn has brought attention to a wide variety of early modern objects, some of which were previously ignored, tucked away in museum storage rooms. A material culture approach to early modern European history has driven significant shifts in professional practices by inspiring collaborative and interdisciplinary approaches to research and pedagogy and by forging new relationships between academic historians and museum curators and conservators, sometimes leading to exhibitions.[2] Historians who have long experienced the special "tangible contact with the past" that comes from working with early modern printed books, manuscripts, and documents in rare books rooms, libraries, and archives are increasingly turning to objects in museum collections that provide another kind of connection to the cultures that produced them.[3]

Before we can begin to use these objects for historical research, though, we must first investigate how the objects and our knowledge of them have been shaped by the collecting practices that have made them accessible to us today.

In the seminal essay "Mind in Matter: An Introduction to Material Culture Theory and Method" (1982), Jules David Prown enjoined scholars to use historical objects "actively as evidence rather than passively as illustrations." Provocatively designating objects "the only historical occurrences that continue to exist in the present," Prown argued that scholars are able to share sensory experiences with their historical subjects by engaging directly with objects from the past.[4] Historians who take Prown up on this heady promise inevitably encounter obstacles in putting material culture theory into practice when working with an object in a museum collection. Objects do not survive the centuries unchanged (as Prown acknowledged); in addition to suffering the deleterious effects of the passage of time, objects evolve continuously as a result of their relationships with people and institutions. These changes are not always readily apparent or visible to the naked eye, and failing to see them can lead to disastrous errors of interpretation. As objects change over time, so does our knowledge about them. This is especially true when it comes to anonymous works and objects in media (such as textiles or polychrome sculpture) that were valued highly by the early modern cultures that produced them but enjoyed less cachet among subsequent generations. Finally, Prown's methodology was fundamentally based on sensory experiences—not just looking, but touching and handling an object as its original owners would have done. But museums typically have strict rules governing who is allowed to handle their objects, and many artworks and artifacts on exhibit may be seen only behind display cases. The vast majority of objects are out of sight in storage; at the Metropolitan Museum of Art in New York City (The Met), for example, approximately 6 percent of the museum's enormous collection of European objects made between 1400 and 1800 is on display at any given time.[5] It is worth noting that Prown, an art historian, was a professor and curator of American art at the Yale University Art Gallery—a position that granted unfettered access to the objects he studied.

This essay explores some of the practical aspects of working with historical objects as a historian of early modern Europe in an American museum collection. Based on years of research and teaching with the collection at The Met, this chapter explores the material histories of three objects that were made or used in early modern Spain: a painted portrait of an unidentified man, a length of brocaded velvet, and a sculpture of the Christ Child that includes a set of garments to dress it. These studies are organized chronologically according to each object's accession by the museum—not by when it was made—in order to highlight the important role that collecting practices have played in the objects' lives and our understanding of them.[6] According to Prown's categorization of object types, the three objects examined below belong to the categories of art or adornment, yet for different reasons none of them is a kind of art that traditionally has been attended to by art historians.[7] The case studies that follow draw upon a variety of museum resources—including departmental research files, curatorial expertise, and conservation studies—and firsthand examination of the objects themselves in order to reconstruct their material histories. While each object has its own unique history, taken together they offer some general insights and lessons for historians about the ways early modern European museum objects and our understanding of them have been shaped by relatively recent trends and practices in collecting that are continuously changing.

I

The first case study begins in The Met's Department of European Paintings, where an anonymous late seventeenth-century Spanish portrait of a man hangs in storage. Measuring about two feet tall, the *Portrait of a Man* is a bust-length likeness of a dark-haired, mustachioed, middle-aged man who wears the stiffly starched white collar and monochrome black garments that were de rigueur for a seventeenth-century Spanish gentleman (see fig. 8.1). The painting has been cut out of a larger canvas that represented a religious scene: the man is positioned in front of a partially visible altar draped with rich fabrics and set with silver ornaments. With its juxtaposition of an austerely dressed gentleman against a richly ornamented altar, the *Portrait of a Man* has potential interest as a source for the study of Counter-Reformation Spanish material culture—but first we have to reconstruct what the original artwork would have looked like. As it turns out, the painting has gone through several major changes that are detailed in the object file maintained by The Met's Department of European Paintings, which includes correspondence, photographs, publications, bibliography, and notes taken by curators and conservators over the decades since the painting entered the collection in 1931.[8] The story of this painting's material history, which is full of strange coincidences and unexpected twists and turns, dramatizes some of the ways that twentieth-century collecting practices have shaped the evidence available to scholars working with museum objects today.

The *Portrait of a Man* looked very different than it does today when it appeared on the London art market and was purchased by Eugen Boross, a banker and art collector in Larchmont, New York, in 1922 (see fig. 8.2). At some prior point, the background of the portrait had been overpainted to hide the fact that it had been cut from a larger work. A red curtain had been added to cover up the altar, and the man's hair had been extended into a lush mane that reached his shoulders. His right hand, from which a rosary dangled, had been covered with black paint to blend into his garment. These changes effectively secularized the portrait—which, in its altered state, resembled the famous self-portrait of the Sevillian painter Bartolomé Esteban Murillo that had been purchased by Henry Clay Frick in 1904.[9]

August Mayer, curator of paintings at the Alte Pinakothek in Munich, examined Boross's *Portrait of a Man* and attributed it to Murillo, whose work was in high demand among American collectors.[10] "Spanish art was in vogue," as historian Richard Kagan put it, and no Spanish artist was more popular than Murillo at this time.[11] Mayer published Boross's *Portrait of a Man* with his Murillo attribution in a short article about Spanish portraits in foreign collections that appeared in the June 1923 *Boletín de la Sociedad Española de Excursiones*, a publication designed to promote Spanish culture and local tourism and to document Spain's monuments in photography.[12] In this same article, Mayer also published a bust-length portrait of an elaborately dressed woman in front of an altar that he confidently attributed to the seventeenth-century Spanish court portraitist Juan Carreño de Miranda, a painting then with the New York dealer Scott & Fowles (see fig. 8.3). A Spanish lawyer named Fidel Pérez Mínguez read Mayer's article in the *Boletín* and recognized the likenesses of the man and woman from a different painting, a family portrait that hung in the Chapel of Cristo de Gracia in the village of Las Navas del Marqués (province of Ávila). The group portrait in the chapel, which included a husband, wife, and two children, had an inscription identifying the man as a merchant named Cristóbal García Segovia (1633–92) and the woman as his wife, Isabel Gabriela de Ingunza (1645–ca. 1700).[13] Pérez Mínguez was certain that the anonymous man and woman he had seen in Mayer's article in the *Boletín* were also portraits of García Segovia and his wife, and he speculated that the paintings that the German art historian had attributed to two different artists might actually have been cut out of the same canvas. He published his findings along with a black-and-white photograph of the chapel family portrait, which was in terrible condition, under the title "Desconocidos descubiertos" (Unknown Persons Revealed), in the September issue of the *Boletín de la Sociedad Española de Excursiones*. The group portrait of the García Segovia family in the chapel was subsequently destroyed during the Spanish Civil War.

Pérez Mínguez's identification of the *Portrait of a Man* as the likeness of Cristóbal García Segovia was ignored when the portrait went back on the market in 1931 through the art dealer Franz Kleinberger. Still being represented as a work by Murillo, the portrait's subject was identified, for no discernible reason, as don Antonio Álvarez, the Spanish ambassador to England. The New York City collector Michael Friedsam bought the portrait in January 1931, just a few months before his death. Friedsam bequeathed his large collection of European paintings—including this recently acquired Spanish portrait—to the Metropolitan Museum of Art.[14] The museum's paintings conservation department X-rayed the *Portrait of a Man* and discovered that the painting was a fragment of a larger canvas and that it had been "made to appear like a complete picture" through the addition of the red curtain covering the background. Conservator Stephen Pichetto "cleaned off the repaints and put on a coat of varnish," according to the conservation report, but it was decided not to restore the work—which, now that its true nature had been revealed, was deemed hardly worth the effort.[15] The cleaned *Portrait of a Man* was exhibited alongside masterpieces like Vermeer's *Allegory of Faith* in a special exhibition of the entire Friedsam collection at the Metropolitan Museum of Art between November 1932 and April 1933. The Spanish portrait was then retired to storage, where it has remained ever since.

The technical study and cleaning of the *Portrait of a Man* revealed that the background of The Met's fragment perfectly matched the altar behind the woman whose portrait August Mayer had published as a work by Carreño de Miranda. The amateur art sleuth, Pérez Mínguez, had been absolutely right: the two paintings had been cut from the same canvas.[16] Meanwhile, the portrait of the woman had been purchased by Paris Singer, heir to the Singer sewing machine fortune, in 1925. His daughter-in-law, Margaret Singer, decided in 1960 to donate the painting to the Metropolitan Museum of Art and wrote to invite the paintings curator Theodore Rousseau to view the picture in her home in Paris on the Place de Vosges. "My half of the Miranda is in much better condition than yours!" she wrote teasingly to the curator, maintaining Mayer's attribution of the picture to Carreño de Miranda. Singer claimed to have sent the painting to the museum, but there is no record that it ever arrived. Its present location is unknown.[17]

It is possible to reconstruct roughly what the original painting would have looked like based on the surviving painting of the man (fig. 8.1) and black-and-white photographs of the missing portrait of his wife (fig. 8.3). The complete original painting, which probably featured full-body portraits of husband and wife kneeling, likely would have been over six feet tall. The woman's off-the-shoulder gown, puffed sleeves, and hair ribbons date the painting to the period between 1660 and 1680. The man and woman formed a prayerful pair, with a book of devotions in her hands and a rosary in his. Their bodies faced each other, and the woman looked directly out at the viewer while the man gazed sidelong at an unspecified point to his right (the viewer's left)—perhaps indicating a devotional object or inscription at the original site where the painting hung, whether a church or a private oratory. The altar behind the couple in the painting was covered by a red and gold frontal and a white lace-trimmed altar cloth. On top of the altar, two silver candlesticks held lit white tapers, above which silver lamps hung from chains. At the center of the altar, a small sculpture of the Virgin was mounted on a silver base with an inscription, "Nuestra Señora de . . ." (Our Lady of . . .). The rest of the inscription, which identifies the image, is illegible in the photographs that document the woman's side of the painting. The sculpture of the Virgin wore a dress with a gold-trimmed hem, which is visible in the surviving fragment of the *Portrait of a Man*. This dress and other adornments in the painting were likely gifts that Cristóbal García Segovia and Isabel Gabriela de Ingunza are documented as having donated to the Chapel of Cristo de Gracia, which included a silver lamp, a satin curtain, and dresses for an image of the Virgin.[18]

The large, sumptuous devotional painting described above was not the sort of Spanish art that American collectors were buying in the first decades of the twentieth century. Thus the painting was dismembered so that its usable parts— the portraits—could be sold. In its disguised state, the *Portrait of a Man* had a brief moment in the sun as a supposed Murillo—but the cleaning of the picture brought an end to that. Today, the unrestored fragment is a ruin that will never join the portraits of austere men in black by Murillo, Velázquez, and El Greco that hang alongside other masterpieces in The Met's galleries of Spanish painting. Even in its present state, though, the surviving fragment of the *Portrait of a Man* offers vivid testimony to the material culture of piety in early modern Spain. In particular, the opulent textiles that occupied much of the square footage of the original picture—the silk and metallic woven fabrics, laces, ribbons, and trims that adorned the man's wife, the altar, and the sculpture of the Virgin—were signs of the couple's wealth and virtue. In the painting's own time, those textiles were much more valuable than the portrait that represented them. The following case study takes a closer look at one surviving luxury textile in order to better understand this phenomenon.

II

According to the hierarchy of the arts in early modern Europe, a sumptuous silk was much more valuable than a painting.[19] Made with consummate craftsmanship, sometimes incorporating precious materials, fabric could be worth more than land in premodern Europe.[20] But later generations, with different tastes and values, placed less stock in textiles, even the richest sort, and many lavish brocades had their silver and gold threads removed (or "drizzled"), the precious metals being worth more as raw materials. The devaluation of textiles in subsequent centuries means that they have been neglected in scholarship compared to other arts from early modern Europe. Very little is known about the provenance of the length of red silk velvet brocaded with gold metallic threads in The Met's Department of European Sculpture and Decorative Arts that is the subject of this second case study (see fig. 8.4).[21] It has a much shorter paper trail than the *Portrait of a Man*, examined above. Nonetheless, the object itself communicates a great deal of information about its role in the material culture of early modern Europe. Fortunately, textiles are among the most accessible objects at The Met, where they can be studied firsthand in the museum's Antonio Ratti Textile Center.[22]

The Met's sixteenth-century velvet belongs to the category of fabrics known today as pomegranate cloths of gold, named after their large stylized fruit motifs (which, confusingly, are not necessarily pomegranates) and brocading with precious metals.[23] These superluxury textiles served an important social and economic function as instruments for the display of wealth and prestige in early modern Europe. Produced primarily in Italy and Spain in the sixteenth century, pomegranate cloths of gold were often used to make ecclesiastical vestments and secular garments that would show off their large designs (such as men's circular capes and women's bodices, sleeves, and skirts). They also served as furnishing fabrics, most notably as cloths of honor that hung behind a dignitary's chair on a special occasion (they frequently appear in this form in paintings depicting the Virgin Mary enthroned). This particular piece of velvet is quite long (it measures over seven feet in length) and survives in extremely good condition with its precious metals largely intact—a testament to its good fortune over the centuries.

Knowledge about the origins of The Met's brocaded velvet has evolved over the decades since it entered the collection. The velvet was cataloged as Venetian, and it was dated to the second half of the fifteenth century, when it was acquired by the museum in 1946. The following year, Rudolph Berliner, of the Rhode Island School of Design Museum, studied the new acquisition at the Metropolitan Museum of Art and suggested that its design was more consistent with textiles produced in Spain in the sixteenth century. Nonetheless, at that time the assumption was very strong that textiles of exceptionally high quality had to be Italian, and The Met's velvet continued to be cataloged as such. This changed in 2010 when Melinda Watt and Milton Sonday conducted a technical analysis that revealed structural idiosyncrasies confirming that the velvet was in fact most likely made in Spain.[24] This Spanish attribution is bolstered by an altar painting in the Seville Cathedral, Ferdinand Storm's *The Mass of St. Gregory*, in which three priests wear vestments made from a pomegranate velvet that is strikingly similar to the one at The Met.[25] The painting proves that this particular design was present in Spain in the mid-sixteenth century, but it does not prove that the fabric was actually made there. The Met's current attribution of the velvet hedges its bets by cataloging the piece as Spanish *or* Italian.

The identities of the artisans who wove this brocaded velvet—whether in Spain or in Italy—are irretrievably lost, but their skill is palpably present in the object that has survived them.[26] The weaver, who would have memorized every row that makes up the large pattern, wove the warp threads over metal rods of different diameters, to form larger and smaller loops, and then sliced them open with unfailing precision. The result is two different heights of velvet pile that appear to the eye as lighter and darker shades of red, which form a subtle floral pattern that snakes across the length of the fabric. This red velvet is the background for a large pattern of a golden rose with seven lobes that is embellished with metal loops called bouclé. Magnification reveals that those loops are made from fine, flat strips of silver gilded with gold that are wrapped around a yellow silk core.

Viewing this velvet in the Antonio Ratti Textile Center makes it possible to appreciate up close the precious materials and skilled craftsmanship that made luxury textiles so valuable and prestigious in their own time. Firsthand experience with the object brings about a visceral understanding of the early modern European hierarchy of the arts, which placed a piece of brocaded velvet like this very near the top. After seeing The Met's brocaded velvet in person, a twenty-first-century viewer is better able to look with a period eye at a sixteenth-century painting representing a similar textile and interpret two-dimensional images more accurately: to see different shades of color as shifts in the height of a velvet pile, and to understand clusters of golden loops as a form of currency.[27] But the historian's experience of the object is by necessity limited. We can only imagine how such fabrics would have come to life when they were shaped into three-dimensional garments and worn by a living person, their golden threads sparkling indoors in candlelight or outside beneath the sun. According to Prown's material culture method, touch is a key aspect of "sensory engagement" with a historical object. Yet historians are limited when it comes to the sensory exploration of early modern textiles at professional facilities like the Antonio Ratti Textile Center, where only curators, conservators, and trained technicians are permitted to handle the fabric.[28] A brocaded velvet of such exquisite quality was, of course, an exclusive possession in the sixteenth century—many more people would have seen and admired them than owned and worn them. Like the regular citizens who once saw such velvets on the backs of local nobles or priests in the sixteenth century, the twenty-first-century historian at the museum is able to observe and admire—but not to touch—these woven treasures.

III

Unlike the velvet brocade examined above, the painted lead sculpture that is the subject of this third and final case study has a much more certain provenance. *The Child Jesus Triumphant* (see fig. 8.5) is one of many near-identical works produced in Seville in the 1620s, made after a prototype by the Andalusian sculptor Juan de Mesa (who trained in the workshop of Juan Martínez Montañés, the Sevillian sculptor who first popularized such pint-sized images of the young Christ).[29] Countless copies of the Mesa sculpture were made in wood, bronze, or lead (wood being almost twice as expensive as lead, with bronze in between), many of which traveled across the ocean to Spanish America.[30]

Figure 3.1. *Contra infantium adustionem (For [Treating] Inflammation in Infants),* in Martín de la Cruz and Juan Badiano, *Libellus de medicinalibus Indorum herbis (Little Book of Indian Medicinal Herbs),* also known as *Codex de la Cruz–Badiano,* Tlatelolco, Mexico, 1552, fol. 61r; iron-gall ink and organic pigments on European paper, 8 × 6 in. (20.6 × 15.2 cm). Biblioteca Nacional de Antropología e Historia, Mexico City

Figure 3.2. Impersonator of Huitzilopochtli and celebrants adorned with flowers, in Bernardino de Sahagún and indigenous artists and scribes, *Historia general de las cosas de Nueva España (General History of the Things of New Spain)*, also known as *Florentine Codex*, Tlatelolco, Mexico, ca. 1577, vol. 1, book 2, fol. 6ov; ink and color on European paper. Biblioteca Medicea Laurenziana, Florence, Ms. Med. Laur. Palat. 118, f. 114v

Figure 3.3. (*opposite below, and below*) Description and illustration of Mexican medicinal herbs, in Bernardino de Sahagún and indigenous artists and scribes, *Historia general de las cosas de Nueva España (General History of the Things of New Spain)*, also known as *Florentine Codex*, Tlatelolco, Mexico, ca. 1577, vol. 3, book 11, fols. 133v–134r, ink and color on European paper.
Biblioteca Medicea Laurenziana, Florence, Ms. Med. Laur. Palat. 220

Figure 3.4. Unknown artist, *Relación geográfica map of Santiago Atitlán*, 1585, ink and watercolor on paper, 24 3/16 × 31 7/8 in. (61.5 × 81 cm). Joaquín García Icazbalceta Manuscript Collection, Nettie Lee Benson Latin American Collection, The University of Texas at Austin, JG xx-10

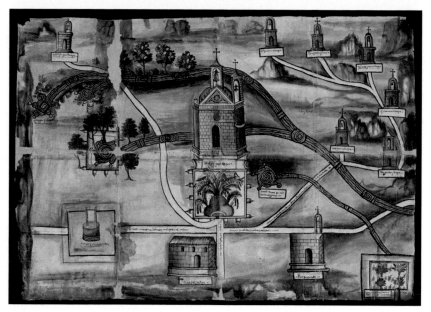

Figure 3.5. Unknown artist, *Relación geográfica map of Guaxtepec*, 1580, ink and watercolor on paper, 24 1/2 × 33 1/2 in. (62 × 85 cm). Joaquín García Icazbalceta Manuscript Collection, Nettie Lee Benson Latin American Collection, The University of Texas at Austin, JG xxiv-3

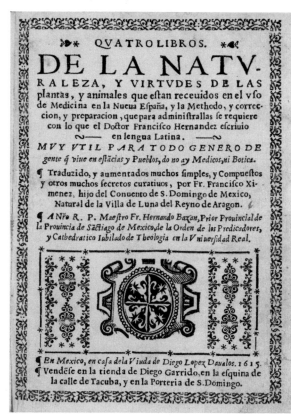

Figure 3.6. Title page of Francisco Ximénez, *Quatro libros: De la naturaleza, y virtudes de las plantas, y animales que estan recevidos en el uso de medicina en la Nueva España (Four Books: On the Nature and Virtues of the Plants and Animals That Are Accepted in Use in Medicine in New Spain*; Mexico, 1615). The Huntington Library, San Marino, California, 9197

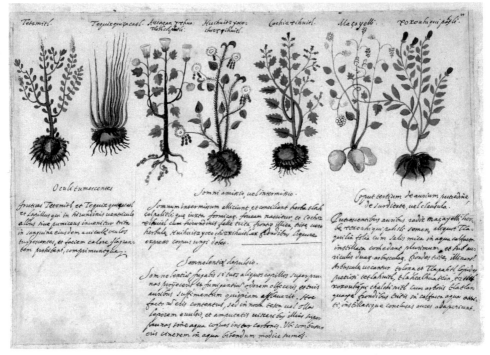

Figure 3.7. Italian artist and scribe, *Copy of an Aztec Herbal (Codex de la Cruz–Badiano)*, 1626, fol. 7, watercolor and bodycolor over black chalk; page, 9 1/3 × 12 1/8 in. (23.7 × 33.4 cm). The Royal Collection / HM Queen Elizabeth II, RCiN 927905. Royal Collection Trust / © Her Majesty Queen Elizabeth II 2019

Figure 3.8. Frontispiece to Francisco Hernández, *Rerum medicarum Novae Hispaniae thesaurus (Treasure of the Medical Things of New Spain*; Rome, 1651), engraving. The Huntington Library, San Marino, California, 34588

MACPALXOCHI QVAHVITL.   MAMAZTLACOTL.

NOTATIONES.

MATLAL XOCHITL.   MACPALXOCHI QVAHVITL.

Arbor eſt . Flor comomano ab Hiſpa-
nis vocata : Folia Mori ſunt . Flos
vndeuis in ramis naſcitur habetq;
figuram quaſi Tulipæ , colore mal-
uaceo ; & pro ſtamine exit quaſi
manus, vel potius pes auis, coloris
ſanguinei : cui à latere luteum
quid coniunctum eſt . Flos admo-
dũ magnus eſt, pro ratione arboris.

MAMAZTLACOTL.

Caulis folioſus & alatus. Flos Chry-
ſanthemi luteus : Folia minutim
inciſa : Radix verò ſubnigricans .

MATLALXOCHITL.

Folia vt Valerianæ græcç, & flos ex
luteo ruber , quinque vt mihi vi-
detur, foliolis conſtans . Stamina
lutea ſunt .

MA-

Figure 3.9. Entries for *Macpalxochi Quahuitl, Mamaztlacotl,* and *Matlal Xochitl*
with woodcut illustrations, in Francisco Hernández, *Rerum medicarum Novae
Hispaniae thesaurus (Treasure of the Medical Things of New Spain*; Rome, 1651), 383.
The Huntington Library, San Marino, California, 34588

Figure 4.1. Dish with flowers and birds, ca. 1660; Dutch (Delft); tin-glazed earthenware with cobalt blue pigment. The blue pigment, shape, and design of this Dutch earthenware plate imitate Chinese kraak porcelain (a type of Chinese export porcelain). Metropolitan Museum of Art, accession no. 30.86.3.

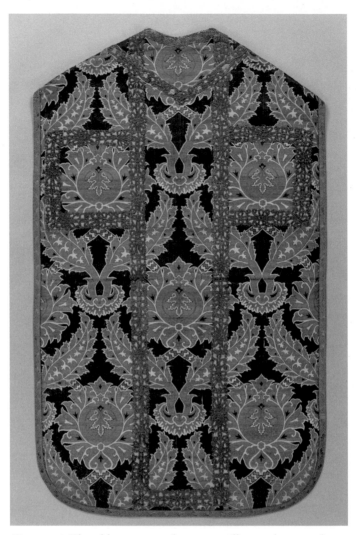

Figure 4.2. Chasuble, seventeenth-century silk, metal-wrapped thread, lampas. This chasuble (the vestment worn to celebrate the Eucharist) is made from Ottoman silk. Both Russian Orthodox and European Catholic clerics owned robes like this one. Metropolitan Museum of Art, accession no. 06.1210

Figure 8.1. Anonymous Spanish artist, *Portrait of a Man*, c. 1660–1680, after removal of later overpaint, revealing its fragmentary status; oil on canvas, 25 × 20 5/8 in. Metropolitan Museum of Art, The Friedsam Collection, Bequest of Michael Friedsam, 1931 (32.100.7)

Figure 8.2. Anonymous Spanish artist, *Portrait of a Man* (see fig. 8.1), before removal of overpaint. Photo courtesy of the Department of European Paintings, Metropolitan Museum of Art

Figure 8.3. Anonymous Spanish artist, *Portrait of a Woman*, c. 1660–1680; oil on canvas. Formerly in the Singer collection; present location unknown. Photo courtesy of the Department of European Paintings, Metropolitan Museum of Art

Figure 8.4. Pile-on-pile cut, voided, and brocaded velvet of silk and gold metallic thread with bouclé details; Spanish or Italian, sixteenth century; 22.5 × 87 in. Metropolitan Museum of Art, Fletcher Fund, 1946 (46.156.120)

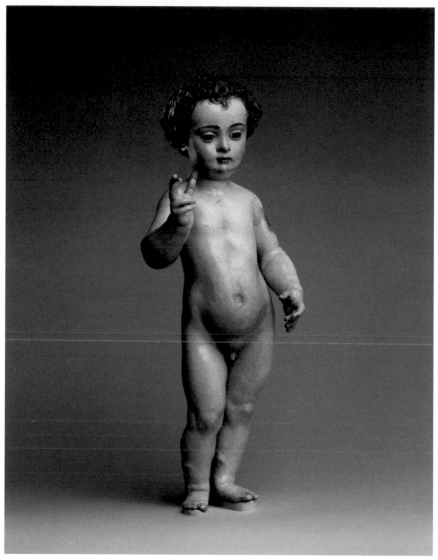

Figure 8.5. Juan de Mesa, *The Child Jesus Triumphant*, ca. 1625; polychromed lead with glass eyes, 16 in. tall. Metropolitan Museum of Art, Gift of Loretta Hines Howard, 1964 (64.164.244a)

Figure 8.6. Juan de Mesa's *Child Jesus Triumphant* (fig. 8.5) dressed in an eighteenth-century silk and silver-gilt lace dress, capelet, and bib, with a silver halo made in the late nineteenth or early twentieth century. Metropolitan Museum of Art, Gift of Loretta Hines Howard, 1964 (64.164.244a, b, c, d, e)

What is special about The Met's copy of this sculpture is that it entered the museum's collection with a complete set of garments and a silver halo.[31] Made to adorn the Sevillian Christ Child generations after it was cast and painted, the sculpture's layette provides evidence of what the anthropologist Arjun Appadurai called the "social life" of the object.[32] Unlike The Met's *Portrait of a Man*, which came from a larger canvas that was dismembered and defaced to meet market demands, *The Child Jesus Triumphant* remained relevant over the centuries through additions that did not fundamentally alter the original object. The material history of *The Child Jesus Triumphant* is a story of transformation through accumulation.

Sculptures of the Christ Child were popular in convents and private homes across Counter-Reformation Catholic Europe. Primarily associated with nuns, these sculptures were also collected by secular women and men. The sculptures played a special role in devotional practices at Christmastime and could also be part of the daily lives of their owners, who might dress and care for these religious dolls as one would for a living infant.[33] Margarita de Austria (also known as Margarita de la Cruz; 1567–1633), daughter of the Habsburg emperor Maximilian II, famously lavished love and attention on her collection of eleven sculptures of the Christ Child at the Convent of Las Descalzas Reales, in Madrid. According to her seventeenth-century biographer, the royal nun dressed the sculptures, talked to them, and put them to bed. She named them after their appearances: there was the Germanic blonde *El Alemán*, the handsome *El Hermoso*, and the severe-looking *El Grave*, among others.[34]

Within the genre of Christ Child sculptures, different types (*advocaciones*) are distinguished by their postures, clothes, and accessories. The Met's sculpture belongs to the "triumphant" category: standing upright, the right hand raised in benediction, this Christ Child triumphs over sin and death. The nudity of the image shows the humanity of Jesus while also celebrating his naming and circumcision.[35] Whether hand-carved from noble wood or cast in economical lead, these sculptures were painted in flesh tones that give them a lifelike appearance. The verisimilitude of The Met's painted lead *Child Jesus Triumphant* is further enhanced by glass eyes. Measuring 16 inches high and weighing 4.7 pounds, the sculpture is notably smaller and lighter than an actual child of flesh and blood. Visible wear and tear provides material evidence of how the object was used. The bumps protruding from the lower legs were formed when rods were inserted through the feet to make the figure stand upright on a carved wooden pedestal, leaving impressions in the soft lead.[36] The holes in the figure's head—five of them—were made intentionally so that a halo could be attached. The flesh-toned paint has worn away in places to reveal the grey metal underneath, most notably on the left upper arm where an owner might have grasped the sculpture to pick it up.

As a sculpture celebrating the circumcision, *The Child Jesus Triumphant* was supposed to be seen in the nude (Counter-Reformation prudery did not apply to images of Christ as a child), but a century or so after the sculpture was cast, a set of elaborate clothes was made to dress it.[37] This is not unusual: it was quite popular in the eighteenth century to dress devotional sculptures of the Christ Child in real clothes that reflected particular moments in the life of Christ, or garments of different religious orders. The outfit for The Met's *Child Jesus Triumphant* consists of three pieces—a gown, a cape, and a bib—made from cream-colored silk brocaded with pink and blue flowers (a pattern that appears outsized on these miniature garments). The three pieces are trimmed in silver-gilt lace and further embellished with sequins, embroidered flowers, and additional flowers made from silk and wire. Sometime in the nineteenth or twentieth century, a silver halo was made to adorn the sculpture as well, probably a replacement for an older one.

The appearance and meaning of the seventeenth-century *Child Jesus Triumphant* are radically transformed when it is dressed in its clothes and halo (see fig. 8.6). The garments, which are layered on top of each other, completely cover the sculpture, leaving exposed only the right hand, raised in blessing, and the head, which is partially obscured by the silver halo. Covering the sculpture's body also changes its significance. No longer representing the circumcision, what does *The Child Jesus Triumphant* mean when it is dressed like this? A clue can be found at the Convent of Las Descalzas Reales in Madrid, where there are sculptures of Christ the King—images of the Christ Child holding a globe with a cross and seated on a diminutive throne—dressed in outfits made from colorful eighteenth-century floral brocades adorned with silver lace.[38] This pairing of dress and image suggests that *The Child Jesus Triumphant*'s outfit may have transformed the sculpture into Christ the King.

The clothes that were made for *The Child Jesus Triumphant* thus served several different purposes. The luxe eighteenth-century fabric modernized—and enriched—the early seventeenth-century sculpture's appearance. At the same time, the new suit of clothes broadened the sculpture's range of functions: it could still represent the circumcision and naming of Christ when nude, but it could also be turned into Christ the King when it was dressed. Finally, the garments gave the sculpture's owner an opportunity to dress and undress the little figure and thus to experience an intimate and interactive relationship with this doll-like work of art. The clothes and halo that were made for *The Child Jesus Triumphant* are material witnesses to the different lives that the sculpture had in the hands of various owners, and to the continuing relevance and dynamism that this devotional image retained in the centuries after its creation.

It is not known who owned and cared for *The Child Jesus Triumphant* prior to the twentieth century, when it entered the collection of the artist Loretta

Hines Howard. As a collector of eighteenth-century Neapolitan crèche figures, Howard must have appreciated this Spanish sculpture's connection to the nativity. In 1964, Howard made a large donation to the Metropolitan Museum of Art that included *The Child Jesus Triumphant* along with more than 140 Neapolitan Christmas figurines.[39] After making the donation, Howard annually decorated the museum's Christmas tree with the crèche figures from her collection, starting a tradition that would continue after her death in 1982.[40] While Howard's Neapolitan crèche figures became the centerpiece of a cherished annual holiday tradition, *The Child Jesus Triumphant* rested dormant in storage and went unseen by the public for more than fifty years.

For much of the twentieth century, painted sculptures like *The Child Jesus Triumphant* were ignored by most Anglophone art historians and museum curators, who dismissed them as popular devotional objects that did not merit serious scholarly study, despite pioneering work on polychrome sculpture by Beatrice Gilman Proske in the late 1960s and Gridley McKim-Smith in the early 1970s.[41] This began to change with the publication in 1998 of Susan Verdi Webster's book on Sevillian processional sculpture, which was followed by two exhibitions in 2008, curated by Ronda Kasl and Xavier Bray, that integrated painted devotional sculpture with paintings on canvas.[42] In the wake of this newfound interest in Spanish polychrome sculpture, The Met's *Child Jesus Triumphant* finally emerged from storage to be included in the 2018 exhibition *Like Life: Sculpture, Color, and the Body*.[43] The Christ Child sculpture was dressed in its elaborate outfit and halo for display in the exhibit—a curatorial decision that deemphasized the seventeenth-century nude sculpture (which was largely obscured) and focused attention on the object's layered multigenerational history instead. The exhibition launched a new, public era for this intimate devotional sculpture, which was conserved and photographed and is now better documented than ever for the next stage in its life cycle.

The material histories of the objects examined in this essay—the portrait of a man, the brocaded velvet, and the Christ Child sculpture—expose how the history of collecting affects the evidence that historians of early modern European material culture may access in museum collections. As these case studies show, the appearance of early modern objects sometimes changes dramatically over time—whether through irreversible destruction (as in the *Portrait of a Man*) or through the accumulation of layers (as in *The Child Jesus Triumphant*). These cases also show how objects in museum collections can disappear from sight and reemerge decades later as trends come and go in the academic and museum worlds, and how our knowledge about objects grows in the particular moments when they happen to be in vogue. As the case of The Met's length of brocaded velvet reveals, knowledge about anonymous objects is always a work in progress, and historians have to tolerate a certain amount of ambiguity and uncertainty when working with them.

When it comes to reconstructing the material histories of museum objects, departmental files—the paper kind, which contain more information than can usually be found online—are essential repositories of knowledge compiled by generations of curators and conservators. The case studies explored in this essay have also highlighted the importance of technical studies that expose aspects of the objects' material histories that are hidden beneath the surface—whether it is the weave structure of a textile under its velvet pile or the original image of a painting covered by a deceptive layer of overpainting. This brings forth a crucial point about the extent to which historians depend on the expertise of curators and conservators, who have specialized knowledge of materials and techniques, a broad range of comparative examples to draw upon, and the latest information on changing attributions. Dialogue with museum experts is an essential component of material culture research. A material culture approach is an inherently collaborative effort, and as such it has the potential to bring historians' attention and appreciation not only to the finished objects that they study but also to the collaborative processes that produced them.[44]

NOTES

Research for this article was made possible by a Jane and Morgan Whitney Fellowship at the Metropolitan Museum of Art (2005–6), a Mellon Grant for New Initiatives in Curatorial Training that supported a CUNY Graduate Center seminar at The Met (2012), and a research fellowship at the Bard Graduate Center (2017). I would like to thank Carmen Ripollès for inviting me to collaborate on a panel about material culture in the early modern Iberian world at the annual meeting of the College Art Association (2017) and for commenting on an early version of this essay.

1. See, for example, Bleichmar and Mancall 2011; Findlen 2013; Richardson, Hamling, and Gaimster 2016; Gerritsen and Riello 2016; and Hamling and Richardson 2016.

2. Major interdisciplinary collaborative projects include the HERA-funded "Fashioning the Early Modern" project, directed by Evelyn Welch (see Welch 2017 and the project website, http://www.fashioningtheearlymodern.ac.uk/), and "The Making and Knowing Project," directed by Pamela Smith (see Bilak et al. 2016; https://www.makingandknowing .org/). For examples of historian-curated exhibitions, see Ulrich 2015; Avery, Calaresu, and Laven 2015.

3. Grafton 1993, 10.

4. Prown 1982, 1, 3.

5. According to the museum's online catalog (http://www.metmuseum.org/art/collection), consulted on January 7, 2020, the museum's collection included 77,223 objects from Europe (1400–1800), of which 4,707 were on display.

6. The history of collecting is a significant and growing field of scholarship; to give but one example related to the subject of this essay, see Reist and Colomer 2012.

7. For a critical analysis of the relationship between art history and material culture, see Yonan 2011.

8. I am very grateful to Patrice Mattia, associate administrator in The Met's Department of European Paintings, for facilitating my access to the *Portrait of a Man* (32.100.7) and its file (hereafter cited as European Paintings object file); to curator emerita Katharine Baetjer for sharing her recollections of the painting's history at The Met; and to curators Stephan Wolohojian and Ronda Kasl for examining the painting with me and sharing their keen observations. Some of the information in the paper file is summarized in the annotated bibliography that accompanies the painting on The Met website: http://www.metmuseum .org/art/collection/search/437728.

9. Salomon 2017, 80, 81.

10. See Kagan 2012. Works by Murillo in American collections before 1925 are cataloged in Cano and Ybarra 2012.

11. Kagan 2019, 231.

12. Mayer 1923, 124. Mayer's attribution of the painting to Murillo was also published in Térey 1923.

13. Pérez Mínguez 1923. On Cristóbal García Segovia, see the two-part study by Quirós Rosado (2008, 2009).

14. An issue of the museum's bulletin devoted to the Friedsam collection did not describe the "Murillo" (or any other Spanish works, except for one by Goya); *Metropolitan Museum of Art Bulletin* 27 (11), pt. 2 (November 1932).

15. Undated X-ray notes and conservation report dated June 16, 1932, in European Paintings object file 32.100.7.

16. The most likely explanation for the striking coincidence that Mayer published two pieces cut from the same painting in his article is that they would have entered the art market and thus come to the curator's attention at the same time.

17. European Paintings object file 60.177.

18. These gifts were enumerated in the inscription to the family portrait in the Chapel of Cristo de Gracia; see Pérez Mínguez 1923, 210.

19. On the hierarchy of the arts in the Renaissance versus subsequent periods, see Belozerskaya 2005.

20. Smail 2016, 60.

21. This case study of The Met's brocaded velvet is based on conversations with Melinda Watt, formerly the supervising curator of the Antonio Ratti Textile Center, to whom I am deeply indebted. I am also grateful to textile conservators Cristina Carr and Giulia Chiostrini, and to the Ratti staff members who made the piece available for study on multiple occasions. The public catalog record and a color image of the velvet are available on The Met's website: http://www.metmuseum.org/art/collection/search/227208.

22. The Antonio Ratti Textile Center stores textiles from the museum's various curatorial departments. There has been a textile study room at the Metropolitan Museum of Art since 1909; the current facility opened in 1995. Montebello 1995–96.

23. Monnas 2008, 258–65. On the materials and techniques that went into making Renaissance velvets, see Monnas 2012.

24. On the production of luxury velvets in Spain, see the exhibition catalog (which includes an English translation of the text), *L'art dels velluters* 2011.

25. See Ferdinand Storm's *Mass of St. Gregory* (https://commons.wikimedia.org/wiki

/File:Misa_de_San_Gregorio_Magno_(Retablo_de_la_capilla_de_los_Evangelistas_de _la_catedral_de_Sevilla).jpg) and its setting at the center of the altarpiece in the Chapel of the Evangelists in the Seville Cathedral (https://commons.wikimedia.org/wiki/File :Retablo_de_la_capilla_de_los_Evangelistas._(Catedral_de_Sevilla).jpg). On the Flemish artist Ferdinand Storm (known as Hernando de Esturmio in Spain), see Serrera Contreras 1983, 96–97.

26. Studies of artisan skill and knowledge include Smith 2004; P. Long 2001; Smith and Schmidt 2007; Klein and Spary 2009.

27. For the original formulation of the "period eye," see Baxandall 1972; for the application of Baxandall's concept to early modern textiles and clothing, see Rublack 2016.

28. Prown 1982, 9.

29. García Sanz 2010, 242–45, 251; Proske 1967, 52.

30. Fons del Museu Frederic Marès 1996, 329–30.

31. Masters student Gabriela Valero Arias brought my attention to these objects, which were the subject of her 2016 seminar paper for a course that I taught in the art history program at the CUNY Graduate Center. I am very grateful to Peter Bell, former curator in the Department of European Sculpture and Decorative Arts at The Met, who made it possible for another class to examine the Christ Child sculpture and its garments and halo together at the Antonio Ratti Textile Center in 2017. I also thank Denny Stone, senior collections manager, for weighing the Christ Child, and curators Denise Allen and Ronda Kasl for generously answering questions about the sculpture's attribution.

32. Appadurai 1986.

33. García Sanz 2010, 150.

34. Palma 1636, f. 225r. On Margarita de Austria's devotion to her sculptures of the Christ Child, see E. Goodman 2001, 160–71; García Sanz 2010, 131–33; and Tiffany 2016.

35. García Sanz 2010, 239–45.

36. The Met's *Child Jesus Triumphant* does not have its original pedestal, but there are numerous examples that do, including those at the Museu Frederic Marès in Barcelona (Fons del Museu Frederic Marès, 1996, cat. no. 291, 329–30) and at the San Diego Museum of Art (http://collection.sdmart.org/Obj29015?sid=1780&x=4833).

37. García Sanz 2010, 244–45, 389.

38. García Sanz 2010, 420–25.

39. Raggio 1965.

40. Obituary, "Loretta Hines Howard, Artist and Crèche Figure Collector," *New York Times* (April 3, 1982), sec. 1, p. 18.

41. Proske 1967 and McKim-Smith 1974.

42. Webster 1998; Kasl 2009; Bray 2009.

43. The catalog inexplicably changes the attribution of the sculpture to "Spanish or Mexican," but The Met continues to recognize it as a work by Juan de Mesa. Syson et al. 2018, cat. no. 72, 190–91, 200.

44. For more on collaboration, see Blair's chapter in this volume.

CHAPTER NINE

New Knowledge Makers

ANN BLAIR

The knowledge transmitted from ancient Greece and Rome also included categories for classifying knowledge, notably into a hierarchical distinction between "liberal" and "mechanical" disciplines. These terms have proved tenaciously long-lived even as the nature of the distinction has taken various forms over time. While cultural historians pride themselves on attending to actors' categories in order to make sense of them in historical context, we should be aware that these categories can also pose an impediment to finding answers to some historical questions. In investigating methods of knowledge making in particular, historians have increasingly looked beyond the theories of knowledge in search of evidence about practices. Since direct evidence is often sparse, historians also glean clues by reading their sources against the grain, that is, for information that they were not designed to convey. In the past three decades we have learned a great deal about early modern working methods. This historiography rejects the long-traditional distinctions between handwork and intellectual work, artisan and scholar, practice and theory, and highlights instead "the complicated mix of knowledge, know-how, and technique" that generated knowledge, in particular of the natural world.[1] Rather than attempt a synthesis of recent research or propose an overarching historical narrative about knowledge making in early modern Europe, I examine how the notion of "mechanical" work has become obsolete and collaboration has instead become a promising new object of analysis.

The Category "Mechanical"

The hierarchical distinction that has elevated some activities as "liberal" over others has taken a few different forms since its ancient origins. Today we contrast a liberal arts education with professional training, emphasizing that the liberal arts broaden the mind, encourage curiosity and discovery across a wide range of fields, and foster personal and moral growth, instead of training students for a specific line of employment. But this dyadic contrast is of recent origin.[2] Plato distinguished the knowledge of crafts as *technê* from the knowledge of eternal truths, or *epistêmê*.[3] With a more concrete set of definitions, Aristotle identified liberal occupations in his *Politics* as those appropriate for the education of a free man, by contrast with illiberal or banausic arts that encompassed "any craft or branch of learning [that] renders the body, the soul, or the thought of free people useless for the uses and actions of virtue." What made an activity banausic was not just the nature of the activity but the purpose for which it was performed. So Aristotle explained: "For what one does for one's own sake, for the sake of friends, or because of virtue is not unfree, but someone who does the same thing because of others would in many cases seem to be acting like a hired laborer or a slave."[4] All activities performed for someone else or for pay were illiberal, including, for example, the work of reading and writing performed by a slave in ancient Greece or Rome.

It is a later opposition that is most familiar to us, dominant for more than a millennium: liberal versus mechanical. The ninth-century Irish philosopher John Scotus Eriugena is credited with coining the term "mechanical arts" to designate fields founded on practical experience rather than intellectual instruction. In the twelfth century, Hugh of St. Victor systematized the seven mechanical arts to match the seven liberal arts: fabric making, armament making, and commerce; agriculture, hunting, medicine, and theatrics.[5] The medieval distinction hinged mostly on the opposition of hand and mind, though the earlier distinction between something done for pay and something done freely still resonated since most products of mechanical arts were paid for whereas knowledge in the medieval period supposedly could not be bought or sold.[6]

During the Renaissance, these ancient and medieval distinctions persisted in theory, but in practice the mechanical arts acquired new prestige. The rediscovery of Vitruvius's *De architectura*, which called for training architects equally in the liberal arts and in practical crafts, inspired emulators such as Leon Battista Alberti (1404–72) who mastered engineering as well as humanist philology.[7] And practitioners of activities traditionally considered mechanical successfully claimed greater recognition in one or more of the following ways: by receiving patronage from noble and royal courts, by publishing books on their art (clearly a "liberal" activity), by proclaiming their unique expertise and explicitly challenging the traditional notion that the theoretical was superior to the practical.[8] Only an elite among artists (painters, sculptors, jewelers, architects) or mathematical practitioners (military advisers, navigators, instrument makers, astronomers) managed this feat of social and cultural upward mobility. Most mechanical arts continued to be practiced without the financial rewards or social prestige associated with such recognition. Nevertheless the mechanical arts were discussed and praised in print, even before Francis Bacon famously emphasized the need to draw on both practical and theoretical skills in natural philosophy.[9]

The Basel polymath Theodor Zwinger composed his *Theatrum humanae vitae* in 1565 as a massive encyclopedic compendium of human experience. He bestowed considerable praise on the mechanical arts for their "ingenious achievements useful and pleasant to daily human life." He upheld the traditional distinction and hierarchy but noted that it was not clear-cut: "The liberal arts indeed rely more on the aid of the mind than of the body and are thus considered more worthy. But no one denies that both theoretical and practical character [*habitus*] are illustrated in the work of mechanical [men]."[10] Zwinger devoted one volume in his *Theatrum* to the mechanical arts, which he expanded well beyond Hugh of St. Victor's seven, including art making (e.g., music, painting, sculpting), the care of the body (from surgery to haircutting, and from wet-nursing to pall bearing), the preparation of food, clothes, and utensils, and the recently developed art of typography.[11] The latter might conjure to us another connotation of "mechanical" as reliant on machines. Although this meaning was also present in antiquity—for example, with the prototype of Archimedes—Zwinger did not emphasize this point.[12] In 1571, Zwinger introduced in his treatment of the topic a section on the "vicarious mechanical arts by which a man does work for another man."[13] Without naming Aristotle, Zwinger returned here to that Aristotelian notion of "illiberal" as work performed for another.

Zwinger divided work that was called mechanical because it was vicarious into tasks pertaining to the mind and tasks pertaining to the body. The first category was the larger of the two, and it included scribes and amanuenses and those hired to balance accounts, to administer, or to mourn for another. The second category of vicarious arts, pertaining to the body, included tasters and mercenaries, for example. Some of these might also have been considered mechanical because they used their hands (scribing certainly was a demanding physical discipline), but they were listed here because they involved working for another person, for pay.[14] In that subordination to the wishes of another, "mechanical" could be contrasted with those tasks requiring judgment performed by the person commissioning the work. The Spanish scholastic Juan Caramuel made a similar distinction in the advice he published in 1664 about how to make an index. He recommended a procedure that was already well known at the time: an amanuensis should copy out the items to be indexed, cut these entries into separate slips of paper, and alphabetize the slips. But he called attention especially to the delegation of this work: "Have this done I say, do not do it yourself. Indeed this is mechanical work [*labor mechanicus*] and does not require you; it is enough to direct others to do it."[15] Caramuel designated this work as mechanical and appropriate for delegation, ostensibly because it did not require judgment—anyone with sufficient literacy could perform these functions. Similarly, Joseph Scaliger dismissed the work he did over two years to index the massive collection of inscriptions by Gruter as a workman's task. And yet the twenty-four separate indexes spanning 228 pages that he devised on a range of unusual themes, from grammatical forms to the kinds of professions mentioned in the inscriptions, clearly resulted from great learned expertise. Regardless of its sophistication, indexing ranked as mechanical in these seventeenth-century contexts, and Scaliger asked Gruter not to name him in the indexes, presumably because he would have considered it degrading to have that work attributed to him.[16]

Much more could be said about early modern acceptations of "mechanical," which variously emphasized reliance on physical rather than mental work, or work performed at the request and direction of another, or work performed by a machine. It is sufficient here to note the pervasiveness of the liberal-mechanical distinction and the hierarchy rooted in it. Although the mechanical arts gained more attention and prestige from some learned men and wealthy patrons in the Renaissance, the traditional hierarchy that subordinated the mechanical to the liberal arts informed centuries of European educational systems. As a result, the critique of the multiple dichotomies opposing handwork and mindwork has been slow to develop, since intellectual historians and historians of science have long focused on individuals noted for their innovative ideas.

Historiographical Developments

Edgar Zilsel offered an early rebuttal to this approach in a 1945 article in which he argued that the decisive innovation of the Scientific Revolution was the crumbling of the "wall which since antiquity had separated the 'liberal' from the 'mechanical' arts."[17] Zilsel identified "superior artisans"—people who combined artisanal skills with the ability to write about them—as crucial sources for the notion of progress: unlike scholars trapped in competitive arguments about abstract concepts, artisans modeled working toward practical goals through collaborations that were essential to the ambitions and the success of early modern science. Zilsel's conclusions were part of a broader theory of the origins of genius that he developed in dialogue with Marxism and in reaction against the logical positivism of fellow members of the Vienna Circle. He favored historical empiricism instead of abstract theories as the basis for analyzing scientific method, but his writings were not well received at the time.[18] Nevertheless Zilsel's "superior artisans" also received attention from a very different quarter. Eva Germaine Rimington (better known as E. G. R.) Taylor, the first woman professor of geography in the United Kingdom, wrote on the history of her discipline throughout her career and, in retirement, composed two landmark volumes on the mathematical practitioners of early modern England, including navigators, surveyors, and instrument makers.[19] Studies of experimentation in the Middle Ages and the early modern period were also among the first to note the significance of practical knowledge.[20]

Nevertheless, the predominant analyses of early modern science of the mid-twentieth century—by Alexandre Koyré, I. B. Cohen, Charles Coulston Gillispie, and A. Rupert Hall, among others—focused on theories and the individuals associated with them, often explicitly denying the contribution of artisans. Others have pointed out how this emphatic focus on scientific advances driven by internal intellectual factors held special appeal in the decades after World War II as a strategy for keeping science out of the hands of politicians.[21] The story of the growth of externalist approaches has been told many times: from early suggestions in the 1950s and 1960s, including Kuhn's *Structure of Scientific Revolutions* (1962), to the rise of the Strong Programme in the Sociology of Scientific Knowledge in the 1980s and 1990s. Attention to artisans was not a major theme during those years but surfaced as a "weak subset of the 'externalist' position" that social context affects scientific ideas.[22] In a remarkable article in 1989, Steven Shapin noted both the constant presence of laboratory assistants in the work of Robert Boyle and their effacement in published accounts of that work.[23]

The pursuit of a more holistic and broadly contextualized portrait of scientific activity led to a great extension of Zilsel's original insight that many kinds of people contributed to scientific knowledge in the early modern period (among other contexts). The 2006 *Cambridge History of Early Modern Science* highlighted this range of contributions by organizing a central section of the book around the different locales in which historians have studied the exchange and creation of knowledge: libraries and lecture halls, but also markets and piazzas, homes and households, courts and academies, anatomy theaters, botanical gardens, natural history collections, laboratories, sites of military science and technology, coffeehouses and print shops, and networks of travel, correspondence, and exchange.[24] Those articles described burgeoning areas of research that have continued to grow since then.

In the intervening years, we have learned a great deal more about the role of markets in stimulating the collection of natural specimens and information and the making of models (including wax preparates), drawings, catalogs of collections, and books.[25] Agents and brokers put buyers and sellers in touch with each other and governments relied on expert mediators to manage public engineering projects.[26] Urban spaces have been portrayed as crucial sites for knowledge making given the exceptional density of people and opportunities for exchange across social groups.[27] Exciting new studies of cross-cultural interactions have highlighted the role of go-betweens, whether informal or professional, who could serve in one or more roles as interpreters, translators, unofficial diplomats, local guides, or informants.[28] Closer attention to households has revealed families in which women tested, gathered, and traded recipes with peers, or contributed to the work of their husbands (often alongside daughters and sons) by writing, drawing, organizing, and analyzing.[29] Servants could also be involved, whether they were specially hired as secretaries or not.[30] Historians have also uncovered the roles of slaves in developing remedies for tropical illnesses, whether from African sources or from contact with Amerindians, and in interaction with Europeans in the Americas.[31] From this work Londa Schiebinger emphasizes that many factors prevented the easy exchange of this knowledge. We can assume, for example, that most of this knowledge, developed and transmitted orally, has been lost. Even the cases we do know about were laden with circumstances that inhibited the recording and preservation of knowledge; in Schiebinger's study the slave healers' desire for secrecy and the European prejudice toward the information were also hampering factors, in addition to the distortions added by the shifts in languages and cultural contexts involved in the transmission.[32]

The picture we now have of knowledge making in early modern Europe is full of many more kinds of people, places, and objects than are visible in any set of categories from the period. Historians are piecing together evidence from multiple sources, often fragmentary, and forging a new path in describing the nature of these knowledge-making activities and relationships. One of the vivid conclusions we can draw is that each of these knowledge makers was an agent with a "strategic field of vision and a rationality," in the words of Sanjay Subrahmanyam, even if those were bounded by various limitations. Regarding go-betweens in colonial settings in particular, he concludes that they were "heavily constrained in their actions and thus caught as it were between a rock and a hard place."[33] Other knowledge makers brought to light in this current historiography were also in positions that were subaltern in some way—due to their lower social status or to their being enslaved, or in the employ of another, or female—and yet they actively participated in the processes of observing, reporting, and interpreting information and experience. To those above them in these hierarchical relationships, their contributions may well have seemed mechanical (as in made for hire or requiring no judgment, or both) and thus not worth mentioning or crediting, but each of these contributors was an agent who participated in shaping the final knowledge product. Historians are often unable to reconstruct the exact nature of these contributions given the fragmentary and indirect nature of the evidence, but we can look forward to learning more in the years ahead about the participation of hitherto marginal figures in early modern knowledge making, in scientific fields but also many others.

An Example: The Agency of the Scribe

An unusual passage in a Latin treatise illustrates how a task delegated as mechanical (in this case, scribing) resulted in a change in the content of the work—and not simply by scribal error but by the scribe's reliance on his judgment and personal experience. In his *Theatrum vitae humanae*, Theodor Zwinger had classified scribing as mechanical because it was vicarious, or performed for another, and, within that category, as pertaining to the mind. Scribes were routinely hired by authors for a number of purposes in early modern Europe, including making a clean copy of a manuscript to be sent to the printer. Such a task was presumably the occasion for the anecdote reported in the 1603 *De recta nominum impositione* (*On the Correct Imposition of Names*), by the French nobleman Pontus de Tyard (1521–1605). Tyard is best known today for his poetry, but he was also seigneur of Bissy-sur-Fley in Burgundy, where the fifteenth-century castle he once inhabited still stands. After a few years in the court of Henri III, he was appointed bishop of Châlon-sur-Saône in 1578, until he resigned in 1594 (in favor of a nephew) and retired to his family lands. There he focused on scholarship, particularly on the etymology of terms from Greek and Hebrew origins, and wrote *On the Correct Imposition of Names*. The section on natural historical terms concluded (after sections on quadrupeds and birds) with gems and rocks. Here Tyard discussed, for example, the magnet, the amethyst (whose power included dispelling drunkenness), and those strange stones that form inside various animals (in modern parlance, gallstones). Among them he mentioned the "perdicite," a stone said to form inside the partridge (hence its name, derived from the Latin *perdix* for partridge).[34] Two printed marginal notes alongside the following passage called attention to the themes of interest in it, "New concerning the perdicite" and "Philippus Robertus":

> When I had retired to my castle of La Salle near the end of the summer 1591, with Philippe Robert, learned jurist and great connoisseur of Greek and Latin letters: and there he spent most of the day on his Demosthenes or Isaeus, while I devoted myself all my studies to Philo, one day, as evening was falling, Caumont, my secretary, came to see us, exulting: "I have (he said looking at me), I have, my master, something that you could please add to your little book on the

right imposition of names which you have given me to transcribe. Here is a *perdicite* which I pulled out from the gullet of one small partridge (which was served to us among the remains of your meal) with my own fingers." And, contemplating this kind of bite-sized crystal, of delicate transparency, ranging from a splendid watery glow to a very weak red, but of such a size and shape that made it difficult to think that it could have been swallowed by such a very small bird as this partridge, or that it could have formed naturally in it, I therefore with surprise and doubt: "Hey Caumont (I said) what game are you playing with us?" And he, blushing a bit, expressing respect: "I beg your indulgence (he said), I assure you, that the truth itself is not more true [than what I said]. If it is not so, master, henceforth do not believe me on any matter divine or human. I split the gullet of that bird, came across this little stone and removed it myself from the neck of the bird which I then ate. Your chaplain [giver of pious alms] over there, and this valet here and all the other domestics who saw this with their own eyes, I can bring them before you if you'd like." I agreed, smiling: "Give me this stone (I said), you have convinced me." And I turned to Philippe Robert: "I am reminded of this passage in Aelian (I said): 'There are partridges near Pisidia in the region of Antioch that eat even stones.' But we are far from Antioch." After a little discussion on this stone between myself and the very learned Robert, I enclosed it in my collection of precious stones and kept it carefully with me. Nonetheless when, some time later, I was writing these lines, it seemed to me that its red tint had completely dissipated (although its crystalline transparency remained). But from this digression we must return to the road.[35]

We cannot confirm this incident from other sources nor get the perspectives of the others involved; nevertheless we can use this account (since it seems unlikely to be completely fictitious) to provide a rare glimpse into the making of a passage in a learned book.

Tyard is writing in the first person to recount an interaction that took place twelve years earlier during a summer retreat to one of his castles, presumably on his lands in Burgundy. The passage is also a tribute to Philippe Robert, as highlighted in the marginal note, who was a magistrate in the *parlement* of Burgundy until his death in 1594.[36] The anecdote is framed by the two learned men exercising learned *otium*, or leisure, during the summer respite from their worldly duties. In a kind of "parallel play," each focused, apparently in the same room, on studying ancient authors of their choice: Philo for Tyard and Demosthenes and Isaeus for Robert—indeed Robert left at his death a manuscript translation of the orations of Isaeus, which had been lost by 1666.[37] The digression ends with the two discussing the ancient references that came to mind concerning the strange stone.

Into that calm and dignified frame burst Caumont, Tyard's *"calligraphus a*

manu," a scribe with a good hand, likely employed for the long term as a secretary. Secretaries were typically employed to take dictation, make copies of
letters and other outgoing documents for record-keeping purposes, and organize their employers' correspondence, notes, and books. We hear about others
in Tyard's large household. The chaplain ("giver of pious alms") and the valet
were perhaps in the room at the time, depending on the valence we give to the
pointing terms "this" (*hic*) and "that" (*iste*), by which he designated them. And
Caumont offered to bring in "many other domestics" to confirm his testimony,
presumably because they were all dining together when Caumont came across
the stone. Caumont is described as copying the text of the book we are reading.
Interestingly, that book was published only much later. Perhaps Tyard composed the passage while reminiscing years later or, more likely in my view, the
passage (or even the whole manuscript) had been completed close to this event
of 1591 but then was left to be published later. The work appeared in 1603,
when Tyard was roughly eighty-two, and just two years from his death. Tyard
may have feared his work would be lost if he did not publish it. Indeed, in a
letter to the printer Roussin in the front matter of the book, Tyard explained
that he wished to publish these "little results of nighttime study" (*lucubratiun-
culae*) so that something of his would survive him.[38] Caumont might have moved
on from Tyard's employ by the time the book was published and another scribe
may have prepared the manuscript for printing. But it is also possible that the
passage was composed not too long after the event (Tyard explained that he was
writing "some time after"), in which case Caumont might have had occasion to
copy out the passage that he had prompted Tyard to compose.

The subordinate position of the secretary is clear: he was eating with the
other servants after his master's meal when he came across the strange stone.
Caumont came to see the learned men on his own initiative, at the end of the
day. Since this was presumably not right after the meal he described, it seems
that Caumont waited until a natural break point in the workday so as not to
interrupt the two learned men. Caumont's excitement is met with amused and
somewhat condescending skepticism at first. The detailed presentation of Caumont's testimony and the conversation about it serve to establish the veracity of
the anecdote (as in the rhetorical strategies designed to create "virtual witnessing" of Royal Society experiments later in the century); at the same time it allows Tyard to abstain from endorsing any claims about the perdicite directly.[39]
He told Caumont he was convinced and took possession of the stone, but the
reader is left to weigh all the factors presented in deciding what to think—the
credibility of the secretary versus the lack of supporting ancient authority and
the later degradation of the original color of the stone.

Tyard portrayed himself as the knowledge maker in this interaction, but
from our perspective Caumont also contributed to the content of the text while

carrying out a task that Tyard likely considered merely mechanical, of copying over a book manuscript either while the composition was in progress or once it was finished to send a clean copy to the printer. In fact, not all copying was considered beneath a learned man's station. Early modern scholars sometimes engaged in scribing themselves, especially when copying authoritative texts from a rare manuscript original and thereby creating a copy that would have special significance in the text's accurate transmission. For example, the English antiquarian Sir Simonds d'Ewes reported taking "extraordinary content" at copying in his own hand the *Fleta* (a medieval legal text) from the exemplar owned by Robert Cotton (which is today the only complete copy and may have been so at the time too, in 1626—in any case it was very rare). But d'Ewes's enthusiasm for the task evaporated suddenly when he learned that "one Ralph Starkie" had already made copies of the text and sold them. At that point the scribing was no longer an act of scholarly transmission but one of mere reproduction, so d'Ewes stopped his efforts and had the rest of the manuscript copied out by an "able librarian" whom he hired.[40] Since scribing was a task that every scholar could also perform (barring illness), decisions about when to delegate the work and when to do it oneself were complex and invite further investigation.[41] But for copying out his recently composed book, whether for further editing or for use by the printer, Tyard's decision to delegate the work to a secretary with good handwriting was likely unproblematic. He presumably also expected Caumont to make a reliable copy without modifications—in a kind of "mechanical" operation.

But a surprise was in store. Caumont was no machine in carrying out his assigned task. He was alert to the discussion of bird stones that he was copying. He understood the Latin text and explicitly asked for his experience to be included in the learned treatise. That experience itself—taking notice of a small stone encountered while eating—was likely triggered by his knowledge of the text he had just been copying. In addition, Caumont may have wished to be recognized in print. When artisans and other "mechanicals" were mentioned in learned works of natural philosophy, for their testimony or expert opinion, they were rarely named. Here Tyard named his amanuensis (although the mere name has not sufficed to identify Caumont further), and Caumont would have been aware of the outcome if he wrote out this passage at some point, either under dictation to compose the text or in making a clean copy. Presumably Caumont was pleased at being named, though we can only speculate about the broader dynamics of his relationship with Tyard.[42] For Tyard, was this a favor happily rendered to a well-liked member of the household? Or a case of humoring a pesky servant? The underlying emotions are inscrutable from such little evidence at this cultural remove.

The takeaway for my purpose here is that scribing—indeed any "mechanical" work, whether vicarious work or handwork—was "mechanical" only from the perspective of the person commissioning the work. The person performing the work drew on their own judgment, skills, and experience, thus shaping the outcome, even if that process is rarely as visible centuries later as in the case of Caumont's contribution to Tyard's learned treatise. Considering that almost all printed works went through the hands of a scribe at least once (to make the clean copy that the compositor would use to set the type), we are left to speculate about the contributions (intentional and unintentional) that these countless nameless individuals had the opportunity to make in their decidedly non-mechanical work.

New Terms for Collaboration

How should we describe Caumont's contribution to this passage in Tyard's *De recta impositione*? Coauthorship is too strong a term, but scribing too weak. And others may have contributed to the text too, without leaving as explicit a trace—either directly, as Caumont did, or indirectly as sources of inspiration for Tyard whether or not he cited them. In the analysis of literary texts, genetic criticism especially has attended to the role of models and other sources of inspiration in the process of composition.[43] Harold Love has described a variety of stages of authorship, from "precursory" to "executive" and "revisionary," at each of which others may have been involved beyond the author declared on the title page.[44] More generally, the field of Renaissance literature has devoted much attention of late to coauthorship, especially among playwrights.[45]

As for printed texts, they all passed through the minds and hands of multiple people at the printing house. Tyard himself alluded to this process when he enjoined Roussin, in the front matter of his *De recta impositione*, to "publish [the work] with your types as correctly as you can."[46] The tremendous growth of book history since 1980 has shed light on the many ways in which the printed book was shaped by those whose work was long overlooked as "mechanical" in each of the three senses of the word—handwork, for pay, using a machine. The printing house was of course full of people, comprising at least a printer-publisher who made the decision to print (and for what market, in what format, at what price point, and so forth), a compositor who set the type and determined the layout, and a corrector who prepared the manuscript for print, often modifying the spelling and punctuation, and who then read the proofs with or without help from the author. The printer or author might hire still others to perform specialized work such as editing, abridging, translating, or adding indexes, commendatory odes, or images. All of these agents contributed their judgments and skills to the final printed book, shaping its impact and reception, so we must guard against viewing a book as having sprung ready-made from the mind of its author.[47]

After we acknowledge these usually silent contributors in principle, the next steps—to examine their work in more detail—are often difficult. Correctors, for example, were not mentioned in print except occasionally when taken to task by angry authors, but Anthony Grafton has traced their activities in vivid detail thanks to the surviving archives of the Plantin press in Antwerp.[48] The relations between authors and printers left more visible traces, including explicit statements in printed paratexts and implicit patterns of behavior. But these kinds of evidence must be analyzed with care since they resulted from the drive to sell books, not to document a relationship transparently for the historian. For example, an author might complain in a preface about a printer acting without authorization and yet elect to work with that same printer soon thereafter. Perhaps the two managed to overcome genuine antagonism, or perhaps that antagonism was more rhetorical than real, staged in the paratexts in order to minimize the author's responsibility for the publication and relieve the stress of a critical reception.[49] Many case studies illustrate the range of relationships between authors and printers—friendly, hostile, or a bit of both, geographically close or distant, controlling (in either direction) or easygoing. One collection of such studies focused on the relations between authors and printers calls the players involved in the production of a printed book "co-elaborators of the text."[50] We are just beginning to develop a vocabulary that is nuanced enough to match the great range of roles involved in the production of a text that is typically assigned just one author.

Illustrated books involved even more contributors, as current work emphasizes. Natural historians in the Renaissance prided themselves on offering "drawings from nature" that required either bringing the specimen to the artist, or the artist to the specimen. Conrad Gessner, for example, thanked Lucas Schan of Strasbourg for making images and providing some descriptions of birds from life. Gessner's praise of him as a man "equally skilled in painting and in hunting for birds" suggests that he was the kind of artist who sought out the specimens.[51] Plants could be more easily brought to the artists, as was the case behind a rare depiction of image making in Leonhardt Fuchs's natural history of plants, *De historia stirpium* (1542). Whereas the book opens with a full length image of the author, it closes with a depiction of the three men responsible for the images: the first made a drawing on paper by observing a plant in a vase, another transferred the drawn image to the woodblock, and a woodcarver produced the block used in printing the image.[52] Some natural historians relied on their daughters trained in drawing to supply that labor, adding a familial layer of complexity to the collaborations.[53] Twenty-five years ago historians of science were at the forefront of studying how social hierarchy shaped the authority of truth claims, for example, in seventeenth-century England; and today they are tracking the nuances of collaboration among naturalists in which men of lower social rank could play a leading role in an intellectual partnership even though at other times they were treated like servants.[54]

Building on studies of the workshops where the master painter or sculptor worked with assistants, art historians have been focusing in recent years on collaboration. They have uncovered many variations in the division of labor in these contexts. Customers commissioning a portrait could specify in the contract who (i.e., the master) would be responsible for the most difficult and important elements, such as the face and hands.[55] In some cases (say, in the workshops of Raphael or Rubens) assistants were given general assignments but otherwise free rein, for example, in painting a scene or a background. Others held their assistants within narrow instructions; Michelangelo famously delegated only when necessary and had everyone report to him directly.[56] Analyzing all these variations in modes of working and their impact is challenging, especially without a set of terms to describe them.

"Collaboration" is the term commonly used today to discuss how people work together, but it too encompasses many kinds of work and interactions often specific to different fields. An information scientist seeking to offer practical advice to managers of programming teams has contrasted dynamic collaboration (in which any member of a team can fill any role on it) with static collaboration (with a more rigid division of labor within a team).[57] In the arts, hierarchies are still present between the artist who conceives the work and the technicians who carry it out and thus contribute to the creative process, but receive much less compensation or credit.[58] By contrast, in films the credits have become ever lengthier and more specific.[59] In the sciences, as single-authored papers have become the exception and the number of coauthors runs into double digits (or even triple; e.g., in experimental physics at CERN), some periodicals require an "author-contribution statement" for each article describing what each of the coauthors contributed to it.[60] Collaboration in the humanities is also growing and expanding beyond the classic model of coauthors or coeditors sharing the same kind of work. In particular, projects in digital humanities involve larger teams that encompass different kinds of skills and levels of time commitment than do traditional print humanities. Our increasing awareness of collaboration today has no doubt played a part (along with the factors mentioned above) in the ongoing attention to the complexity of knowledge making in the past, which was so often collaborative, as it still is today.

NOTES

1. P. Long 2011, 128, citing L. Roberts, Schaffer, and Dear 2007, xvii.
2. For an entry into this topic, see Grafton 2010b.
3. Plato, *Republic*, 428b–d, and, more generally, Parry 2020.

4. Aristotle 2017, 190.

5. Whitney 1990, chs. 1 and 3, esp. 70–73.

6. Post, Giocarinis, and Kay 1955.

7. Grafton 2000.

8. The literature on this development and its origins is massive. For a point of entry, see Smith 2004.

9. Pérez-Ramos 1988.

10. "[In artibus mechanicis] Effectio ingeniosa, hominum vitae quotidiane utilis et iucunda, proper quam Periti et Industrij ab ope ferenda Opifices nomen trahunt, commendetur. . . . Nam quo quaeque animi potius quam corporis nituntur praesidio, sic hominis excellentia digniores censentur. Neque vero quicquam prohibet, Theoricos et Practicos habitus Mechanicorum opera illustrari, et vicissim Mechanicos Theoricorum praeceptis dirigi et animari." Zwinger 1571, 3182.

11. See the more than sixty mechanical arts listed, most conveniently, in Zwinger 1565, 36.

12. The *Oxford English Dictionary* dates the first use of this meaning in English to 1567 (meaning II.5.a, "acting, worked or produced by a machine or mechanism").

13. "Mechanicae vicariae, quibus homo pro homine operam subit, vel opus efficit." Zwinger 1571, 267. The section was further expanded in the third and last edition, under the title "Mechanicae operae vicariae, pro homine agentes, ea quae ab illo fieri debebant." Zwinger 1586, 3669–76.

14. "Quo ad animum. Huc pertinent Amanuenses, Scribae, Notarij, Epistolarum magistri, Aquam subministrantes in iudicijs." Zwinger (1571), 3267–68. In 1586 the section was longer and included: "qui . . . rationes subducunt, explorant, lugent, lamentantur." Zwinger 1586, 3670–71.

15. "*Jube* dico, non *Fac*: nam labor iste est mechanicus, nec te indiget: sufficit enim, ut alios dirigas." Caramuel y Lobkowitz 1664, 191 (articulus VIII, section 3223; italics for emphasis in the original); reprinted in Romani 1988, 1–73 (quotation on 30).

16. "Adhuc Gruterus ab amicis plusquam trecentas habuit Inscriptiones, quae magno illi accuduntur Operi. Indicem misi, mei mentionem fieri prohibui. Editio enim tota Gruteri est; ego mediastinus operas tantum contuli." Scaliger to Casaubon, Aug. 6, 1602, in Scaliger 1627, letter 73, 224. See also Gruterus 1603 and Grafton 1975.

17. Zilsel 1945, 342.

18. Raven and Krohn 2000, xxi–xxiii, xliv–xlvii; P. Long 2011, ch. 1.

19. Taylor 1954, 1966. On Taylor, see de Clercq 2007, 4.

20. Rossi 1970, citing Leonardo Olschki among others. For an entry into alchemy, which has featured prominently in this line of inquiry, see Newman 2004.

21. Cook, Smith, and Meyers 2014, 3–5.

22. Cook, Smith, and Meyers 2014, 5. On the sociology of scientific knowledge, see Golinski 1998.

23. Shapin 1989.

24. Park and Daston 2006, part 2: "Personae and Sites of Natural Knowledge."

25. Smith and Findlen 2002; Cook 2007; Margócsy 2014.

26. Cools, Keblusek, and Noldus 2006; Egmond 2008; P. Miller 2015; Ash 2017.

27. Harkness 2007; Marr 2011.

28. See, for example, Bleichmar 2012; Ghobrial 2013; and E. Rothman 2012.

29. See, for example, Rankin 2013 and M. Roberts 2016. On the permeability of the scholarly study, see Algazi 2003, 27–28.

30. Krajewski 2018; Blair 2019.

31. Schiebinger 2017; Parrish 2006; Sweet 2013; Gómez 2017.

32. Schiebinger 2017, 158–64.

33. Subrahmanyam 2009, 440.

34. The term is rare, not found in the early modern French dictionaries on the website of The ARTFL Project (https://artfl-project.uchicago.edu/); but (as a Google search reveals) it appears as an English word in Meissner 1847, 790 (s.v. "Rebhuhn"), where it is defined somewhat differently as "a kind of stone marked like the breast of a partridge," with no source provided.

35. "Quum in Sulano meo Castro sub finem aestatis anni 1591, una cum Philippo Roberto doctissimo Iuris consulto, literarum omnium Latinarum et Graecarum eruditissimo, secessissem: et ibidem illo, suo Demostheni, vel Isaeo, diei meliorem partem daret: ego vero in Philonem omnia mea studia conferrem, die quadam cum iam advesperasceret, Calmontius Calligraphus a manu meus, gestiens, ad nos: habeo (inquit me intuens) habeo, mi Here, quod addere possis si lubet libello tuo de Recta nominum impositione, quem mihi exscribendum dedisti: Ecce perdiciten, quam ego e sinu gutturis unius pulli perdicis (qui inter reliquias coenae tuae nobis appositus est) his ipsis meis digitis extraxi. Ego vero id crystallinum tanquam frustulum conspiciens, eleganti pelluciditate, ab aqueo splendido nitore, in dilutissimum ruborem emicante, sed ea figura et magnitudine, ut vix crederes, hunc a pullo parvissimo, qualis erat hic περδικιδεύς, devoratum fuisse, ac innatum putares: quare miratus, et emiratus, dubiusque, Ohe (inquam) Calmonti, quibus facetiis nobiscum ludis? Ille subpudens, honorem praefatus, Bona (inquit) tua venia dixerim: Nae, hac mea assertione, ipsa certe verior non est veritas. Ni sit ita, mi Here, posthac neque divini, neque humani mihi quicquam accredas: sinum discidi gutturis pulli, lapillum hunc offendi, evulsi ab ipsa ingluvie, quam esitavi: Tuum istum piae stipis erogatorem, huncque cubicularium, caeteros etiam alteros ex domesticis quosdam oculatos testes, te coram produco, si lubet: Acquievi, subridens: et cedo igitur lapillum (inquam) fidem impetrasti. Tunc ad Robertum conversus: mentem meam subiit istud Aeliani, inquam. [Greek passage from Aelian omitted] *Syproperdix circa Antiochiam Pisidiae nascitur, qui etiam lapides vorat.* Sed longo ab Antiochia distamus intervallo. Postquam hunc super, inter doctissimum Robertum, et me, aliquantulum dissertum esset, eum in dactylotheca inclusi, quem diligenter apud me servatum habeo. Verum dum haec, aliquandiu post, scriberem, color ille rubescens, mihi penitus (perspicuitate tamen crystallina remanente) obliteratus et elutus esse visus est. Sed ab hoc diverticulo, ad viam redeundum." Tyard 1603, 80–81. For a modern edition and French translation, see Tyard 2007, 198–99, and notes on 442–43. I am most grateful to Jean Céard for this reference. Tyard is accurately citing Aelian, *On the Nature of Animals*, XVI.7. On Tyard, see Nicéron 1745, 292–302, and Kushner 2018.

36. On Robert, see Kushner 2018, 81–82.

37. Muteau and Garnier 1858–60, 3:61.

38. "Mihi enitendum esse putavi quantum possim, ut post meam mortem, etiam mihi superstes esse videar. . . . Has igitur lucubratiunculas ad te mitto." Tyard 1603, sig. A2r.

39. On virtual witnessing, see Shapin 1984.

40. Woudhuysen 1996, 128. Warm thanks to Arnold Hunt for this reference.

41. For some discussion, see Blair 2016.

42. For another example of an amanuensis who asked to make an appearance in his employer's publication, see Blair 2019, 29–31.

43. For an introduction to this predominantly French approach, see Deppman, Ferrer, and Groden 2004.

44. Love 2002, 40–50.

45. For an entry into this field, see Knapp 2005.

46. "Has igitur lucubratiunculas ad te mitto, ut eas tuis typis quam correcte poteris, publico mandes." Tyard 1603, sig. A2r.

47. Chartier 2014.

48. Grafton 2011b.

49. For one example of such a ploy, see Blair 2013, 146–47; for a typology of various ploys that discuss publication, see Vanautgaerden 2012, 500.

50. Furno 2009; Keller-Rahbé 2010; Ouvry-Vial and Réach-Ngô 2010.

51. "Lucas Schan pictor Argentoratensis aves plurimas ad vivum nobis expressit et quarundam historias quoque addidit, vir picturae simul et aucupii peritus." Gessner 1551, sig. [γ1]v.

52. For an entry into illustrated books, see Kusukawa 2012 and Egmond 2017.

53. Roos 2019.

54. Shapin 1994. On the complex relations between John Ray and Francis Willughby, for example, see Yale 2019, 285–86.

55. Baxandall 1988, 6, 23.

56. See, for example, Alpers 1988, 59; Wallace 1994, 38; or the many collaborators described in Wunder 2017.

57. Nielsen 2011.

58. Santana-Acuña 2016.

59. M. Murphy 2017. Murphy attributes this growth to the shift from celluloid to digital, which much reduced the cost of long credits; he reports as an example that *Ironman 3* (2013) listed 37,000 names.

60. "Authorship Policies" 2009.

CHAPTER TEN

History, Historians, and the Production of Societies in the Past and Future

YUEN-GEN LIANG

Ancient scholars searched the skies not for tools but for portents:
for astronomical events that not only located but prophesied the
great events near which they fell. Research into the past, after all,
must often have been linked to efforts to predict the future.

Anthony Grafton, Joseph Scaliger: A Study in the History
of Classical Scholarship

In *Joseph Scaliger,* Anthony Grafton studied Renaissance humanists working to identify and eliminate accretions to ancient texts. The methods of philological analysis they developed to assess the provenance of sources, trace the genealogy of manuscripts, and compare different versions of texts have since become essential for modern historical craft. These techniques made it possible to divine a history of texts and to expose them as windows onto different times. Yet, despite critical contributions to the practice of history, humanists were not content to let the past remain in the past. In a world that lacked standard measures and consistent memories, Renaissance scholars obsessed over the dates of ancient and medieval events in these textual reconstructions. Discovering these dates was also crucial for charting chronological patterns that would in turn shed light on the future, including the course of God's teleological creation and ultimately the arrival of the End of Times. In Renaissance scholarship, the past and future went hand in hand. Scholars who focused on the distant past also devoted themselves to ascertaining the future.

The essays in this volume propose new horizons in the study of Renaissance and early modern history. Like research on chronology, they start by assessing prior knowledge in order to then suggest future directions, particularly in the field of intellectual history. Continuing down this path, my contribution nonetheless takes a distinct turn toward the social. Examining how historians engage with contemporary societies is of keen importance as today's cultures are ever more engrossed by the present and the future and seem likewise disinterested in the past.[1] Every day, sleek and shiny devices that enhance the ease of our personal lives dominate attention spans and render alien the predigital age. Robotics, machine learning, and genetic engineering are reshaping the structure of economies, the ways humans interact with and trust one another, and how we and other organisms reproduce. These dynamic interminglings of technology, capital, and entrepreneurship foment transformations at a breathtaking pace and win resounding acclaim from popular culture, government policy, and university budgets, all the while thrusting societies toward increasingly portentous horizons.

Under these circumstances, the study of the past faces the challenges of articulating orientations toward the future. This essay examines how practices and practitioners of history contribute to an unfolding future. Four topics help bring insight to this investigation: a case study of the settlement of the Knights of St. John on the Mediterranean island of Malta in the sixteenth century and the reliance on precedents to secure survival in an uncertain future; the role historians play in passing on knowledge, civic values, and professional skills to our students; the development of digital humanities as ways to experiment with technology and make the past more present and accessible; and the place of historians in public affairs. It is uncommon for this seemingly disparate set of subjects to be treated together. The way they combine here expands awareness of the interconnected contributions we as historians make to society. This discussion also addresses diverse forms of historical practice and gives voice to different practitioners of history. Finally, these topics reach out to a broader audience including our students, who will take whatever interest they have concerning the past into the future.

Precedents That Shape the Future: Charles V and the Knights Hospitallers

Humans often look to the past for experiences that can be applied to help envision, condition, and enact a future in the process of unfolding. A case from early modern European history illustrates this point. In June 1530, the Knights of St. John of Jerusalem, also known as the Knights Hospitallers, landed on Malta to establish a new home base.[2] In time, they would erect cities such as Valletta and Mdina that brimmed with massive fortifications and Baroque splendor. They would also launch a corsair fleet that targeted Muslim shipping in the Mediterranean. The order's very survival, however, seemed unlikely at its incipient moment in the 1530s. Ottoman power was surging into the Mediterranean, and Malta, located at the center of the sea basin, was a prime target. To establish a new outpost in the face of such peril, the emperor Charles V and the knights turned to legal precedents. Charters and decrees drew on past rights and privileges to outline the framework of the settlement, including the parameters for supplying the outpost. The emperor, knights, and other officials deployed the past in order to generate a future.

The knights arrived on Malta as Ottoman power was expanding in the sea basin. In 1480, Ottoman forces conquered and briefly occupied Otranto, a town on the southeastern Italian peninsula. The sultan Selim "the Grim" next smashed through Mamluk defenses and took Syria and Egypt in 1516–17. At the same time, the captain Oruç Reis "Barbarossa" seized Algiers and turned it into an Ottoman base in the western Mediterranean. The forces of the Sublime Porte then turned to the island of Rhodes in 1522 and expelled the Hospitallers from their home of two hundred years. Forced into errancy for eight years, the knights did not find a permanent base until reaching Malta, a new fief granted by Charles V.

Though strategic, Malta's position was exposed. The island commanded a key passageway connecting the eastern and western halves of the sea and stood in the path of Ottoman expansion westward. In order to resist the Porte's might, the Knights had to build, arm, and man imposing military installations. Yet fortifying Malta made the Hospitaller base even more of a target for periodic raids by the Ottomans, including a major attack in 1551 and an epic siege in 1565. Apart from these campaigns, the unpredictability of enemy movements or rumors thereof imposed states of fear and anxious anticipation. Furthermore, as an island, Malta had to be provisioned with essential victuals, munitions, and supplies from Italy and elsewhere. Survival for the new Hospitaller community was precarious, and arriving in Malta after peripatetic exile, the knights were not certain if they would even have enough to eat.

The specter of scarcity can be gleaned in the documents preserved by the Università, the island's local governing council. These records are now bundled together as the first manuscript of the Archivio dell'Università housed in the National Library of Malta.[3] Taking pride of place as the first bound folio is a charter of privilege granted by the emperor Charles V to the grand master of the knights. This vital concession permitted the knights to secure quantities of grain and other provisions from Sicily annually and transport them without payment of ordinary duties (*derechos ordinarios*) or new taxes (*nuevos impuestos*) to Malta.[4] The particular importance of this document is signaled by the invocation of Charles's many titles, its solemn composition in Latin, and the sovereign's imperious signature "Yo, el Rey" (I, the King) in a grandiose hand. Issued on April 22, 1531, in Ghent, a city closely associated with the emperor's childhood, the document's ceremonial magnificence is further amplified by the stately vellum used to carry his words. The emperor's will, manifested in this very regal form, bespeaks an early modern notarial culture that, like its patrons, produced such a lasting object in order to transmit the past into the future.

While Charles bestowed the charter as a demonstration of his patronage and majesty,[5] the knights no doubt understood that access to victuals and supplies from Sicily helped ensure their survival. The sparsely populated Maltese islands were home to just around twelve thousand native inhabitants, who were predominantly Christians speaking a variant of Arabic. This population fluctuated, however, as periodic Ottoman raids carried off large numbers of residents into captivity. The islands themselves are rocky, with scanty, uneven rainfall further limiting agricultural cultivation. Under these challenging demographic, military, and ecological circumstances, the privilege granted the knights a number of provisions. Each year, it provided an allowance of "4,000 salmas of wheat, 1,000 salmas of oats for Malta and another 1,000 salmas of wheat for the maintenance of Tripoli."[6] The knights' meager existence was supplemented by "wines, livestock and cured meat, legumes, and other things to eat and munitions for the use and service of this order [*religion*]."[7] Despite the royal grant, the Hospitallers nevertheless had to constantly appeal for the fulfillment of these promises. At times the knights could not acquire the allotted wheat for Tripoli, so they had to provision the North African outpost from Malta's share.[8] At other times, drought rendered the fields of Sicily "sterile" so that the grand master had to send a special representative to plead with the emperor to honor the terms of his agreement.[9]

Focused on food, the records looked to precedents to justify provisioning the knights. The emperor granted the privilege "according to the past when [the knights] were based in Rhodes and what they were accustomed to take."[10] He further justified the allowance according to the favor that his predecessor King Ferdinand of Aragon had shown the knights decades earlier. In this early modern era of global expansion, Europeans often relied on legal precedents to impose order on novel circumstances and encounters that, in many ways, were disorientingly new. Under these conditions, the principle of conservation, exemplified by the commonplace precept "may there be no innovation in this matter" often framed attitudes, positions, and decisions.[11] In the case of the knights, the crown's policy toward provisioning Malta defaulted to a *status quo ante.* If the Hospitallers had received subsidies in the past, they were entitled to and should continue to receive assistance in the present and future. It is important to note that the thinking here was not to constrain the future based on the past but to avoid causing harm to rights and privileges that had been accorded the Hospitallers. Rather than a norm to be prescribed, the past was a resource that could be summoned to help the knights to manage their new situation on Malta.

In addition to applying precedent to accommodate projected needs, language in foundational documents also connected the past and future in terms of lineage. Charles V pointed to Ferdinand as more than just a predecessor; he specifically addressed the Aragonese king as "my lord and grandfather." To further emphasize the affective ties, Charles attested to his "happy memory" (*felice memoria*) of his progenitor.[12] The allusion to lineage served particular purposes. Understood in broad terms, "lineage" at once invokes connections between the past, present, and future. When referring to blood ties, this word also conjures an homage to ancestors; social status in the present that comes from ancestral identity and deeds; and hope for progeny to carry on the line. Drawing attention to blood ties with the Aragonese dynasty that had acquired Malta centuries earlier further enhanced Charles's legitimacy in mandating the Hospitaller settlement in a landscape of competing French, papal, and Ottoman interests. A generation later, upon the death of Charles V and the succession of Philip II, the new king reconfirmed the Hospitallers' privileges by once again deferring to his ancestors' prior practices. The concept of royal lineage, particularly its binding of past, present, and future, served as a guarantor of the Hospitallers' survival on Malta.

Critically, deployment of the past was not an attempt to forestall change. Although the monarchs and the knights drew on precedent and lineage, the future could not resemble the past. Novel circumstances such as the Holy League's defeat of the Ottoman navy at the Battle of Lepanto in 1571 resulted in the rise of the knights as fearsome corsairs of the sea. As Malta's stature grew, it became a key inflection point in the sea basin and an outpost with an international flavor. The Arabic-speaking local Christian population already exemplified its location at a crossroads. The knights hailed from the four corners of Catholic Europe and the men were arranged into eight "tongues" (*langues*): Auvergne, the Crown of Aragon, the Crown of Castile, England, France, the Holy Roman Empire, Italy, and Provence. As the archetypal Christian corsairs in the Mediterranean, the Hospitallers preyed on Ottoman shipping and the Muslims on board. They also seized merchandise carried by Greek Orthodox subjects of the sultanate, forcing these merchants to appear in Maltese courts to defend their identity and legal status as fellow Christians and to sue for the restitution of their goods.[13] Although Malta had been established as a Catholic bastion, its transimperial development featured connective and globalizing dimensions (addressed in Alexander Bevilacqua's chapter in this volume). Charles V and the knights may not have expected some of these eventualities, yet the legal precedents and terms of lineage that gave sustenance to the Hospitallers ultimately enabled this future to emerge. Invocations of the past were not constraints but rather anchored the emperor and the knights themselves in their ability to navigate new developments.

In his book *Past Futures: The Impossible Necessity of History*, the historian of Canada Ged Martin investigates decision-making moments as markers of significance in the interpretation of history.[14] Analyzing these instances, he first describes how an individual made choices based on an assessment of past events that led up to the moment of decision. Just as important, however, Martin stresses that decision-making processes also appraised the possible future outcomes that the different choices might precipitate. Placing decision making in a continuum of past-present-future considerations, he lucidly demonstrates how the creation of the future is inextricably tied to examinations of the past. Indeed, the case study of Malta argues Martin's point: envisioning the scenario of the Hospitallers' settlement on the island, Charles V decided to effectuate this situation based on his grandfather's prior support of the knights and to apply precedents to secure such a reality.

After laying out this schema, Martin goes on to suggest that historians have often neglected to analyze how thinking about the future affected an individual's decision-making process, faulting our own ambivalent attitudes toward making projections about a time that has not yet come to pass.[15] Martin recommends instead taking into consideration possible alternative outcomes in order to mitigate a teleological assumption, of the kind that William Bulman identifies, for example, when historians treat certain sociopolitical formations as "normative," in particular "Western, free-enterprise, electoral democracies."[16] But the implications of Martin's ideas were not clear at the time he was writing—the decade after the demise of the Soviet Union, when others were heralding the victory of the "free world" as the apogee of human political development and consequently the "end of history."[17] Yet the return of illiberal strongmen to prominent political stages highlights anew Martin's admonition that historians cannot shirk the challenges of the future by taking liberal systems as an inevitable course of human decision making. Instead, we must actively wield knowledge of the past to help shape decision making in ever more precarious futures, as I discuss in the following section, on contemporary historical practice.

Lessons about the Past as Fostering Civic and Professional Skills

The uncertain future that the Hospitallers faced evokes the challenges that Renaissance and early modern European historians are confronting today.[18] Along with the rest of the humanities, the discipline is grappling with technological, economic, and cultural changes as large sectors of societies increasingly worship at the altar of algorithms, heed the language of utilitarian outcomes, and monetize the value of human processes including learning, growth, communication, and creativity. Practitioners of history, as well as citizens all over the world, are being buffeted by the deployment of "alternative facts" that deliberately warp memory and history to advance pernicious ideological objectives. To address these struggles, it is essential to articulate ways that the study of the past and scholars who pursue that study offer value propositions to twenty-first-century society. Just as Ann Blair expands research on early modern knowledge by incorporating the evidence of mechanical arts and their practitioners, I present the testimony of historians living, breathing, and engaging with societies around us in order to extend the evidence that our work as constructive citizens uncovers. I now discuss what we do as teachers, digital humanities scholars, and participants in collaboration and partnership with the public sphere.

An explanation of our relevance to society is grounded, of course, in the intrinsic interest of history itself. As a burgeoning new age of global encounters, political upheaval, intellectual and artistic exploration, and "self-fashioning," the early modern period invites contemporary audiences whose lives are undergoing similar and rapid changes to connect with a transformative historical epoch. The multiethnic and transnational nature of Malta's population, the crossing of what is now often thought of as hard and durable boundaries between "Europe-Christendom" and the "Middle East–Islam," and the many acts of creation required of the knights settling on Malta, all of these dynamics are pertinent to our world today. As with Charles V and the Hospitallers' recall and application of precedents, a historian's role in this case is to preserve knowledge of the past, analyze its importance, and authoritatively interpret it with active reference to his or her own societal context.[19] In addition, through narrative skills, we share these histories with audiences for whom stories have personal meaning as well as entertainment value. Indeed, storytelling—the building of narrative arcs with beginnings, middles, and ends; the highlighting of details that flesh out the story; the situating of details in sufficient context to assist understanding; and the painting of scenes through rich descriptions—these techniques have long been our forte. In a competitive marketplace of media and ideas, it is incumbent on us to demonstrate our ability to communicate the interest of history to a variety of audiences.[20]

Besides telling illuminating narratives of the past, we can communicate and reproduce the advantages of our knowledge by other means. Charles V and the Hospitallers invoked precedents to provide sustenance to the knights and ensure the reproduction of their community. Historians also foster social reproduction by passing on our knowledge and contributing to the personal and professional formation of new generations of students. The basis of what we teach students may remain the content of what happened in the past. But we also convey vital and marketable skills beyond the abilities discussed above that are useful in a variety of contexts.[21] It would be shortsighted to think of education in solely utilitarian or vocational ways, and exposure to the arts and humanities can impact an individual's quality of life, notably by fostering a sense of empathy or enhancing one's own happiness. However, it would shortchange our achievements not to recognize the array of skills we practice in our classrooms. These include critical thinking, active listening, communication (written, oral, and visual presentation), information management (research, data absorption, analysis, contextualization, and organization), argumentation (identifying degrees of importance through assessment, situation, and interpretation), and leadership (persuasion, teamwork, and project management). Careers in the information economy demand these skills and they are also essential to building a responsible information economy.[22]

The training we impart may serve to produce budding historians, but more often our students fruitfully implement this training in nonacademic settings. The applications are multitudinous.[23] Our classrooms can serve both as interactive communities and as semiprofessional laboratories where students model sociability and experiment with ideas. The discussions that we proctor bring into dialogue different perspectives that mirror communication among groups of individuals possessing distinct ideologies, expertise, and backgrounds in a pluralistic society. The articulation of arguments and the marshaling of evidence model how leaders both in everyday communities and in work teams persuade associates to arrive at shared conclusions and invest in common goals. The ability to gather and discern data is no doubt important to functioning in an information economy, but it has proven to be just as critical for constructive citizenship in an age of mendacious public figures and "fake news."

Enhancing students' capabilities ultimately contributes to the future of the profession. Students of history are some of the best ambassadors for the discipline, not to mention potentially lifelong consumers of knowledge. Our students, most of whom pursue careers outside traditional academic paths, can show others the value of studying history. History and the broader humanities make vital social contributions by fostering both constructive citizenship and professional skills. Helping form students into more conscientious citizens and professionals expands the notion of a history education, and thus enhances our value as teachers of history.

The Experimentation of Digital Humanities

As new technologies arise, historians, like others, adopt the innovations that are dominating daily life. Over the past twenty years or more, digital humanities have emerged as approaches, platforms, and modes for using electronic technology to undertake an array of scholarly endeavors.[24] Digital humanities are expanding access to sources, data, and information while developing analytical approaches, collaborations, presentation platforms, and other creative possibilities. Although the field has its critics, it is undeniable that nearly all scholars make extensive use of electronic devices to access the virtual world.[25] Still, digital humanities are continuing to undergo experimentation and there is much to consider about how to use them as vehicles to the future. The digital world dissolves many boundaries erected by academics in previous generations by melding old genres, media, interfaces, and audiences and inventing new ones capable of changing the nature of categories, creators, and consumers. Although the following discussion focuses mainly on digital humanities in a scholarly realm, digital humanities are practiced by many people and come in diverse forms, most of which are not intentionally scholarly.

Assessing the technological and informational changes taking place today, some digital humanities scholars see these pioneering times as an opportunity: "Ours is an era," Anne Burdick observes, "in which the humanities have the potential to play a vastly expanded creative role in public life."[26] Indeed, as everyday users of the internet, academics are cognizant of the nature and power of digital platforms. Even if we are not deliberately working on a digital humanities project, the ability to access materials such as the latest journal articles or centuries-old documents has altered the ways we do research. Over the years, digital humanities have come to encompass many practices. In addition to the massive output of a variety of digital media, other practices include the use of computer programs and microprocessors to analyze data and the presentation of content that combines text, sound, and image in simultaneous and networked ways.[27]

Much of the production in digital humanities reflects the kind of work historians do, and these projects also have the potential of expanding our engagement with other knowledge and skill sets. Historians are expert content creators and curators. We gather information that we then process through analysis, contextualization, and interpretation. To previously disparate and incomplete data, we apply narrative structures in order to create intelligible stories. However, some aspects of digital production are likely to be unfamiliar to some humanities scholars, at least at the outset. Technical proficiency in coding, managing large data sets, creating multimedia presentations, writing for broader audiences, marketing products, and other tasks may require assistance from other experts. Digital humanities may encourage us to work in different ways than before. Collaboration here is perhaps one of the hallmarks of creating in digital humanities, and it also opens doors to interdisciplinary and interprofessional activities. Digital humanities have the potential to weave us into virtual and physical communities, ones that are immediate, collaborative, and sociable.

The formulation of digital humanities projects enables new creativity. Called by some the "generative humanities," they are opportunities to interact with text, image, sound, sequencing, simultaneity, hyperlinking-networking, performance, software, platforms, people, and audiences. Though requiring a variety of proficiencies, these processes may give creators greater hands-on control over their products than publishers would normally be able to provide in conventional publications. Consequently, digital humanities create room for experimentation. As Burdick has written, "Digital Humanities infrastructures encourage PROTOTYPING, generating new projects, beta-testing them with audiences both sympathetic and skeptical, and then actually looking at the results. Building on a key aspect of design innovation, Digital Humanities must have, and even encourage FAILURES."[28] Digital humanities offer opportunities for scholars to develop new skills related to evolving technologies, and these attempts and even stumbles help shape the direction that some of these technologies take.

Indeed, experimentation is required for scholars to take advantage of virtual reality and its ability to reach an audience the size of the World Wide Web. The massive online publication of sources has enabled a variety of projects that catalog, render searchable, analyze, interpret, hyperlink, and geolocate information and will no doubt continue to spawn new ways to use and present records. These endeavors serve important heuristic purposes, yet scholars could also both envision broader audiences or target distinct ones for particular projects. Conceiving projects as products that meet market needs or create them (e.g., "killer apps")—and borrowing from business strategy and analysis of competitive advantage—could motivate digital humanities scholars to reach out beyond the academy. We need especially to promote products that have both educational and popular appeal.

For over twenty years already, digital humanities have enabled scholars to make use of technology to develop their work. Still, the impacts of digital techniques on the humanities in institutionalized form are unclear, and a number of questions have emerged that are worth considering for the future of the humanities. These include: How effectively do computational analyses represent and interpret human subjectivities? How does programming that is more linear in trajectory compare with the free-associational capabilities of the human scholar? Do the significant sums disbursed for digital humanities reflect the increasing corporatization of universities? How do the uncertainties of long-term financial support, institutional structures, and technological durability affect the sustainability of projects? Does computing even constitute a practice of the humanities? Perhaps the most salient question of all is whether digital humanities techniques advance a new way of *thinking* in the humanities or merely augment the way we would otherwise work.

Digital technology massively influences the structure and minutiae of lives on both individual and societal levels. Consequently, it also raises fundamental questions about human agency. As scholars who study "the world created by humans," this is a crucial subject to address.[29] Machine learning, artificial intelligence, and robotics are automating many parts of the economy and daily life, such as manufacturing, automobile driving, plane piloting, the review of legal documents, stock transactions, and surgery. The digital processes may be programmed by human individuals, but do these activities still count as human? Regarding the study of history, how do researchers engage with tools of production that are no longer easily and immediately manipulable like pen and paper? How much do the abilities and limits of machines and computational analyses drive research questions? How much do computational methods such as modeling and statistical analysis impact researchers' close reading, critical thinking, and subjective interpretation? Are researchers on the inside or the outside of analytical processes? Will artificial intelligence render "cyborg" the study of history, just as the addictive attachment to smartphones have led some to suggest that electronic appendages have already attached onto humans?

Ultimately, the digital world, including digital humanities, augurs changes in academic categories and structures.[30] Digital technology has massively broadened and multiplied the production, dissemination, and consumption of information and content, as websites, blogs, YouTube, Facebook, Twitter, Instagram, and any number of other digital platforms have amply shown. As a corollary, the number of people participating in discussing, analyzing, and interpreting such information and content has also exploded. Masses of individuals interact with, are vested in, take ownership of, and are authorities over information as never before. The production and consumption of information are also changing, making these processes incredibly frenetic, fleeting, and monetized. Although most of these practices are not formally academic, it is undeniable that many parallel what humanities scholars do. Digital technology does not make the humanities obsolete. Instead, it expansively decentralizes and popularizes the practice of engaging with "the world created by humans." Within this changing landscape, scholars of Renaissance and early modern Europe find and cultivate allies among these social forces.

Lived and Living Lives

At the 89th Academy Awards on February 26, 2017, the actress Viola Davis testified to the value of historical inquiry. Accepting the Oscar for best supporting actress, she declared with almost uncontained urgency: "You know there's one place that all the people with the greatest potential are gathered—one place—and that's the graveyard. People ask me all the time, 'What kind of stories do you want to tell, Viola?' And I say, exhume those bodies, exhume those stories, the stories of the people who dreamed—big—and never saw those dreams to fruition, people who fell in love and lost." As if this were not enough of a tribute to history and the humanities, she continued: "I became an artist and thank God I did because we are the only profession that celebrates what it means to live a life."[31] Almost instantly after these words were uttered, Davis's speech—already broadcast on television—was disseminated across the internet. Although she is not a professional historian, the actress's words nevertheless described the practice and value of the study of history to a vast audience. She did so by personifying history as lived life, an immediately accessible explanation that can activate viewers' intellectual and emotional sensibilities.

Without having to say it, Davis associated lived life with poignancy, majesty, tragedy, the quotidian, absurdity, and humor. Her invocation of lived life appealed to the audience's intuitive understanding of the past as grasped through the passage of their own lives or through experiencing other peoples' lives.[32] Carried further, these experiences implicate our habitation of spaces, development of relationships, growth and maturation, work and labor, joy and suffering, consumption of goods—essentially the processes that form part of human lives. Davis ultimately spoke of a history that appeals to an audience far broader than what academics usually reach, and she did it in such a way as to assume people would instinctively grasp the value of what she was referencing. Davis's example underscores the fact that professional historians in no way monopolize the craft of history or knowledge of the past, a point reinforced by Frederic Clark's essay in this volume. History, which is present everywhere around us, belongs to no one person, and everyone has the potential to engage with it. In an age when the public already fashions and consumes histories on its own terms, what roles do Renaissance and early modern European historians play? In this final section, I heed Davis's call, not so much to "exhume" bodies as to cultivate lived or living lives of historians as examples. In many ways, our mentors, colleagues, and students are already leading the way with "plan A."[33]

Some Renaissance and early modern European historians have distinguished themselves in the service of broad public concerns. For example, Lisa Jardine, a British scholar whose books on early modern science have been issued by mass-market presses like HarperCollins and Macmillan, served two terms on Britain's Human Fertilisation and Embryology Authority. As chair, she oversaw high-profile public consultations on mitochondrial DNA replacement in 2012. This in-vitro fertilization process substitutes faulty mitochondrial DNA in one woman's egg with healthy mitochondrial DNA from a third-party egg. When doctors started performing this controversial procedure in 2015, newspapers printed headlines such as "Britain's House of Lords Approves Conception of Three-Person Babies" and the American television series *The Good Fight* featured it in one of its first episodes.[34] Jardine also served in many other capacities as a school governor, a trustee of the Victoria and Albert Museum, a council member of the Royal Institution, and a judge of the Man, Booker, and Whitbread literary prizes, and she even hosted *Seven Ages of Science*, a show on BBC Radio 4.

Political, social, and cultural turmoil following the 2016 US elections has also amplified public interest in history itself and consequently prompted historians to raise their voices. As the Trump administration has broken long-held norms of governance and conduct, politicians, journalists, and the public have looked to historical figures, events, and precedents in order to contextualize and critique Trump's deeds. Just in the first months of 2017, Trump's barring of Muslim refugees from Syria was compared to the turning away of Jews on the MS *St. Louis* in 1939; his firing of Acting US Attorney General Sally Yates was juxtaposed to the "Saturday Night Massacre" in 1973; a hypothetical Muslim registry was analogized to the internment of Americans of Japanese descent in 1942–46; and the figure of Mussolini was invoked as warning against the rise of neofascists in the United States.[35] This recourse to history captured the imagination of a variety of publics. A *Washington Post* article noted that a cross-section of American society was invoking history "to praise or condemn Trump," including a comedic writer for the television series *The Office* and *Thirty Rock*, "a guy named Eric from Ohio," and a Human Rights Campaign blog.[36]

Professional historians have engaged in these intense public debates, often using digital platforms to share their knowledge of the past. On January 28, 2017, the day after Trump signed Executive Order 13769, barring citizens of seven predominantly Muslim countries from entering the United States, Heather Cox Richardson made her "shock event post" on Facebook.[37] The professor of US history argued that the executive order served as an event designed to shock, sow discord, and divide the public. The first affront, however, serves only as a distraction from an ulterior purpose that, when revealed, would encounter less resistance by an already fragmented public. Richardson then recalled that Confederate militants used a shock event, the firing on Fort Sumter in Charleston Harbor, to provoke southern states to leave the Union and initiate the US Civil War. She ended the post by explaining how Abraham Lincoln assembled previously splintered factions to form a new Republican Party and urging readers to find ways to unite. Within a month, the post had been shared 82,000 times, received more than 50,000 "Likes," and garnered some 6,600 comments.[38] A *Boston Globe* article reporting on the post took the title "Our Chance to Write History."[39]

Remarking on the explosion of references to the past in early 2017, the historian of early modern France David Bell marveled: "I've never seen so many people desperate to refer to historical examples. . . . Everyone seems to have an example." The professor of US history Jill Lepore reasoned: "These moments from the past are enticing because of the depth of uncertainty. . . . People are reaching out for whatever twig is streaming by to give some meaning to what they are seeing."[40] This flourishing interest in historical precedents also exposed the powerful, latent dangers in misremembering and misusing history and memory. Michael Rosenwald, author of the *Washington Post* article cited above, cautioned: "Professional historians worry that specious and cherry-picked comparisons will reverberate through social networks as gospel, deepening the country's division."[41] US historian Moshik Temkin and others discounted facile analogies and argued that it is necessary to focus on the unique factors that led to the current political predicament in order to accurately diagnose and respond to it.[42] Despite these concerns, historians are continuing to voice their knowledge of the past in a number of media outlets for the purpose of interpreting the present and suggesting future actions.

In this context, some scholars are engaging in activism by undertaking fundamentally historical tasks such as document preservation. Apprehensive that Trump's appointees in the Environmental Protection Agency and the National Oceanic and Atmospheric Administration would downplay climate change by removing key scientific studies, teams of scientists in 2017 downloaded reports published on the agencies' websites to store on alternative platforms.[43] One effort was spearheaded by Bethany Wiggin, the scholar of early modern German literature and the history of the book who also founded the Penn Program in Environmental Humanities.[44] Regardless of disciplinary background, the actions of these scholars recognize the importance of maintaining a historical database. Their intention of passing these collections onward highlights, once again, the important ways that history and its practitioners inform and shape the future.

Early modern historians are educators, activists, public administrators, and interpreters of knowledge; a few of them also play senior roles in the US federal government. In 2012, Alexander Bick completed a dissertation at Princeton University on decision making within the board of directors of the Dutch West India Company in the seventeenth century.[45] Soon afterward, he joined the US Department of State's Policy Planning Staff and then the White House's National Security Council as director for Syria.[46] In an interview on March 3, 2017, Bick acknowledged the many differences between the historical training he received in graduate school and the tools required for government service. Nevertheless, analyzing early modern institutions and negotiations helped him think about ways to approach contemporary problems, including the multiple crises emanating from the Arab Spring. On a daily basis, Bick applied research, writing, and editing skills to crafting policy memoranda and recommendations for the secretary of state and later the president.[47] Moreover, historians recognize that even copious information is incomplete, and Bick used this understanding to interpret and draw conclusions from data as a White House staff member. Bick further noted that he was hardly the only humanities doctorate in the offices where he served, and that practices often considered "scholarly" were both common and respected throughout the government, including the Department of Defense, where senior officers disseminate extensive reading lists, and the intelligence agencies, where the ability to sift, corroborate, and analyze information is of paramount importance.

Back to the Future

Alexander Bick's experience takes this essay full circle back to the relationship between history and the future. US secretary of state Dean Acheson once pointed out that the task of the Policy Planning Staff was to "anticipate the emerging form of things to come."[48] Indeed, Bick's work on staff sought to devise policy to create a future; in our interview, he described how the skills of historical practice helped him move toward this goal. Fundamentally, historians are accustomed to think over longer time spans and to perceive the continuum, sometimes punctuated by ruptures, between the past, present, and future.[49] Historians also understand that human development does not occur in a linear, point-to-point fashion and that contingent and unexpected factors also shape circumstances. Events are not self-contained, isolated incidents but have repercussions over time and space. These sensibilities seem second nature to historians, especially scholars of the premodern age. Nonetheless, it is worth keeping in mind that this long-term perspective is necessary to complement the shorter-term, everyday concerns and memories that are more readily accessed by the general public.[50]

These skills of historical practice equip historians with valuable abilities to shape the future, but of what use is studying the past if interpretations and conclusions about history are themselves provisional? The historian of revolutionary France Lynn Hunt is careful to acknowledge that "even when historical interpretation is based on true facts, it is logically coherent, and as complete as a scholar can make it, the truth of that interpretation remains provisional. New facts can be discovered, and the benchmark for completeness shifts over time."[51] This aspect of historical practice resembles, significantly, characteristics of the future and the uncertainty of its unfolding. Herein lies history's special contribution. Meaning derived from studying the past is tied to the situational context of the interpreter. In this way, the practice of history will continue to change and adapt in relation to emerging conditions. That such a future is public—in the public domain, owned by no one, and yet of interest to everyone—further reinforces historians' role. Recall Ged Martin's argument that the future is created through decisions that consider the past. Through our selection of facts, analyses, interpretations, and narrations, historians are principal creators of that past. We reanimate history, enabling the public to know how the world arrived at its present state and, with this knowledge, to make decisions about the future. Although "there is no public office of the long term that you can call for answers" about epochal changes,[52] historians can provide such service to the citizenry and to policymakers. In closing this essay, I return to the epigraph: research into the past is essential to efforts to create the future.

The "world created by humans" is everywhere around us. Even in the virtual age, humans and societies still have pasts. In an era of massive amounts of information, interest in and engagement with visual, audio, and textual data have increased exponentially. Everyone participates in history, and the humanities remain as relevant as ever. We are creators of content, with the transformational age of the Renaissance and early modernity as particularly salient points of reference for contemporary culture. Our knowledge and skills train students in civic values and professional abilities. We adapt technology to the needs of our analysis, presentation, and dissemination. We collaborate with a public that also shares intuitive feelings for the past. We do all of this and much more. How historians shape the future ultimately distills down to the questions: How do we want to relate to society? How do we want to reproduce and advance our practice? And who are we as historians and, just as importantly, as citizens contributing to the common good? Ultimately, whoever we want to be, we offer our learning of the past and our whole selves to shape the future.

NOTES

I am grateful to Ann Blair and Nicholas Popper for inviting me to join this volume and for their thoughtful feedback and suggestions.

Epigraph: Grafton 1993, 261.

1. Other historians, including early modern Europeanists, are also undertaking such an inquiry. See, for example, Hunt 2018; Guldi and Armitage 2014; and Staley 2007.

2. For histories of the knights and Malta, see Sire 1994; Atauz 2008; and Luttrell 1975.

3. National Library of Malta, Archivio dell'Università (hereafter NLM, AdU); vol. 1 contains documents from the reigns of Charles V and Philip II.

4. Charles V to the Grand Master of St. John of Jerusalem, Ghent, April 22, 1531, vol. 1, f. 2, NLM, AdU. Though numbered f. 2, it is the first bound document in the manuscript. Preceding this folio is an unbound loose-leaf letter folded four times and numbered f. 1.

5. Charles V to Viceroy of Sicily, Ratisbon, July 2, 1532, vol. 1, f. 3; Charles V to Viceroy of Sicily, Innsbruck, May 4, 1552, vol. 1, f. 52, both in NLM, AdU.

6. A salma is a measure of weight comparable to today's metric ton (1,000 kilograms). In 1530, Charles V also granted Tripoli in Libya to the knights as a fief.

7. "Los vinos, carnes bivas y saladas, legumbres y otras cosas de comer y municiones para uso y seruicio de la dicha religion." Charles V to Viceroy of Sicily, Ratisbon, July 2, 1532, vol. 1, f. 3, NLM, AdU.

8. Grand Master of St. John to Charles V, Malta, September 14, 1545, vol. 1, f. 47, NLM, AdU.

9. Charles V to Viceroy of Naples, Innsbruck, May 4, 1552, vol. 1, f. 54, NLM, AdU.

10. "Segun que en lo pasado hallandose en Rodas y despues las acostumbraron sacar." Charles V to Viceroy of Sicily, Ratisbon, July 2, 1532, vol. 1, f. 3, NLM, AdU.

11. "Que no se les haga innouacion alguna." Charles V to Viceroy of Sicily, Ratisbon, July 2, 1532, vol. 1, f. 3, NLM, AdU.

12. Charles V to Viceroy of Sicily, Ratisbon, July 2, 1532, vol. 1, f. 3, NLM, AdU. Growing up at the Burgundian court, Charles never met his Spanish grandfather. This reference is even more poignant as a memory of an unknown, though exceedingly illustrious ancestor.

13. See Greene 2010.

14. G. Martin 2004.

15. This hesitance results in the insistence that "the past now has no causal or structural relationship to the inherently provisional present," to which Martin empathically responds that rather "it is the present that has no independent existence, and to deny its relationship with the past leaves us not only bereft of a sense of our location in time but dangerously exposed to the delusion." G. Martin 2004, 6.

16. G. Martin 2004, 6.

17. Fukuyama 1989, 3–18.

18. This section builds upon Anthony Grafton and James Grossman's call for graduate programs to prepare students for nontraditional careers. Grafton and Grossman 2011.

19. I use the term "authoritatively" here to mean having the necessary breadth and depth of knowledge, the ability to contextualize and draw appropriate comparisons, and the skill to narrate the history in a balanced and effective way.

20. From Queen Elizabeth I to the Medici, from the life of William Shakespeare to lost paintings of Leonardo da Vinci, from witches to Vlad Dracula, the individuals and figures that populate early modern European history still live large in the imaginations of popular culture today. One early modern historian who has novelized her research is Deborah Harkness in her All Souls Trilogy.

21. Touba Ghadessi and I discuss the application of humanities skills in professional settings in Liang and Ghadessi 2013, 40–41. Also see the mission statement of the Wheaton Institute for the Interdisciplinary Humanities (www.wheatoncollege.edu/wiih).

22. Here I outline some of the more practical skills. The American Historical Association's Tuning Project has also compiled a description of what students can achieve in history courses and degree programs. See American Historical Association 2018.

23. I treat the two simultaneously to emphasize the overlap between citizenship and work.

24. For an introduction that lays out the many components of this field, see Gardiner and Musto 2015.

25. Gardiner and Musto (2015) contend that digital humanities are "an intrinsic part of our culture now" and have "become a part of the air we breathe" (x).

26. Burdick 2012, vii.

27. It should be noted that the cliometrics of the 1960s and 1970s also undertook quantitative analyses.

28. Burdick 2012, 21. Also see Anthony Grafton's remarks on digital history in Grafton 2011e, ix–xix.

29. Gardiner and Musto 2015, 31.

30. Gardiner and Musto 2015, ch. 8, esp. 135–39.

31. Viola Davis, speech at the 89th Academy Awards, February 27, 2017.

32. Hobsbawm (1972) offered a more expansive explanation: "All human beings are conscious of the past . . . by virtue of living with people older than themselves. All societies likely to concern the historian have a past, for even the most innovatory colonies are populated by people who come from some society with an already long history. To be a member of any human community is to situate oneself with regard to one's (its) past, if only by rejecting it. The past is therefore a permanent dimension of the human consciousness" (3).

33. Grafton and Grossman 2011.

34. Devlin 2015; *The Good Fight*, season 1, episode 4.

35. Rosenwald 2017.

36. Rosenwald 2017.

37. Richardson 2017.

38. As of February 28, 2017. Richardson 2017.

39. Abraham 2017.

40. David Bell and Jill Lepore, quoted respectively in Rosenwald 2017.

41. Rosenwald 2017.

42. Temkin 2017.

43. Dennis 2016; Schlanger 2017.

44. Penn Program in Environmental Humanities, Datarefuge, accessed February 23, 2017, http://www.ppehlab.org/datarefuge.

45. Bick 2012.

46. Alexander Bick in discussion with author, March 3, 2017. As a former IIE-Fulbright scholar in Syria and the son of a refugee, I took a particular interest in Bick's work.

47. US Department of State, "Policy Planning Staff," accessed September 7, 2020, https://www.state.gov/bureaus-offices/bureaus-and-offices-reporting-directly-to-the-secretary/policy-planning-staff/.

48. Dean Acheson quoted in US Department of State, "About Us — Policy Planning Staff," accessed September 7, 2020, https://www.state.gov/about-us-policy-planning-staff/.

49. See Guldi and Armitage 2014, ch. 1.

50. For discussion of the tensions between memory and history and also between an intensive "presentism" and a longer-term historical perspective, see Hartog 2003.

51. Hunt 2018, 54–55.

52. Guldi and Armitage 2014, 1.

Where the Reasoning Beings Were

ANTHONY GRAFTON

In 1583, Joseph Scaliger published his *Opus novum de emendatione temporum*, a treatise on calendars and dates from many times, places, and cultures—a work of polyglot erudition and technical ingenuity. In the second book, Scaliger offered a dense and detailed treatment of "the lunar cycle of the ancient Saxons." Tacitus and Caesar, he noted, had recorded that the ancient Gauls and Germans used a lunar calendar. The ancient Saxons and Danes, however, had created something much more remarkable: "a marvelous disposition of the year, based on the states of the ocean, on whose shores they lived."[1] They divided the year into two halves, in accordance with the high tides (*malinae*), which reached their peak near the two equinoxes. And they correlated their months, which Scaliger listed, to the tides as well, connecting these to the movements of the moon. Scaliger named his source for this information, an Indigenous writer of the eighth century: "We owe the names of the Saxon months to Bede, who, as a Saxon by nation, set out much that is worth knowing about them. You may draw the rest from there."[2] He also made clear how impressive he found the Anglo-Saxon calendar, as it is now known. If the Saxons' year was solar and their moons were lunar, it followed that their calendar was governed by a luni-solar cycle. The Greek historian Diodorus Siculus recorded that Apollo visited the Hyperboreans, inhabitants of a northern island, and played his cithara for them every nineteenth year (2.47.7).[3] Evidently, then, the cycle had lasted nine-teen years, corresponding to the Greek nineteen-year cycle traditionally ascribed to the Athenian astronomer Meton.[4]

In this case, however, the creators of the calendar had followed not the chang-ing shadows cast by gnomons but "the remarkable swellings of the Ocean."[5] True, there was nothing marvelous about the fact that a calendar cycle based on the tides matched that based on astronomical observation: "for if the changes of the ocean follow the moon, and the movement of the moon follows the nineteen-year cycle, then the changes of the ocean undoubtedly also fol-low the same nineteen-year cycle."[6] But Scaliger found this calendar striking for another reason: "I am surprised not that the year works this way, but rather that those men observed it."[7] Sharpening his nib, he asked readers if they did not find the technical proficiency of this ancient Germanic people surprising: "The Roman people, conquerors of the world, used a completely worthless calendar until the time of Julius Caesar. By contrast, this people, whom the Romans saw as barbarous, not only ordered their lunar year by the rule of na-ture, that is, the ocean, but also mastered the course of the solar year."[8] Scaliger was not the only European writer in this period to insist that not all the peoples whom Europeans had traditionally classified as ignorant barbarians deserved this title.[9] Montaigne, the first volume of whose *Essays* had appeared in 1580, described the complex, sophisticated practices of the Indigenous inhabitants of Brazil in his essay "On Cannibals." He argued that they were more civilized, in crucial ways, than the Europeans of his own time. Cannibalism, after all, inflicted no pain on its victims, unlike the grisly forms of torture and execution practiced in Europe: "I think it is more barbarous to eat a man alive than to eat him dead."[10] The practices that interested Scaliger were more technical and less appealing than the aspects of diet, warfare, and dress—or undress—that Mon-taigne described. But he drew a related conclusion: "The Chaldeans and the East did not have a monopoly on knowledge. The inhabitants of the West or the North were also reasoning beings."[11] Scaliger undermined any notion of western superiority even more sharply than Montaigne. Following a widespread tradition rooted in ancient sources, he took it for granted that the ancient Chal-deans and other Eastern peoples had been more proficient at astronomy and many other studies than the Romans.[12] When he added the supposed barbar-ians of the far North and West to the ranks of the better-known, Near Eastern sages who had cracked the secrets of nature, and made clear that the rulers of the world had been ignorant of them, he showed how ready he was to find pro-found knowledge of nature in unexpected historical places. Montaigne, by con-trast, never abandoned his faith in the supreme value of the Greek and Latin classics.

The history of past scholarship devoted to the calendar, as Scaliger reconstructed it, was immense and complex. And the world of contemporary writers whose reports he collated and analyzed was as diverse as it was populous. Even in the first edition of the *De emendatione temporum*, his modern sources included texts by Ethiopians and Jews, a Portuguese Jesuit and a Syrian Christian with whom he corresponded—as well as a wide range of works by earlier experts on calendars and chronology. His bibliography became more crowded with each subsequent edition of the book. Scaliger's resources grew rapidly after he became a professor at the newly founded Leiden University in 1593. His network of scholarly correspondents and informants expanded and he benefited from direct contact with Dutch trading companies. But many of his most stunning discoveries—like his rediscovery of Hellenistic Judaism—depended on his ability to recognize fragments of genuine ancient texts and forms of learning in books that scholars had read and edited before him, often with the help of the insights and practices they had developed. His response to Bede was typical of his approach to his subject.

Scaliger's assessment of the material provided by Bede, though informed and shrewd, was necessarily incomplete. He was right to infer that the knowledge of the tides on which the "Saxon" calendar rested had been compiled by shore dwellers, through direct observation. But he did not have access to the full range of texts on the calendar that chronologers would begin to explore in the seventeenth century. Hence he could not have worked out that Bede's knowledge actually came from Irish scholars who combined mastery of the computus with experience of coastal saltworks.[13] Although Scaliger read Bede with care and ingenuity, moreover, he seems not to have realized that the library in Jarrow where Bede worked offered him rich textual resources. Bede's lists of month names impressed Scaliger, as they had earlier scholars—some of whom added further lists to their copies of *On the Reckoning of Time*. But Scaliger missed one vital clue to the origins of Bede's information. When Bede named the "Greek" (Macedonian) months in chapter 24 of *De ratione temporum*, he wrote that they had been "recently transmitted to us from Rome."[14] His new source was a short text *On the Year*, itself drawn from the prologue of a calendar assembled by a Roman official, Polemius Silvius, in the middle years of the fifth century.[15] Here as elsewhere, Bede melded the traditions of Christian scholarship, as he knew them from an ever-expanding range of literary works, with knowledge drawn from local and nearby sources. Although Scaliger sensed something of Bede's intellectual situation, he lacked the evidence he would have needed to trace every stone in Bede's mosaic back to its original source.

In attending to peoples outside the Roman world as well as those inside it, in dwelling on technical subjects like chronology and the calendar, in taking an interest in ancient languages other than Latin, Greek, and Hebrew, Scaliger seems—or would have seemed a generation ago—a typical scholar of the later

Renaissance. As traditionally defined, this period began roughly in the middle of the sixteenth century when many learned Europeans began to treat the entire past—including the barbarous Middle Ages—as worthy of study, to master new languages, from Hebrew and Arabic to Anglo-Saxon and Turkish, and to formulate plans for more comprehensive cosmographies, geographies, and histories. Historians have often connected these innovations with the Protestant Reformation, assuming that the most innovative thinkers were either Protestants or irenic Catholics, and that these intellectual innovations meant the abandonment of traditional humanism, with its exclusive emphasis on the languages and experiences of ancient Greece and Rome.

In fact, when Scaliger praised the calendrical skills of barbarous peoples, he was not innovating. The defects of the ecclesiastical calendar had attracted widespread attention in the later Middle Ages. They stimulated the production of detailed proposals for reform.[16] But they also stimulated the development of new forms of scholarship, as attention turned to the possibility that studying the early history of the Christian calendar might help to rebuild it. The Erfurt-trained humanist Johannes Sichard and the Cologne-trained astronomer Johannes Bronckhorst both collected and edited Bede's works on the calendar, long before Scaliger.[17] Bronckhorst worked from two very old manuscripts from the Cologne Cathedral Library, Dombibliothek 102 (tenth and eleventh centuries) and 103 (ca. 795 CE), as well as others not yet identified.[18] His careful edition, splendidly illustrated with images adapted from the manuscripts, appeared in 1537. Bronckhorst made clear that as he mastered his material, he came to see his edition as a historical, rather than a practical, enterprise. Computing the date of Easter, which had greatly occupied Bede, "is considered a trivial problem now, but formerly, as we can see from many very famous and ancient synods, it was an effort of great importance."[19] Discussing the various ways of carrying out calendrical computations on one's fingers, he took care to justify the effort he had made to preserve the treatise on this curious subject ascribed to Bede: "Before this book of Bede was published, these things were unknown, and I think that this antiquity owes its preservation to him. Whatever it is worth, then, you owe it all to Bede."[20] Bronckhorst printed an old commentary on Bede's work from one of his manuscripts word for word, "for the sake both of the age of the exemplar, which deserves a certain veneration in itself, and of their learning."[21] At another point, he described a passage from this commentary as "learned, and drawn from learned writers."[22] Like Scaliger, Bronckhorst was surprised to encounter erudition in the works and world of an early medieval Anglo-Saxon. He noted that some of "these texts were written with too much erudition to seem to be the work of that Bede who is called the Venerable"—but he too accepted what his evidence showed him, remarking that "these matters can only be treated in an erudite way."[23] And even when he added to Bede's work, drawing up detailed tables of the months of many peoples to supplement Bede's sparse lists, he built on a precedent set by the early medieval author. Instead of praising these texts exuberantly, Bronckhorst set them into what he saw as their context, the long-term development of calendrical skills. More clearly than Scaliger, he saw that Bede had drawn much of his material from older texts, some of them not otherwise attested. But he also realized that reasoning beings had lived on the bleak shores of northern England in the seventh and eighth centuries.

Bronckhorst converted to Protestantism, and his son Everard became a colleague of Scaliger's at Leiden. The Swedish cleric Olaus Magnus, author of the *History of the Northern Peoples*, whose rich text and lurid illustrations fascinated readers across Europe, remained a faithful Catholic and died in the monastery of St. Brigitta at Rome. Yet he too knew the fascination of ancient calendrical practices: in his case, the sticks as tall as a person, "inscribed with Gothic characters," that the early inhabitants of Scandinavia had used to tell time and observe the stars in an age when "books were not yet in use," or were at least very rare.[24] Olaus admitted that the techniques embodied in the rune sticks were relatively primitive and that the more cultivated Swedes who read books had developed far more precise methods. But he also made clear, in detail, that even in his own day rustics and peasants "persevered immovably" in using and teaching these practices "as they had learned them from their elders."[25] And he seems to have admired the peasants "who are so skillful that they can predict, on one day, what the golden number is or will be, and the dominical letter, the leap year, the intervals, the moveable feasts, and even the phases of the moon, after ten, or six hundred, or a thousand years."[26] Like Scaliger—and decades earlier—Olaus Magnus found calendrical techniques a key to the cultural level of those who practiced them. More remarkably, he traced remnants of these ancient practices among the peasants of his own time—and treated them both as historical survivals of an earlier world and as rich with ethnographic interest.

Evidently, there was nothing peculiar to the late Renaissance, or to Protestant learning, in Scaliger's treatment of Bede. The enthusiasm with which he praised the tidal calendar cycle of the Saxons was distinctive. In other respects, from his interest in early calendars to his historical study of early medieval culture, Scaliger followed leads left him by much earlier scholars. A single example is enough to reveal the fragility of our old stories about the intellectual history of Renaissance Europe. The articles collected in this volume batter them into fragments—and leave us looking forward to a wide range of innovative studies that will eventually replace them.

As the authors in this collection note, older constructions of the past were largely swept away by a wave of brilliant revisionist books and articles in the 1970s, 1980s, and 1990s—works that made the history of early modern Europe a field of deep interest to many nonspecialists and that shaped research in other disciplines and periods. Few scholars any longer believe in Jacob Burckhardt's construction of the Renaissance as the origin of modernity—or in the constructions that succeeded it, from Hans Baron's effort to show that the core of Renaissance thought was a new, civic version of humanism that took shape in early fifteenth-century Florence, to Paul Oskar Kristeller's effort to show that the core of Renaissance thought was an effort to revive a set of ancient literary disciplines. Few join Theodor Mommsen and Erwin Panofsky in tracing the origins of historicism to humanist students of the ancient world, or Max Weber in tracing the origins of capitalism to the Protestant Reformation. Their books remain and continue to be read, but the larger assumptions that inspired them have lost credibility over time. The revisionist works of Natalie Zemon Davis and Carlo Ginzburg, Robert Darnton and Stephen Greenblatt, Michael Baxandall and Lisa Jardine have revealed an early modern Europe far too complex and rich in local variations to be subsumed under the great theses of earlier generations, and they in turn have inspired newer generations of scholars. Current ways of studying the early modern period—many of them represented, as well as described, by these chapters—yield complicated, colorful mosaics rather than the dramatic frescoes and portraits produced by earlier generations. Yet they have a richness, and even a drama, of their own.

The early modern intellectual world that scholarship is now revealing to us was, in the first place, global. From the fourteenth century on, Latin Christians engaged with people of radically diverse origins and cultures: from the Jews and Muslims who were, for the most part, either expelled from the Latin West or forced to convert to Christianity, to the Ethiopians and sub-Saharan Africans, Ottomans and Arabs, Persians and Indians, Mexicans and Peruvians with whom exploration, trade, warfare, and diplomacy brought them into more and more regular contact. Ideas and people, as well as ships laden with slaves and spices, traveled in all directions. European understandings of Judaism and Islam; the religions and the addictive stimulants of the New World; and the dress, music, and eating habits of Asians were all shaped not only by Europeans themselves but by a vast range of informants and intermediaries, people of mixed origin and multiple languages, whose expert knowledge and distinctive perspectives had a deep impact. Scaliger and Montaigne, for example, both learned from the same Syrian Jacobite, Ignatius Nimatallah, who informed Scaliger about the Chinese twelve-year animal cycle and gave Montaigne a remedy for the stone.[27] The paths that knowledge took were often crooked, sometimes perversely so, and some of the most exciting articles in this volume make clear how much Europeans learned from others—and how hard it can be to establish the exact details of their knowledge transactions.

The early modern was also a world in which scholars continued to rely on and engage with traditions. Yet these, too, were more numerous and varied than older histories suggested. Many scholars joined Scaliger in looking back to medieval Christian sources for information of every kind—including histories of the ways their own lands had been settled and accounts of their ancient laws and constitutions, which often proved highly relevant in the present. When Giovanni Pico della Mirandola needed a foundation for the speech he planned to deliver at the start of a great disputation in Rome, he looked to the Jewish Kabbalah.[28] Others ransacked the traditions of Arabic scholarship and science. Still others mounted elaborate inquiries into the natural knowledge and ritual practices of the inhabitants of New Spain and the Andes. The new forms of textual knowledge that became available to early modern Europeans often proved as challenging to existing assumptions and beliefs as the testimony of live human informants.

The new histories that encompass these developments have been opened up, in part, by applying new methods. Recent scholarship on the intellectual history of early modern Europe takes many forms—but much of it has concentrated, in the past three decades, on a linked set of fields: the history of books and readers, of information, and of knowledge-making practices. Specialists in these fields cover an immense amount of territory: everything from the way in which books were made by authors, scribes, and printers to the ways reports about foreign powers were composed, circulated, and archived by diplomats. But all of them share a deep interest in the ways scholars and others actually created knowledge: how they analyzed texts and compiled notebooks, worked with informants and secretaries, scribes and compositors, and ventured into fields not traditionally seen as falling among their interests. Humanists, these articles show, were as interested in ecclesiastical history and church governance as they were in the rules of classical Latinity. They used their skills to climb the ladder of preferment in papal Rome and to criticize the wealth and corruption of the Roman church (and sometimes did both at once). They devised new ways, both philological and quantitative, to study history, and composed histories of everything from philosophy and the arts to Christianity itself. The forms of inquiry they cultivated were not uniformly modern—and certainly not uniformly secular. Yet they were powerfully influential. The humanists' studies of the history of the church and the operations of nature provided many of the methods and materials that went into the new religious settlements of the period after 1700. Their editions and commentaries, translations and collections of fragments transformed the study of ancient philosophy. They collaborated with helpers very different from themselves, from the divers "more like fish than men" who helped Alberti study the sunken ships in Lake Nemi to the secretaries who—even if they ate with the servants—sometimes had well-informed opinions of their own about their employers' theories.[29] And for all their errors and exaggerations, their studies of the languages, habits, religions, and customs of the world's peoples were often richly edifying.

Historians of early modern Europe, moreover, have realized over the past three generations that intellectual histories cannot be restricted to the study of formal texts. Possessions—especially highly valued possessions—offer a new key to the thought and work of many whose own words are not accessible to us. Inventories of books, coins, and other valuables identify humanists whose written works have not survived. Household furnishings of every kind, from portraits to pottery, illuminate the tastes and interests of aristocrats and merchants, both in Europe and around the world. Understanding what material objects have to tell us is anything but easy and straightforward. They are often hard to read, and even more often—as the antiquarians of the early modern period could have told us—they have been subject to the ravening tooth of time. Historians who hope to draw inferences from New World codices and Old World clothing must read the work of connoisseurs. Like the humanists, too, they must collaborate with others whose expertise complements theirs, including curators, archivists, and librarians. Still, the richness of the results that these inquiries yield is clear. They place us in a past that is three-dimensional and brightly colored—and which, in its unsuspected richness and variety, echoes that of the larger worlds, European and global, in which it came to pass.

Like the new histories composed in early modern Europe, our new intellectual histories are eclectic and generative. That makes them impossible to summarize in anything like a capsule. But it also makes them appropriate to our own time, when new forms of archiving and publication have taken shape alongside the traditional ones, and when new ways of understanding societies and cultures and their interconnections are also slowly coming into view. If, as Yuen-Gen Liang argues in his essay, history can still offer precedents and guidance, as it did in the past, it must be a history like this: a history that works not to offer new comprehensive formulas to replace those of Burckhardt and Weber, formative and influential though they were, but to respond, flexibly and openly, to the immense richness and complexity of the past that is now being revealed. The essays collected here invite new generations to embrace this task and craft new histories—histories that try to do justice, as Scaliger did, to the universal presence and multiple interactions of reasoning beings.

NOTES

1. "Veterum Gallorum & Germanorum annum Lunarem fuisse ex Caesare & Cor. Tacito cognoscimus [Scaliger here quotes from Tacitus *Germania* 11]. . . . At veterum Saxonum & Danorum mira anni ordinatio fuit ex affectibus Oceani, cuius littora accolebant illi." Scaliger 1583, 111.

2. "Nomina mensium Saxonicorum Bedae accepta referimus, qui, ut Saxo genere, de illis satis multa & cognitu digna retulit. Reliqua inde haurias licet." Scaliger 1583, 111.

3. Quoted in Scaliger 1583, 112.

4. This cycle, widely known and used in the ancient Near East, rests on the fact that 235 lunar months and 19 solar years are almost identical. If seven months are intercalated into 19 lunar years, a lunar calendar can be cyclically coordinated with the movement of the sun.

5. Scaliger 1583, 111.

6. "Nam si Oceani mutationes secundum Lunam: Lunae autem progressus secundum ἐννεαδεκαετηρίδα: proculdubio et Oceani mutationes sequuntur ἐννεαδεκαετηρίδα ipsam." Scaliger 1583, 111–12.

7. "Nam id quidem potius ab illis hominibus observatum miror, quam ita accidere." Scaliger 1583, 111.

8. "Non miraris, gentem Romanam orbis victricem & dominam ineptissima anni forma ad tempora usque Iulii Caesaris usam: eam vero, quae Romanis videbatur barbara, non solum annum Lunae ad naturae regulam, Oceanum dico, direxisse, sed etiam anni Solaris cursum tenuisse?" Scaliger 1583, 112.

9. On earlier discussions and their complexities, see Ginzburg 2017.

10. "Je pense qu'il y a plus de barbarie à manger un homme vivant qu'à le manger mort." Michel de Montaigne, *Essais* 1.31, ed. P. Villey and V.-L. Saulnier, online ed. by P. Desan at The Montaigne Project (https://artflsrv03.uchicago.edu/philologic4/montessaisvilley/navigate/1/3/32/; consulted February 22, 2020).

11. "Quare non omnis sapientia penes Chaldaeos & Orientem fuit. Etiam Occidentis aut Septentrionis homines fuerunt λογικὰ ζῶα." Scaliger 1583, 112.

12. On these traditions, see the classic works of Iversen (1993), Yates (2002), and Walker (1972) and the more recent studies of B. Curran (2007), Popper (2012), Stolzenberg (2013), and Levitin (2015).

13. Smyth 1996, 241–62; Bede 2004, lxvi–lxvii, 306–12.

14. "Quo illos ordine annum observare vel menses, et nuper transmissus ad nos de roma computus eorum annalis ostendit; et canones qui dicuntur apostolorum idem antiquioribus literis edocuere." Bede, *De ratione temporum*, ch. 14, in Bede 1943, 210. See also Bede 2004, 52 and the editor's commentary, 282–84, 285–87.

15. C. Jones 1934.

16. See, most expertly, Nothaft 2018.

17. For Sichard's edition, see Bede 1529. On Sichard and his work as an editor, see Lehmann 1911.

18. Bede 1943, vii, 151, 169, 368.

19. "Quae tametsi levis observatio hodie iudicatur, olim tamen (ut ex multis celeberrimis & antiquissimis synodis est videre) praecipui laboris fuit." Bede 1537, sig. a^v.

20. "Est igitur istius computationis cognitio ad pleraque veterum intelligenda necessaria: quod propterea longius egi ne frivolum mox iudicetur quod prima fronte non blanditur. Priusquam hic Bedae liber vulgaretur, incognita ista erant, & sola ut opinor haec vetustas per eum servata est. Quare quidquid huius est id, Bedae in universum debes." Bede 1537, sig. f ij^r.

21. "SCHOLIA IN caput XIII. Sunt huius libri & sequentis loca quaedam vetustis commentarijs incerti autoris explicata, quae quia in vetusto exemplari capitibus hisce ad-

scripta fuerant, tum propter vetustatem exemplaris quae ex se habet aliquid venerationis, tum propter eruditionem quoque ad verbum, quemadmodum ibi erant scripta huc transtuli, adscripto titulo veteris commentarij, ut nostra ab illis essent divisa." Bede 1537, sig. [a vi^v].

22. "Quae sequuntur sic in veteri commentario invenimus, quae quoniam erudita sunt, & ab eruditis scriptoribus sumpta, non existimabam esse omittenda." Bede 1537, sig. d^r.

23. "Horum omnium quaedam sunt illustrata doctissimis annotationibus, in quibus quoniam magnam atque venerandam vetustatem invenimus, non putavi esse praetermittenda: item lepidissima quaedam minimeque vulgaria schemata, quibus res ipsa oculis subijcitur. Cuius autem hoc sit sive Bedae sive alterius, isto loco non habeo dicere, censuram tamen nostram singulis locis apponemus. Sunt omnes hi libri maiore eruditione scripti quam ut videantur eius esse Bedae qui cognominatur Venerabilis, sed eius fidei invenimus necessaria argumenta, & huiusmodi res non nisi erudite tradi possunt." Bede 1537, sig. a^v.

24. Magnus 1555, 53. Describing the illustration on the same page, he writes: "Cernitur hic homo senex atque adolescens, baculum Gothicis characteribus insignitum habentes, tali ratione insculptum, ut videatur, quibus instrumentis vetustissimo tempore, dum librorum usus non esset, lunae solisve et caeterorum syderum virtutes & influentias infallibili eventu cognoverint, prout hoc tempore fere incolae omnes agnoscunt. Baculus itaque humana longitudine formatus est." Gabriel Harvey's copy of this very popular book is Princeton University Library EX Oversize DL45 .O438 1555q, digitized by the Archaeology of Reading project (https://archaeologyofreading.org/). On Olaus Magnus and his work, see Johannesson 1991 and Émion 2018. For the reception of the *Historia*, see Maxwell 2004 and R. Lewis 2018.

25. "Verum de vulgo haec consideratio habetur, quod sicuti a senioribus traditam astronomiae scientiam & practicam in praeostensis baculis & characteribus acceperat: ita immobiliter in eadem accipenda tradendaque, etiam post sacrae fidei susceptionem, perseverat." Magnus 1555, 54.

26. "ita ut rustici seu villani adeo periti reperiantur, & sint, ut die una praedicere possint, quotus quisque aureus numerus sit, literaeque Dominicalis, annus bissextilis, intervalla, festa mobilia, & ipsae lunares mutationes, post decem, vel sexcentos, aut mille annos fient aut erunt." Magnus 1555, 54.

27. Grafton 1983–93, 2:107–9; Levi della Vida 1948, 22–25.

28. Copenhaver 2019.

29. See McManamon 2016.

ALEXANDER BEVILACQUA, Department of History, Williams College

ANN BLAIR, Department of History, Harvard University

DANIELA BLEICHMAR, Departments of Art History and History, University of Southern California

WILLIAM J. BULMAN, Department of History, Lehigh University

FREDERIC CLARK, Department of Classics, University of Southern California

ANTHONY GRAFTON, Department of History, Princeton University

JILL KRAYE, Warburg Institute, London (emerita)

YUEN-GEN LIANG, Institute of History and Philology, Academia Sinica, Taiwan

ELIZABETH MCCAHILL, Department of History, University of Massachusetts Boston

NICHOLAS POPPER, Department of History, College of William and Mary

AMANDA WUNDER, Department of History, Lehman College; doctoral faculty in Art History and History, The Graduate Center, City University of New York

ACKNOWLEDGMENTS

This volume emerged from a conference in honor of Anthony Grafton held at Princeton University in May 2015 entitled "Proofreaders and Polymaths," coorganized by Daniela Bleichmar, Elizabeth McCahill, and the two of us (program available at https://graftoniana.princeton.edu/). In the years since then, the present essays have evolved to consider the cultural and intellectual history of early modern Europe beyond the already broad horizons of the conference.

We are indebted to the remarkable ensemble of speakers at the event and the extraordinary audience, including many who traveled from afar to attend. But even beyond that group, we wish to express our profound appreciation to the community of scholars, teachers, students, librarians, archivists, editors, friends, and others whose work, ideas, conversations, support, labor, and much else have made the Graftonian republic of letters such a joyous place to inhabit over the years. And, above all, we wish to thank Tony for his unfailing acumen, warmth, humor, patience, friendship, and collaborative spirit.

We are grateful to Matt McAdam and Will Krause for all their support of this project, to Felice Whittum for compiling the bibliography, to Steven Baker for copyediting, to Shannon Li for indexing, and to the whole production team at Johns Hopkins University Press. Special thanks to the two anonymous readers for their incisive feedback and criticism.

Ann Blair and Nick Popper

Abraham, Yvonne. 2017. "Our Chance to Write History." *Boston Globe* (Feb. 4).

Acuña, René, ed. 1981–88. *Relaciones geográficas del siglo XVI*. Mexico City: UNAM.

Adams, Robyn, and Rosanna Cox, eds. 2011. *Diplomacy and Early Modern Culture*. New York: Palgrave.

Adorno, Rolena. 1992. "The Discursive Encounter of Spain and America: The Authority of Eyewitness Testimony in the Writing of History." *William and Mary Quarterly* 49 (2): 210–28.

Afanador Llach, María José. 2011. "Nombrar y representar: Escritura y naturaleza en el *Códice de la Cruz-Badiano*, 1552." *Fronteras de la Historia* 16 (1): 13–41.

Aguilar-Moreno, Manuel. 2006. *Handbook to Life in the Aztec World*. New York: Facts on File.

Alam, Muzaffar, and Sanjay Subrahmanyam. 2007. *Indo-Persian Travel in the Age of Discoveries, 1400–1800*. Cambridge: Cambridge University Press.

Alberti, Leon Battista. 2011. *On Painting: A New Translation and Critical Edition*. Edited and translated by Rocco Sinisgalli. Cambridge: Cambridge University Press.

Alcántara Rojas, Berenice. 2011. "In Nepapan Xochitl: The Power of Flowers in the Works of Sahagún." In *Colors between Two Worlds: The Florentine Codex of Bernardino de Sahagún,* edited by Gerhard Wolf and Joseph Connors, with Louis A. Waldman, 106–32. Florence: Kunsthistorisches Institut; Villa I Tatti.

Algazi, Gadi. 2003. "Scholars in Household: Refiguring the Learned Habitus, 1480–1550." *Science in Context* 16: 9–42.

Alpers, Svetlana. 1988. *Rembrandt's Enterprise*. Chicago: University of Chicago Press.

———. 1993. *The Art of Describing: Dutch Art in the Seventeenth Century*. Chicago: University of Chicago Press.

Amato, Lorenzo. 2014. "Le arguzie dei curiali e la Roma dei papi nella prima metà del Quattrocento." *Roma nel Rinascimento*: 59–80.

American Historical Association. 2018. "AHA Tuning Project: 2016 History Discipline Core." Accessed March 4. https://www.historians.org/teaching-and-learning/tuning-the -history-discipline/2016-history-discipline-core.

American Historical Review Forum. 2012. "Forum: Historiographic 'Turns' in Critical Perspective. *American Historical Review* 117: 698–813.

Amos, N. Scott. 2015. *Bucer, Ephesians, and Biblical Humanism: The Exegete as Theologian*. Cham, Switzerland: Springer.

Anderson, Jennifer L. 2012. *Mahogany: The Costs of Luxury in Early America*. Cambridge, MA: Harvard University Press.

Andrews, Walter G., and Mehmet Kalpakli. 2005. *The Age of Beloveds: Love and the Beloved in Early-Modern Ottoman and European Culture and Society*. Durham, NC: Duke University Press.

App, Urs. 2010. *The Birth of Orientalism*. Philadelphia: University of Pennsylvania Press.

Appadurai, Arjun, ed. 1986. *The Social Life of Things: Commodities in Cultural Perspective*. Cambridge: Cambridge University Press.

Ariew, Roger. 2011. *Descartes among the Scholastics*. Leiden, Netherlands: Brill.

Ariew, Roger, and Alan Gabbey. 1998. "The Scholastic Background." In *The Cambridge History of Seventeenth-Century Philosophy*, vol. 1, edited by Daniel Garber and Michael Ayers, 425–53. Cambridge: Cambridge University Press.

Aristotle. 1476. *De animalibus*. Translated by Theodore Gaza. Venice: Johannes de Colonia and Johannes Manthen.

———. 2017. *Politics*. Translated by C. D. C. Reeve. Indianapolis: Hackett.

L'art dels velluters: Sedería de los siglos XV–XVI. 2011. Valencia: Generalitat Valenciana. Exhibition catalog.

Ascoli, Albert, and Unn Falkeid, eds. 2015. *The Cambridge Companion to Petrarch*. Cambridge: Cambridge University Press.

Àsh, Eric H. 2017. *The Draining of the Fens: Projectors, Popular Politics, and State Building in Early Modern England*. Baltimore: Johns Hopkins University Press.

Atasoy, Nurhan, Walter B. Denny, Louise W. Mackie, Hülya Tezcan, and Şerife Atlıhan. 2001. "Trade in Ottoman Silks in the Balkans, Poland, and Russia." In *Ipek: The Crescent and the Rose; Imperial Ottoman Silks and Velvets*, edited by Julian Raby and Alison Effeny, 176–81. London: Azimuth Editions.

Atauz, Ayşe Devrim. 2008. *Eight Thousand Years of Maltese Maritime History: Trade, Piracy, and Naval Warfare in the Central Mediterranean*. Gainesville: University Press of Florida.

Augustine. 2009. "De dialectica," chap. 6. In *Medieval Grammar and Rhetoric: Language Arts and Literary Theory, AD 300–1475*, edited by Rita Copeland and Ineke Sluiter, 344–49. Oxford: Oxford University Press.

"Authorship Policies." 2009. *Nature* 458 (April 30): 1078.

Avcioğlu, Nebahat. 2010. *Turquerie and the Politics of Representation, 1728–1876*. Farnham, UK: Routledge.

Avcioğlu, Nebahat, and Finbarr Barry Flood, eds. 2010. *Globalizing Cultures: Art and Mobility in the Eighteenth Century*. Ars orientalis 39. Washington, DC: Smithsonian.

Avery, Victoria J., Melissa Calaresu, and Mary Laven, eds. 2015. *Treasured Possessions: From the Renaissance to the Enlightenment*. London: Philip Wilson.

Ávila Blomberg, Alejandro de. 2009. "The Codex Cruz-Badianus: Directions for Future Research." In *Flora: The Aztec Herbal; The Paper Museum of Cassiano dal Pozzo*, edited by Martin Clayton, Luigi Guerrini, and Alejandro de Ávila Blomberg, 45–50. London: Royal Collection; Harvey Miller.

Ayers, Michael. 2004. "Popkin's Revised Scepticism." *British Journal of the History of Philosophy* 12: 319–32.

Azzolini, Monica. 2013. *The Duke and the Stars: Astrology and Politics in Renaissance Milan*. Cambridge, MA: Harvard University Press.

Backus, Irena. 2004. *Historical Method and Confessional Identity in the Era of the Reformation, 1378–1615.* Leiden, Netherlands: Brill.

Bacon, Francis. (1623) 2011. *De augmentis scientiarum.* In *The Works of Francis Bacon,* Vol. 1: *Philosophical Works I,* edited by James Spedding, Robert Leslie Ellis, and Douglas Denon Heath. Cambridge: Cambridge University Press.

Baker, Keith. 1994. "Enlightenment and the Institution of Society: Notes for a Conceptual History." In *Main Trends in Cultural History: Ten Essays,* edited by Willem Melching and Wyger Velema, 114–17. Amsterdam: Rodopi.

———. 2001. "Epistémologie et politique: Pourquoi l'*Encyclopédie* est-elle un dictionnaire?" In *L'"Encyclopédie": Du réseau au livre et du livre au réseau,* edited by Robert Morrissey and Philippe Roger, 51–58. Paris: Champion.

Baker, Nicholas, and Brian Maxson, eds. 2015. *After Civic Humanism: Learning and Politics in Renaissance Italy.* Toronto: Centre for Reformation and Renaissance Studies.

Barbaro, Ermolao. (1492–93) 1973–79. *Castigationes Plinianae et in Pomponium Melam.* Rome: Eucharius Silber. Modern edition, edited by Giovanni Pozzi. 4 vols. Padua: Antenore.

———. 1545. *Compendium scientiae naturalis ex Aristotele.* Venice: Cominus de Tridino.

Baron, Hans. 1955. *The Crisis of the Early Italian Renaissance: Civic Humanism and Republican Liberty in an Age of Classicism and Tyranny.* Princeton, NJ: Princeton University.

———. 1988. *In Search of Florentine Civic Humanism: Essays on the Transition from Medieval to Modern Thought.* Princeton, NJ: Princeton University Press.

Barrera-Osorio, Antonio. *Experiencing Nature: The Spanish American Empire and the Early Scientific Revolution.* Austin: University of Texas Press, 2006.

Barret-Kriegel, Blandine. 1988. *Les historiens et la monarchie.* 4 vols. Paris: PUF.

Baudouin, François. 1561. *De institutione historiæ universæ et ejus cum jurisprudentia conjunctione prolegomenon, libri II.* Paris: Andreas Wechel.

Bauer, Ralph, and Marcy Norton. 2017. "Introduction: Entangled Trajectories; Indigenous and European Histories." *Colonial Latin American Review* 26 (1): 1–17.

Baxandall, Michael. 1972. *Painting and Experience in Fifteenth-Century Italy: A Primer in the Social History of Pictorial Style.* Oxford: Clarendon Press.

———. 1988. *Painting and Experience in Fifteenth-Century Italy.* 2nd ed. Oxford: Oxford University Press.

Bayle, Pierre. 1705. *Continuation des pensées diverses.* Rotterdam: Leers.

Bayly, C. A. 2002. "'Archaic' and 'Modern' Globalization in the Eurasian and African Arena, c. 1750–1850." In *Globalization in World History,* edited by A. G. Hopkins, 45–72. New York: W. W. Norton.

Bede, Venerable. 1529. *Bedae presbyteri Anglosaxonis viri eruditissimi De natura rerum et temporum ratione libri duo.* Edited by Johannes Sichardus. Basel: Henricus Petrus.

———. 1537. *Bedae presbyteri Anglosaxonis, monachi Benedictini, viri literatissimi Opuscula cumplura de temporum ratione.* Edited by Jan van Bronckhorst. Cologne: Quentel.

———. 1943. *Bedae Opera de temporibus.* Edited by C. W. Jones. Cambridge, MA: Mediaeval Academy of America.

———. 2004. *The Reckoning of Time.* Edited and translated by Faith Wallis. Liverpool: Liverpool University Press.

Belozerskaya, Marina. 2005. *Luxury Arts of the Renaissance.* Los Angeles: J. Paul Getty Museum.

Bentley, Jerry H., Sanjay Subrahmanyam, and Merry E. Wiesner-Hanks, eds. 2015. *The Construction of a Global World, 1400–1800 CE*, parts 1 and 2. Vol. 6 of *The Cambridge World History*. Cambridge: Cambridge University Press.

Berg, Maxine. 2005. *Luxury and Pleasure in Eighteenth-Century Britain*. Oxford: Oxford University Press.

Bernard, Jacques. 1714. *De l'excellence de la religion*. Amsterdam: Wetstein.

Berry, Jessica N. 2004. "The Pyrrhonian Revival in Montaigne and Nietzsche." *Journal of the History of Ideas* 65: 497–514.

Bertelli, Silvio. 2001. *The King's Body: The Sacred Rituals of Power in Medieval and Early Modern Europe*. University Park: Pennsylvania State University Press.

Bevilacqua, Alexander. 2016. "How to Organise the Orient: D'Herbelot and the *Bibliothèque Orientale*." *Journal of the Warburg and Courtauld Institutes* 79: 213–61.

———. 2018. *The Republic of Arabic Letters: Islam and the European Enlightenment*. Cambridge, MA: Harvard University Press.

Bevilacqua, Alexander, and Helen Pfeifer. 2013. "Turquerie: Culture in Motion, 1650–1750." *Past and Present* 221: 75–113.

Biagioli, Mario. 1993. *Galileo, Courtier: The Practice of Science in the Culture of Absolutism*. Chicago: University of Chicago Press.

Biagioli, Mario, and Jessica Riskin, eds. 2012. *Nature Engaged: Science in Practice from the Renaissance to the Present*. New York: Palgrave Macmillan.

Bianchi, Luca. 2004. "Fra Ermolao Barbaro e Ludovico Boccadiferro: Qualche considerazione sulle trasformazioni della 'fisica medievale' nel Rinascimento italiano." *Medioevo* 29: 341–78.

Bick, Alexander. 2012. "Governing the Free Sea: The Dutch West India Company and Commercial Politics, 1618–1645." PhD diss., Princeton University.

Bietenholz, Peter. 1994. *Historia and Fabula: Myths and Legends in Historical Thought from Antiquity to the Modern Age*. Leiden, Netherlands: Brill.

Bilak, Donna, Jenny Boulboullé, Joel Klein, and Pamela H. Smith. 2016. "The Making and Knowing Project: Reflections, Methods, and New Directions." *West 86th: A Journal of Decorative Arts, Design History, and Material Culture* 23 (1): 35–55.

Bittel, Carla, Elaine Leong, and Christina Von Oertzen. 2019. *Working with Paper: Gendered Practices in the History of Knowledge*. Pittsburgh: University of Pittsburgh Press.

Black, Robert. 1995. "The Donation of Constantine: A New Source for the Concept of the Renaissance." In *Language and Images of Medieval Italy*, edited by Alison Brown, 51–85. Oxford: Clarendon Press.

———. 1998. "Humanism." In *The New Cambridge Medieval History*, vol. 7, c. *1415–1500*, ed. Christopher Allmand, 243–77. Cambridge: Cambridge University Press.

———. 2001. *Humanism and Education in Renaissance Italy: Tradition and Innovation in Latin Schools from the Twelfth to the Fifteenth Century*. Cambridge: Cambridge University Press.

Blackwell, Constance. 2001. "Vocabulary as a Critique of Knowledge: Zabarella and Keckermann." In *Philologie und Erkenntnis: Beiträge zu Begriff und Problem frühneuzeitlicher "Philologie,"* edited by Ralph Häfner, 131–49. Tübingen, Germany: Max Niemeyer.

Blair, Ann. 1997. *The Theater of Nature: Jean Bodin and Renaissance Science*. Princeton, NJ: Princeton University Press.

———. 2003. "Reading Strategies for Coping with Information Overload ca. 1550–1700." *Journal of the History of Ideas* 64 (1): 11–28.

———. 2005. "*Historia* in Theodor Zwinger's *Theatrum Vitae Humanae*." In *Historia: Empiricism and Erudition in Early Modern Europe*, edited by Gianna Pomata and Nancy G. Siraisi, 269–96. Cambridge, MA: MIT Press.

———. 2008. "Disciplinary Distinctions before the 'Two Cultures.'" *The European Legacy: Toward New Paradigms* 13: 577–88.

———. 2010. *Too Much to Know: Managing Scholarly Information before the Modern Age.* New Haven, CT: Yale University Press.

———. 2013. "Authorial Strategies in Jean Bodin." In *The Reception of Bodin*, edited by Howell A. Lloyd, 137–56. Leiden, Netherlands: Brill.

———. 2016. "Early Modern Attitudes toward the Delegation of Copying and Note-Taking." In *Forgetting Machines: Knowledge Management Evolution in Early Modern Europe*, edited by Alberto Cevolini, 265–85. Leiden, Netherlands: Brill.

———. 2019. "Erasmus and His Amanuenses." *Erasmus Studies* 39 (1): 22–49.

Bleichmar, Daniela. 2012. *Visible Empire: Botanical Expeditions and Visual Culture in the Hispanic Enlightenment.* Chicago: University of Chicago Press.

———. 2017. *Visual Voyages: Images of Latin American Nature from Columbus to Darwin.* New Haven, CT: Yale University Press.

Bleichmar, Daniela, and Peter C. Mancall, eds. 2011. *Collecting across Cultures: Material Exchanges in the Early Modern Atlantic World.* Philadelphia: University of Pennsylvania Press.

Bleichmar, Daniela, and Meredith Martin, eds. 2016. *Objects in Motion in the Early Modern World.* Chichester, UK: John Wiley & Sons.

Bloom, Jonathan. 2001. *Paper before Print: The History and Impact of Paper in the Islamic World.* New Haven, CT: Yale University Press.

Boardman, John. 1994. *The Diffusion of Classical Art in Antiquity.* Princeton, NJ: Princeton University Press.

Bod, Rens. 2013. *A New History of the Humanities: The Search for Principles and Patterns from Antiquity to the Present.* Oxford: Oxford University Press.

Bod, Rens, and Julia Kursell, eds. 2015. "The History of Humanities and the History of Science" (Focus section). *Isis* 106 (2): 337–90.

Bod, Rens, Jaap Maat, and Thijs Weststeijn, eds. 2010–14. *The Making of the Humanities.* 3 vols. Amsterdam: Amsterdam University Press.

Bod, Rens, Julia Kursell, Jaap Maat, and Thijs Weststeijn, eds. 2016. *History of Humanities* 1:2, 211–301.

Bödeker, Hans Erich, and Georg G. Iggers, eds. 1986. *Aufklärung und Geschichte: Studien zur deutschen Geschichtswissenschaft in 18. Jahrhundert.* Göttingen: Vandenhoeck & Ruprecht.

Bodin, Jean. 1566. *Methodus ad facilem historiarum cognitionem.* Paris: Martinum Juvenem.

———. 1606. *The Six Books of a Common-weale.* Translated by Richard Knolles. London: Impensis G. Bishop.

———. 1945. *Method for the Easy Comprehension of History.* Translated by Beatrice Reynolds. New York: Columbia University Press.

Bolgar, R. R. 1954. *The Classical Heritage and Its Beneficiaries.* Cambridge: Cambridge University Press.

Bollbuck, Harald. 2014. *Wahrheitszeugnis, Gottes Auftrag und Zeitkritik: Die Kirchenges-chichte der Magdeburger Zenturien und ihre Arbeitstechniken.* Wiesbaden: Harrassowitz.

Boone, Elizabeth Hill. 1998. "Pictorial Documents and Visual Thinking in Postconquest Mexico." In *Native Traditions in the Postconquest World,* edited by Elizabeth Hill Boone and Thomas Cummins, 149–99. Washington, DC: Dumbarton Oaks.

———. 2000. *Stories in Red and Black: Pictorial Histories of the Aztecs and Mixtecs.* Austin: University of Texas Press.

Borghero, Carlo. 1983. *La certezza e la storia: cartesianismo, pirronismo, e conoscenza storica.* Milan: Angeli.

Boros, Gábor, ed. *Der Einfluss des Hellenismus auf die Philosophie der Frühen Neuzeit.* Wiesbaden: Harrassowitz, 2005.

Botley, Paul. 2004. *Latin Translation in the Renaissance: The Theory and Practice of Leonardo Bruni, Giannozzo Manetti, and Desiderius Erasmus.* Cambridge: Cambridge University Press.

Bots, Hans, and Françoise Waquet. 1997. *La République des Lettres.* Brussels: De Boeck.

Bourdieu, Pierre. 1977. *Outline of a Theory of Practice.* Translated by Richard Nice. Cambridge: Cambridge University Press.

Bouterse, Jeroen, and Bart Karstens. 2015. "A Diversity of Divisions: Tracing the History of the Demarcation between the Sciences and the Humanities." *Isis* 106: 341–52.

Boyle, Marjorie O. 1981. *Christening Pagan Mysteries: Erasmus in Pursuit of Wisdom.* Toronto: University of Toronto Press.

Bray, Xavier, ed. 2009. *The Sacred Made Real: Spanish Painting and Sculpture, 1600–1700.* London: National Gallery; distributed by Yale University Press.

Bredekamp, Horst. 1995. *The Lure of Antiquity and the Cult of the Machine: The Kunst-kammer and the Evolution of Nature, Art, and Technology.* Translated by Allison Brown. Princeton, NJ: Princeton University Press.

Breen, Benjamin. 2013. "No Man Is an Island: Early Modern Globalization, Knowledge Networks, and George Psalmanazar's Formosa." *Journal of Early Modern History* 17: 391–417.

Brockey, Liam. 2007. *Journey to the East: The Jesuit Mission to China, 1579–1724.* Cambridge, MA: Harvard University Press.

Brockliss, Laurence. 2002. *Calvet's Web: Enlightenment and the Republic of Letters in Eighteenth-Century France.* Oxford: Oxford University Press.

Brook, Timothy. 2008. *Vermeer's Hat: The Seventeenth Century and the Dawn of the Global World.* London: Bloomsbury Press.

Brotton, Jerry. 2002. *The Renaissance Bazaar: From the Silk Road to Michelangelo.* Oxford: Oxford University Press.

Brown, Alison. 1990. "Hans Baron's Renaissance." *Historical Journal* 33: 441–48.

Buchwald, Jed, and Mordechai Feingold. 2012. *Newton and the Origin of Civilization.* Princeton, NJ: Princeton University Press.

Bulman, William J. 2015. *Anglican Enlightenment: Orientalism, Religion, and Politics in England and Its Empire, 1648–1715.* Cambridge: Cambridge University Press.

———. 2016. "Hobbes's Publisher and the Political Business of Enlightenment." *Historical Journal* 59 (2): 339–64.

Bulman, William J., and Robert G. Ingram, eds. 2016. *God in the Enlightenment.* Oxford: Oxford University Press.

Burckhardt, Jacob. (1860) 1990. *The Civilization of the Renaissance in Italy*. Translated by S. G. C. Middlemore. London: Penguin.

Burdick, Anne. 2012. "Preface." In *Digital Humanities*, edited by Anne Burdick, Johanna Drucker, Peter Lunenfeld, Todd Presner, and Jeffrey Schnapp. Cambridge, MA: MIT Press.

Burke, Peter. 1991. "Tacitism, Scepticism, and Reason of State." In *The Cambridge History of Political Thought 1450–1700*, edited by J. H. Burns and Mark Goldie, 477–98. Cambridge: Cambridge University Press.

———. 2000. *A Social History of Knowledge: From Gutenberg to Diderot*. Cambridge: Polity.

———. 2015a. *The French Historical Revolution: The Annales School, 1929–2014*. Revised ed. Cambridge: Polity Press.

———. 2015b. *What Is the History of Knowledge?* Cambridge: Polity Press.

Burke, Peter, and R. Po-chia Hsia, eds. 2007. *Cultural Translation in Early Modern Europe*. Cambridge: Cambridge University Press.

Burman, Thomas E. 2007. *Translating the Qur'an in Latin Christendom, 1140–1560*. Philadelphia: University of Pennsylvania Press.

Burns, Kathryn. 2010. *Into the Archive: Writing and Power in Colonial Peru*. Durham, NC: Duke University Press.

Burns, William. 2016. *The Scientific Regime in Global Perspective*. New York: Oxford University Press.

Butcher, John, Andrea Czortek, and Matteo Martelli, eds. 2017. *Gregorio e Lilio: Due Tifernati protagonisti dell'Umanesimo italiano*. Umbertide, Italy: University Book.

Buteo, Johannes [Jean Borrel]. 1554. *Opera Geometrica*. Lyon, France: Th. Bertellus.

Butterfield, David. 2019. "Critical Method in Lambin's Lucretius: Collation and Interpolation." In *The Marriage of Philology and Scepticism: Uncertainty and Conjecture in Early Modern Scholarship and Thought*, edited by Gian Mario Cao, Anthony Grafton, and Jill Kraye. London: Warburg Institute.

Caferro, William. 2011. *Contesting the Renaissance*. Malden, MA: Wiley-Blackwell.

Caffiero, Marina. 2011. *Forced Baptisms: Histories of Jews, Christians, and Converts in Papal Rome*. Translated by L. Cochrane. Berkeley: University of California Press.

Cajetan, Tommaso De Vio. 1575. "Tractatus de conceptione beatae Mariae Virginis ad Leonem X." In *Opuscula Omnia*, vol. 2, tract 1, 137–42. Lyon, France: Sumptibus Philippi Tinghi Florentini.

Calvo, Hortensia. 2003. "The Politics of Print: The Historiography of the Book in Early Spanish America." *Book History* 6 (1): 277–305.

Calzona, Arturo, ed. 2009. *Leon Battista Alberti: architetture e committenti: atti dei convegni internazionali del Comitato Nazionale VI centenario della nascita di Leon Battista Alberti: Firenze, Rimini, Mantova, 12–16 ottobre 2004*. Florence: L. S. Olschki.

Campana, Augusto. 1946. "The Origin of the Word 'Humanist.'" *Journal of the Warburg and Courtauld Institutes* 9: 60–73.

Campi, Emidio, Simone De Angelis, Anja-Silvia Goeing, and Anthony T. Grafton, eds. 2008. *Scholarly Knowledge: Textbooks in Early Modern Europe*. Geneva: Librairie Droz.

Candido, Igor. 2010. "The Role of the Philosopher in Late *Quattrocento* Florence: Poliziano's *Lamia* and the Legacy of the Pico-Barbaro Epistolary Controversy." In Angelo Poliziano, *Lamia: Text, Translation, and Introductory Studies*, edited by Christopher S. Celenza, 95–129. Leiden, Netherlands: Brill.

Cano, Ignacio, and Casilda Ybarra. 2012. "Early Collecting of Bartolomé Esteban Murillo's Paintings in the United States (1800–1925)." In *Collecting Spanish Art: Spain's Golden Age and America's Gilded Age*, edited by Inge Reist and José Luis Colomer, 297–323. New York: The Frick Collection.

Cao, Gian Mario. 2001. "The Prehistory of Modern Scepticism: Sextus Empiricus in Fifteenth-Century Italy." *Journal of the Warburg and Courtauld Institutes* 64: 229–79.

Caradonna, Jeremy. 2012. *The Enlightenment in Practice: Academic Prize Contests and Intellectual Culture in France, 1670–1794*. Ithaca, NY: Cornell University Press.

Caramuel y Lobkowitz, Juan. 1664. *Theologia praeterintentionalis . . . est theologiae fundamentalis tomus IV*. Lyon, France: Borde, Arnaud, Borde & Barbier.

Carhart, Michael C. 2007. "*Historia Literaria* and Cultural History from Mylaeus to Eichhorn." In *Momigliano and Antiquarianism: Foundations of the Modern Cultural Sciences*, edited by Peter N. Miller, 184–206. Toronto: University of Toronto Press.

Carlebach, Elisheva. 2011. *Palaces of Time: Jewish Calendar and Culture in Early Modern Europe*. Cambridge, MA: Belknap Press of Harvard University Press.

Carrillo Castillo, Jesús. 2003. "Naming Difference: The Politics of Naming in Fernández de Oviedo's *Historia general y natural de las Indias*." *Science in Context* 16 (4): 489–504.

———. 2004. *Naturaleza e imperio: La representación del mundo en la "Historia general y natural de las Indias" de Gonzalo Fernández de Oviedo*. Madrid: Fundación Carolina/Doce Calles.

———. 2008. "The Eyes of the New Pliny: The Uses of Images in Gonzalo Fernández de Oviedo's *Historia general y natural de las Indias*." ·In *The Art of Natural History: Illustrated Treatises and Botanical Paintings, 1400–1850*, edited by Therese O'Malley and Amy R. W. Meyers. Washington, DC: National Gallery of Art.

Céard, Jean, Judit Kecskéméti, B. Boudou, and H. Cazès, eds. 2003. *La France des humanistes: Henri II Estienne, éditeur et écrivain*. Turnhout, Belgium: Brepols.

Celenza, Christopher. 2004. *The Lost Italian Renaissance: Humanists, Historians, and Latin's Legacy*. Baltimore: Johns Hopkins University Press.

———. 2017. *The Intellectual World of the Italian Renaissance: Language, Philosophy, and the Search for Meaning*. New York: Cambridge University Press.

Chabrán, Rafael, and Simon Varey. 2000. "The Hernández Texts." In *The Mexican Treasury: The Writings of Dr. Francisco Hernández*, edited by Simon Varey and translated by Rafael Chabrán, Cynthia L. Chamberlin, and Simon Varey, 3–25. Stanford, CA: Stanford University Press.

Champion, Justin. 1992. *The Pillars of Priestcraft Shaken: The Church of England and Its Enemies, 1660–1730*. Cambridge: Cambridge University Press.

Charney, Noah. 2013. "Anthony Grafton: How I Write." *Daily Beast* (July 17).

Chartier, Roger. 1989. "Texts, Printings, Readings." In *The New Cultural History*, edited by Lynn Hunt, 154–75. Berkeley: University of California Press.

———. 2014. *The Author's Hand and the Printer's Mind: Transformations of the Written Word in Early Modern Europe*. Translated by Lydia Cochrane. Cambridge: Polity Press.

Chew, Samuel C. 1937. *The Crescent and the Rose: Islam and England during the Renaissance*. Oxford: Oxford University Press.

Chiabò, Myriam, Rocco Ronzani, and Angelo M. Vitale, eds. 2014. *Egidio da Viterbo cardinal agostiniano tra Roma e l'Europa del Rinascimento*. Rome: Centro Culturale Agostiniano Roma nel Rinascimento.

Chocano Mena, Magdalena. 1997. "Colonial Printing and Metropolitan Books: Printed Texts and the Shaping of Scholarly Culture in New Spain, 1539–1700." *Colonial Latin American Historical Review* 6 (1): 69–90.

Christ-von Wedel, Christine. 2013. *Erasmus of Rotterdam: Advocate of a New Christianity.* Toronto: University of Toronto Press.

Clark, Frederic. 2017. "Universal History and the Origin Narrative of European Modernity: The Leiden Lectures of Jacob Perizonius (1651–1715) on *Historia Universalis.*" *Erudition and the Republic of Letters* 2: 359–95.

———. 2018. "Reading the Life Cycle: History, Antiquity, and *Fides* in Lambarde's *Perambulation* and Beyond." *Journal of the Warburg and Courtauld Institutes* 81: 191–208.

Clayton, Martin, Luigi Guerrini, and Alejandro de Ávila Blomberg, eds. 2009. *Flora: The Aztec Herbal.* Series B, Natural History, pt. 8 of *The Paper Museum of Cassiano dal Pozzo.* London: Royal Collection; Harvey Miller.

Clendinnen, Inga. 1991. *Aztecs: An Interpretation.* Cambridge: Cambridge University Press.

Cochrane, Eric W. 1976. "Science and Humanism in the Italian Renaissance." *American Historical Review* 81 (5): 1039–57.

———. 1981. *Historians and Historiography in the Italian Renaissance.* Chicago: University of Chicago Press.

Cohen, H. Floris. 1994. *The Scientific Revolution: A Historiographical Inquiry.* Chicago: University of Chicago Press.

———. 2010. *How Modern Science Came into the World: Four Civilizations, One 17th-Century Breakthrough.* Amsterdam: Amsterdam University Press.

Cohen, Jeremy. 1982. *The Friars and the Jews: The Evolution of Medieval Anti-Judaism.* Ithaca, NY: Cornell University Press.

Coleman, Janet. 1997. "Structural Realities of Power: The Theory and Practice of Monarchies and Republics in Relation to Personal and Collective Liberty." In *The Propagation of Power in the Medieval West,* edited by Martin Gosman, Arjo Vanderjagt, and Jan Veenstra, 207–30. Groningen, Netherlands: Egbert Forsten.

Coller, Ian. 2011. *Arab France: Islam and the Making of Modern Europe, 1798–1831.* Berkeley: University of California Press.

Colley, Linda. 2007. *The Ordeal of Elizabeth Marsh: A Woman in World History.* New York: HarperPress.

Collins, David. 2008. *Reforming Saints: Saints' Lives and Their Authors in Germany, 1470–1530.* Oxford: Oxford University Press.

Colombero, Carlo. 1982. "Colonna, Pietro." In *Dizionario biografico degli Italiani,* vol. 27, pp. 402–4. Rome: Istituto della Enciclopedia Italiana.

Connors, Joseph. 2011. "Foreword." In *Colors between Two Worlds: The Florentine Codex of Bernardino de Sahagún,* edited by Gerhard Wolf and Joseph Connors with Louis A. Waldman, xi–xv. Florence: Kunsthistorisches Institut/Villa I Tatti.

Conrad, Sebastian. 2016. *What Is Global History?* Princeton, NJ: Princeton University Press.

Cook, Harold J. 2007. *Matters of Exchange: Commerce, Medicine, and Science in the Dutch Golden Age.* New Haven, CT: Yale University Press.

———. 2014. "Creative Misunderstandings: Chinese Medicine in Seventeenth-Century Europe." In *Cultures in Motion,* edited by Daniel T. Rodgers, Bhavani Raman, and Helmut Reimitz, 215–40. Princeton, NJ: Princeton University Press.

Cook, Harold J., Pamela Smith, and Amy Meyers. 2014. "Introduction." In *Ways of Making and Knowing: The Material Culture of Empirical Knowledge*, edited by Pamela Smith, Amy Meyers, and Harold J. Cook. Ann Arbor: University of Michigan Press.

Cools, Hans, Marika Keblusek, and Badeloch Noldus, eds. 2006. *Your Humble Servant: Agents in Early Modern Europe*. Hilversum, Netherlands: Iutgeverij Verloren.

Cooper, Frederick. 2001. "What Is the Concept of Globalization Good For? An African Historian's Perspective." *African Affairs* 100: 189–213.

Cooper, John M. 2004. "Justus Lipsius and the Revival of Stoicism in Late Sixteenth-Century Europe." In *New Essays on the History of Autonomy: A Collection Honoring J. B. Schneewind*, edited by N. Brender and L. Krasnoff, 7–29. Cambridge: Cambridge University Press.

Copenhaver, Brian. 2019. *Magic and the Dignity of Man: Pico della Mirandola and His Oration in Modern Memory*. Cambridge, MA: Harvard University Press.

Copson, Andrew, and A. C. Grayling, eds. 2015. *The Wiley Blackwell Companion to Humanism*. Hoboken, NJ: Wiley-Blackwell.

Corens, Liesbeth, Kate Peters, and Alexandra Walsham, eds. 2016. "The Social History of the Archive: Record-Keeping in Early Modern Europe." Special issue, *Past & Present*, supplement 1.

———, eds. 2018. *Archives and Information in the Early Modern World*. Proceedings of the British Academy 212. Oxford: Oxford University Press.

Coroleu, Alejandro. 2014. *Printing and Reading Italian Latin Humanism in Renaissance Europe (ca. 1470–ca. 1540)*. Newcastle upon Tyne: Cambridge Scholars.

Couzinet, Marie-Dominique. 1996. *Histoire et méthode à la Renaissance: Une lecture de la Methodus ad facilem historiarum cognitionem de Jean Bodin*. Paris: J. Vrin.

Cowan, Brian. 2005. *The Social Life of Coffee: The Emergence of the British Coffee House*. New Haven, CT: Yale University Press.

Craig, John. 1964. "Craig's Rules of Historical Evidence." *History and Theory*, Beiheft 4. The Hague: Mouton.

Cristóbal Dominguez, Freddy. 2011. " 'We Must Fight with Paper and Pens': Spanish Elizabethan Polemics, 1585–1598." PhD diss., Princeton University.

Croce, Benedetto. 1921. *Theory and History of Historiography*. London: Harrap.

Crosby, Alfred W. 1972. *The Columbian Exchange: Biological and Cultural Consequences of 1492*. Westport, CT: Greenwood Press.

———. 1986. *Ecological Imperialism: The Biological Expansion of Europe, 900–1900*. New York: Cambridge University Press.

Cruz, Martín de la, and Juan Badiano. (1964) 1991. *Libellus de medicinalibus Indorum herbis: Manuscrito azteca de 1552, según traducción latina de Juan Badiano; Versión española con estudios y comentarios por diversos autores*. Mexico City: IMSS.

Cudworth, Ralph. 1678. *The True Intellectual System of the Universe*. London: Richard Royston.

Cummings, Anthony. 2012. *The Lion's Ear: Pope Leo X, the Renaissance Papacy, and Music*. Ann Arbor: University of Michigan Press.

Curran, Andrew S. 2011. *The Anatomy of Blackness: Science and Slavery in an Age of Enlightenment*. Baltimore: Johns Hopkins University Press.

Curran, Brian. 2007. *The Egyptian Renaissance: The Afterlife of Ancient Egypt in Early Modern Italy*. Chicago: University of Chicago Press.

Curran, Mark. 2012. *Atheism, Religion, and Enlightenment in Pre-revolutionary Europe.* Woodbridge, UK: Boydell.

Dadlani, Chanchal B. 2019. *From Stone to Paper: Architecture as History in the Late Mughal Empire.* New Haven, CT: Yale University Press.

Darnton, Robert. 1982. "What Is the History of Books?" *Daedalus* 111 (3): 65–83.

———. 1984. *The Great Cat Massacre: And Other Episodes of French Cultural History.* New York: Basic Books.

———. 2007. " 'What Is the History of Books?' Revisited." *Modern Intellectual History* 4: 495–508.

Daston, Lorraine. 2014. "Objectivity and Impartiality: Epistemic Virtue in the Humanities." In *The Modern Humanities*, vol. 3 of *The Making of the Humanities*, edited by Rens Bod, Jaap Maat, and Thijs Weststeijn, 27–41. Amsterdam: Amsterdam University Press.

Daston, Lorraine, and Peter Galison. 2007. *Objectivity.* New York: Zone Books.

Daston, Lorraine, and Elizabeth Lunbeck, eds. 2011. *Histories of Scientific Observation.* Chicago: University of Chicago Press.

Daston, Lorraine, and Glenn Most. 2015. "History of Science and History of Philologies." *Isis* 106: 378–90.

Davies, Martin. 1995. "Making Sense of Pliny in the Quattrocento." *Renaissance Studies* 9: 240–57.

Davies, Surekha. 2016. *Renaissance Ethnography and the Invention of the Human: New Worlds, Maps, and Monsters.* Cambridge: Cambridge University Press.

Davis, Natalie Z. 1983. *The Return of Martin Guerre.* Cambridge, MA: Harvard University Press.

———. 2006. *Trickster Travels: A Sixteenth-Century Muslim between Worlds.* New York: Hill and Wang.

———. 2013. "Foreword." In "Forum: Revisiting Joan Kelly's 'Did Women Have a Renaissance?'" *Early Modern Women: An Interdisciplinary Journal* 8: 241–47.

Dear, Peter. 1985. "*Totius in Verba*: Rhetoric and Authority in the Early Royal Society." *Isis* 76 (2): 144–61.

———. 1991. *Mersenne and the Learning of the Schools.* Ithaca, NY: Cornell University Press.

———. 1995. *Discipline and Experience: The Mathematical Way in the Scientific Revolution.* Chicago: University of Chicago Press.

De Clercq, Peter. 2007. "The Life and Work of E. G. R. Taylor (1879–1966)." *Journal of the Hakluyt Society* (Feb.).

De Jonge, Henk Jan. 1975. "The Study of the New Testament: The New Testament among Theologians and Philologists—A General Sketch." In *Leiden University in the Seventeenth Century: An Exchange of Learning*, edited by Theodoor Herman Lusingh Scheurleer and G. H. M. Posthumus Meyjes, 65–109. Leiden, Netherlands: Brill.

Delbourgo, James, and Nicholas Dew, eds. 2008. *Science and Empire in the Atlantic World.* New York: Routledge.

D'Elia, Anthony. 2007. "Stefano Porcari's Conspiracy against Pope Nicholas V in 1453 and Republican Culture in Papal Rome." *Journal of the History of Ideas* 68 (2): 207–31.

———. 2009. *A Sudden Terror: The Plot to Murder the Pope in Renaissance Rome.* Cambridge, MA: Harvard University Press.

———. 2016. *Pagan Virtue in a Christian World: Sigismondo Malatesta and the Italian Renaissance.* Cambridge, MA: Harvard University Press.

Dennis, Brady. 2016. "Scientists Are Frantically Copying U.S. Climate Data, Fearing It Might Vanish under Trump." *Washington Post* (Dec. 13).

Deppman, Jed, Daniel Ferrer, and Michael Groden, eds. *Genetic Criticism: Texts and Avant-textes.* Philadelphia: University of Pennsylvania Press, 2004.

De Ridder-Symoens, Hilde, ed. 1996. *Universities in Early Modern Europe (1500–1800).* Vol. 2 of *A History of the University in Europe.* Cambridge: Cambridge University Press.

Des Chene, Dennis. 1996. *Physiologia: Natural Philosophy in Late Aristotelian and Cartesian Thought.* Ithaca, NY: Cornell University Press.

Des Maizeaux, Pierre ed. 1740. *Scaligerana, Thuana, Perroniana, Pithoeana, et Colomesi-ana.* 2 vols. Amsterdam: Covens & Mortier.

De Smet, Rudolf, and Karin Verelst. 2001. "Newton's *Scholium generale*: The Platonic and Stoic Legacy—Philo, Justus Lipsius, and the Cambridge Platonist." *History of Science* 39: 1–30.

Devlin, Hannah. 2015. "Britain's House of Lords Approves Conception of Three-Person Babies." *Guardian* (Feb. 24).

De Vries, Jan. 2008. *The Industrious Revolution: Consumer Behavior and the Household Economy, 1650 to the Present.* Cambridge: Cambridge University Press.

Dibon, Paul. 1990. *Regards sur la Hollande du siècle d'or.* Naples: Vivarium.

Ditchfield, Simon. 1995. *Liturgy, Sanctity, and History in Tridentine Italy: Pietro Maria Campi and the Preservation of the Particular.* Cambridge: Cambridge University Press.

———. 2010. "Decentering the Catholic Reformation: Papacy and Peoples in the Early Modern World." *Archiv für Religionsgeschichte* 101: 186–208.

Dobbs, Betty J. T. 1988. "Newton's Alchemy and His 'Active Principle' of Gravitation." In *Newton's Scientific and Philosophical Legacy,* edited by Paul B. Scheurer and G. Debrock, 50–80. Dordrecht, Netherlands: Kluwer.

———. 1991. *The Janus Faces of Genius: The Role of Alchemy in Newton's Thought.* Cambridge: Cambridge University Press.

Dobie, Madeline. 2010. *Trading Places: Colonization and Slavery in Eighteenth-Century French Culture.* Ithaca, NY: Cornell University Press.

Drayton, Richard. 2000. *Nature's Government: Science, Imperial Britain, and the "Improvement" of the World.* New Haven, CT: Yale University Press.

Duarte, Rocío. 2008. "The Colegio Imperial de Santa Cruz de Tlatelolco and Its Aftermath: Nahua Intellectuals and the Spiritual Conquest of Mexico." In *A Companion to Latin American Literature and Culture,* edited by Sarah Castro-Klarén, 86–105. Malden, MA: Blackwell.

Dubcovsky, Alejandra. 2016. *Informed Power: Communication in the Early American South.* Cambridge, MA: Harvard University Press.

Duhem, Pierre. 1931–59. 10 vols. *Le système du monde: Histoire des doctrines cosmologiques de Platon à Copernic.* Paris: A. Hermann.

Duindam, Jeroen. 2016. *Dynasties: A Global History of Power, 1300–1800.* Cambridge: Cambridge University Press.

Dunkelgrün, Theodor. 2017. "The Christian Study of Judaism in Early Modern Europe." In *The Early Modern World, 1500–1815,* edited by Jonathan Karp and Adam Sutcliffe, 316–48. Vol. 7 of *The Cambridge History of Judaism.* Cambridge: Cambridge University Press.

Eacott, Jonathan. 2016. *Selling Empire: India in the Making of Britain and America, 1600–1830*. Chapel Hill: University of North Carolina Press.

Eberhard, Johann August. 1788. *Allgemeine Geschichte der Philosophie*. Halle, Germany: Hemmerdesche Buchhandlung.

Edgerton, Samuel Y. 2001. *Theaters of Conversion: Religious Architecture and Indian Artisans in Colonial Mexico*. Albuquerque: University of New Mexico Press.

Egmond, Florike. 2008. "Apothecaries as Experts and Brokers in the Sixteenth-Century Network of the Naturalist Carolus Clusius." *History of Universities* 23 (2): 59–91.

———. 2010. *The World of Carolus Clusius: Natural History in the Making, 1550–1610*. London: Pickering & Chatto.

———. 2017. *Eye for Detail: Images of Plants and Animals in Art and Science, 1500–1630*. London: Reaktion Books.

Eisenstein, Elizabeth. 1979. *The Printing Press as an Agent of Change*. 2 vols. Cambridge: Cambridge University Press.

Elias, Norbert. 2000. *The Civilizing Process: Sociogenic and Psychogenetic Investigations*. Translated by Edmund Jephcott. Oxford: Blackwell.

Ellis, Markman. 2004. *The Coffee House: A Cultural History*. London: Weidenfeld & Nicolson.

Elman, Benjamin A. 2005. *On Their Own Terms: Science in China, 1550–1900*. Cambridge, MA: Harvard University Press.

———. 2006. *A Cultural History of Modern Science in China*. Cambridge, MA: Harvard University Press.

Émion, François. 2018. "Le nord selon Olaus Magnus." *Études Germaniques* 73 (2): 193–214.

Emmart, Emily W. 1940. *The Badianus Manuscript, Codex Barberini, Latin 241, Vatican Library: An Aztec Herbal of 1552*. Baltimore: Johns Hopkins University Press.

Enterline, Lynn. 2012. *Shakespeare's Schoolroom: Rhetoric, Discipline, Emotion*. Philadelphia: University of Pennsylvania Press.

Euben, Roxanne L. 2006. *Journeys to the Other Shore: Muslim and Western Travelers in Search of Knowledge*. Princeton, NJ: Princeton University Press.

Eytzinger, Michael. 1579. *Pentaplus Regnorum Mundi*. Antwerp: Christophorus Plantinus.

Fasolt, Constantin. 2004. *The Limits of History*. Chicago: University of Chicago Press.

Febvre, Lucien, and Henri-Jean Martin. (1958) 1976. *The Coming of the Book*. Translated by David Gerard. London: N.L.B.

Feiner, Shmuel. 2002. *The Jewish Enlightenment*. Translated by C. Naor. Philadelphia: University of Pennsylvania.

Feingold, Mordechai, ed. 1990. *Before Newton: The Life and Times of Isaac Barrow*. Cambridge: Cambridge University Press.

Feldhay, Rivka, and F. Jamil Ragep, eds. 2017. *Before Copernicus: The Cultures and Contexts of Scientific Learning in the Fifteenth Century*. Montreal: McGill-Queen's University Press.

Ferguson, Wallace. 1948. *The Renaissance in Historical Thought: Five Centuries of Interpretation*. Cambridge, MA: Riverside Press.

Fernández Christlieb, Federico, and Pedro Sergio Urquijo. 2006. "Los espacios del pueblo de indios tras el proceso de congregación, 1550–1625." *Investigaciones Geográficas* 60: 145–58.

Fernández de Oviedo, Gonzalo. (1526) 2010. *Sumario de la natural historia de las Indias.* Barcelona: Lingkua Digital.

Ferreri, Luigi, ed. 2014. *Marco Musuro.* Vol. 1 of *L'Italia degli umanisti.* Turnhout, Belgium: Brepols.

Findlen, Paula. 1994. *Possessing Nature: Museums, Collecting, and Scientific Culture in Early Modern Italy.* Berkeley: University of California Press.

———. 1997. "Francis Bacon and the Reform of Natural History." In *History and the Disciplines: The Reclassification of Knowledge in Early Modern Europe,* edited by Donald R. Kelley, 239–60. Rochester, NY: University of Rochester Press.

———, ed. 2013. *Early Modern Things: Objects and Their Histories, 1500–1800.* New York: Routledge.

Findlen, Paula, Michelle Fontaine, and Duane J. Osheim, eds. 2003. *Beyond Florence: The Contours of Medieval and Early Modern Italy.* Stanford, CA: Stanford University Press.

Finlay, Robert. 2010. *The Pilgrim Art: Cultures of Porcelain in World History.* Berkeley: University of California Press.

Flood, Finbarr B. 2009. *Objects of Translation: Material Culture and Medieval "Hindu-Muslim" Encounter.* Princeton, NJ: Princeton University Press.

Floridi, Luciano. 2002. *Sextus Empiricus: The Transmission and Recovery of Pyrrhonism.* New York: Oxford University Press.

———. 2010. "The Rediscovery and Posthumous Influence of Scepticism." In *The Cambridge Companion to Ancient Scepticism,* edited by Richard Bett, 267–87. Cambridge: Cambridge University Press.

Flow, Christian. 2015. "Thesaurus Matters? Frame for the Study of Latin Lexicography." In *Classics in Practice: Studies in the History of Scholarship,* edited by Christopher Stray, 33–73. London: Institute of Classical Studies.

Fons del Museu Frederic Marès, 1996. Vol. 3: *Catàleg d'escultura i pintura dels segles XVI, XVII i XVIII: Època del Renaixement i el barroc.* Barcelona: Ajuntament de Barcelona.

Ford, Philip. 2008. *The Montaigne Library of Gilbert de Botton.* Cambridge: Cambridge University Library.

Forsterus, Valentinus. 1565. *De historia juris civilis Romani.* Basel: Oporinus and Hervagius.

Franklin, Julian. 1963. *Jean Bodin and the Sixteenth-Century Revolution in the Methodology of Law and History.* New York: Columbia University Press.

Frazier, Alison. 2005. *Possible Lives: Authors and Saints in Renaissance Italy.* New York: Columbia University Press.

Freedman, Paul. 2008. *Out of the East: Spices and the Medieval Imagination.* New Haven, CT: Yale University Press.

Freigius, Johannes Thomas. 1583. *Mosaicus.* Basel: Per Sebast. Henricpetri.

Fromont, Cécile. 2014. *The Art of Conversion: Christian Visual Culture in the Kingdom of Kongo.* Chapel Hill: University of North Carolina Press.

Fubini, Riccardo. 2003. *Humanism and Secularization from Petrarch to Valla.* Translated by Margaret King. Durham, NC: Duke University Press.

Fuentes, Marisa. 2016. *Dispossessed Lives: Enslaved Women, Violence, and the Archive.* Philadelphia: University of Pennsylvania Press.

Fukuyama, Francis. 1989. "The End of History?" *National Interest* 16: 3–18.

Furey, Constance. 2006. *Erasmus, Contarini, and the Religious Republic of Letters.* Cambridge: Cambridge University Press.

Furno, Martine, ed. 2009. *Qui écrit? Figures de l'auteur et des co-élaborateurs du texte, XVe–XVIIIe siècle.* Lyon, France: ENS Éditions.

Fussner, F. Smith. 1962. *The Historical Revolution: English Historical Writing and Thought, 1580–1640.* London: Routledge & Kegan Paul.

Gaisser, Julia Haig. 1999. *Pierio Valeriano on the Ill Fortune of Learned Men: A Renaissance Humanist and His World.* Ann Arbor: University of Michigan Press.

Galison, Peter. 1997. *Image and Logic: A Material Culture of Microphysics.* Chicago: University of Chicago Press.

Gallagher, Catherine, and Stephen Greenblatt. 2000. *Practicing New Historicism.* Chicago: University of Chicago Press.

Games, Alison. 2008. *The Web of Empire: English Cosmopolitans in an Age of Expansion, 1560–1660.* Oxford: Oxford University Press.

Garber, Daniel. 2016. "Why the Scientific Revolution Wasn't a Scientific Revolution, and Why it Matters." In *Kuhn's Structure of Scientific Revolutions at Fifty,* edited by Robert Richards and Lorraine Daston, 133–48. Chicago: University of Chicago Press.

García-Arenal, Mercedes, and Fernando Rodriguez Mediano. (2010) 2013. *The Orient in Spain: Converted Muslims, the Forged Lead Books of Granada, and the Rise of Orientalism.* Translated by C. Lopez-Morillas. Leiden, Netherlands: Brill.

García-Arenal, Mercedes, and Gerard Wiegers. (1999) 2003. *A Man of Three Worlds: Samuel Pallache, a Moroccan Jew in Catholic and Protestant Europe.* Translated by M. Beagles. Baltimore: Johns Hopkins University Press.

García Sanz, Ana. 2010. *El Niño Jesús en el Monasterio de las Descalzas Reales de Madrid.* Madrid: Patrimonio Nacional.

Gardiner, Eileen, and Ronald G. Musto. 2015. *The Digital Humanities: A Primer for Students and Scholars.* Cambridge: Cambridge University Press.

Garfagnini, Gian Carlo, ed. 1986. *Marsilio Ficino e il ritorno di Platone: Studi e documenti.* 2 vols. Florence: Olschki.

Garibay, Ángel María. 1964. "Introducción." In *Libellus de medicinalibus Indorum herbis: Manuscrito azteca de 1552, según traducción latina de Juan Badiano; Versión española con estudios y comentarios por diversos autores,* by Martín de la Cruz and Juan Badiano, 3–8. Mexico City: IMSS.

Garin, Eugenio. 1972. "Il pensiero di Leon Battista Alberti: caratteri e contrasti." *Rinascimento* 12: 3–20.

———. 2009. *Interpretazioni del Rinascimento.* Edited by Michele Ciliberto. 2 vols. Rome: Edizioni di Storia e Letteratura.

Garnier, Jean. 1678. *Systema bibliothecae collegii Parisiensis Societatis Jesu.* Paris: Sebastianus Mabre-Cramoisy.

Garone Gravier, Marina. 2011. "Sahagún's Codex and Book Design in the Indigenous Context." In *Colors between Two Worlds: The Florentine Codex of Bernardino de Sahagún,* edited by Gerhard Wolf and Joseph Connors, with Louis A. Waldman, 157–97. Florence: Kunsthistorisches Institut; Villa I Tatti.

Garrod, Raphaële. 2014. "Aristotelianism and Scholasticism." In *Brill's Encyclopaedia of the Neo-Latin World,* edited by Philip Ford, Jan Bloemendal, and Chares Fantazzi, vol. 1, 589–602. Leiden, Netherlands: Brill.

Gascoigne, John. 2009. "The Royal Society, Natural History and the Peoples of the 'New World(s).'" *British Journal for the History of Science* 42 (4): 539–62.

Gaukroger, Stephen. 2006. *The Emergence of a Scientific Culture: Science and the Shaping of Modernity, 1210–1685*. Oxford: Oxford University Press.

Gay, Peter. 1966–69. *The Enlightenment: An Interpretation*. 2 vols. New York: Knopf.

Gentile, Sebastiano. 1997. "Pico filologo." In *Giovanni Pico della Mirandola: Convegno internazionale di studi nel cinquecentesimo anniversario della morte (1494–1994), Mirandola, 4–8 ottobre 1994*, vol. 2., edited by Gian Carlo Garfagnini, 465–90. Florence: Olschki.

———. 2013. "Ficino, Epicuro e Lucrezio." In *The Rebirth of Platonic Theology: Proceedings of a Conference . . . (Florence, 26–27 April 2007)*, edited by James Hankins and Fabrizio Meroi, 119–35. Florence: Olschki.

Gerbi, Antonello. 1985. *Nature in the New World: From Christopher Columbus to Gonzalo Fernández de Oviedo*. Translated by Jeremy Moyle. Pittsburgh: University of Pittsburgh Press.

Gerritsen, Anne, and Giorgio Riello. 2016. *The Global Lives of Things: The Material Culture of Connections in the Early Modern World*. New York: Routledge.

Gessner, Conrad. 1551. *Historia animalium*. Zurich: Froschauer.

Ghobrial, John-Paul. 2013. *The Whispers of Cities: Information Flows in Istanbul, London, and Paris in the Age of William Trumbull*. Oxford: Oxford University Press.

———. 2014. "The Secret Life of Elias of Babylon and the Uses of Global Microhistory." *Past & Present*, no. 222: 51–93.

———. 2016. "The Archive of Orientalism and Its Keepers: Re-imagining the Histories of Arabic Manuscripts in Early Modern Europe." *Past & Present*, supplement 11: 90–111.

Gierl, Martin. 1997. *Pietismus und Aufklärung: Theologische Polemik und die Kommunikationsreform der Wissenschaft am Ende des 17. Jahrhunderts*. Göttingen: Vandenhoeck & Ruprecht.

Gigante, Marcello. 1988. "Ambrogio Traversari interprete di Diogene Laerzio." In *Ambrogio Traversari nel VI centenario della nascita: Convegno internazionale di studi (Camaldoli—Firenze, 15–18 settembre 1986)*, edited by Gian Carlo Garfagnini, 367–59. Florence: Olschki.

Giglioni, Guido. 2015. "Philosophy." In *The Oxford Handbook of Neo-Latin*, edited by Sarah Knight and Stefan Tilg, 249–62. New York: Oxford University Press.

Gilbert, Felix. 1965. *Machiavelli and Guicciardini: Politics and History in Sixteenth-Century Florence*. Princeton, NJ: Princeton University Press.

Gimmel, Millie. 2008a. "Hacia una reconsideración del *Códice de la Cruz Badiano*: Nuevas propuestas para el estudio de la medicina indígena en el período colonial." *Colonial Latin American Review* 17 (2): 273–83.

———. 2008b. "Reading Medicine in the *Codex de la Cruz Badiano*." *Journal of the History of Ideas* 69 (2): 169–92.

Ginzburg, Carlo. (1976) 1980. *The Cheese and the Worms: The Cosmos of a Sixteenth-Century Miller*. Translated by John and Anne Tedeschi. Baltimore: Johns Hopkins University Press.

———. (1993) 2012. "Microhistory: Two or Three Things That I Know about It." *Critical Inquiry* 20 (1993): 10–35. Reprinted in Carlo Ginzburg, *Threads and Traces: True False Fictive*, 193–214. Berkeley: University of California Press, 2012.

———. 1999. *History, Rhetoric, and Proof*. Hanover, NH: University Press of New England.

———. 2015. "Microhistory and World History." In *Patterns of Change*, edited by Jerry H.

Bentley, Sanjay Subrahmanyam, and Merry E. Wiesner-Hanks, 446–73. Vol. 6, part 2, of *The Cambridge World History*. Cambridge: Cambridge University Press.

———. 2017. "Civilization and Barbarism." *Σημειωτική Sign System Studies* 45: 249–62.

Girard, Aurélien. 2011. "Le christianisme oriental (XVIIe–XVIIIe siècles): Essor de l'orientalisme catholique en Europe et construction des identités confessionnelles au Proche-Orient." PhD diss., École Pratique des Hautes Études, Paris.

Glucker, John. 1964. "Casaubon's Aristotle." *Classica et mediaevalia* 25: 274–96.

Goeing, Anja-Silvia. 2017. *Storing, Archiving, Organizing: The Changing Dynamics of Scholarly Information Management in Post-Reformation Zurich*. Leiden, Netherlands: Brill.

Goitein, S. D. 1967–93. *A Mediterranean Society: The Jewish Communities of the Arab World as Portrayed in the Documents of the Cairo Geniza*. 5 vols. Berkeley: University of California Press.

Goldgar, Anne. 1995. *Impolite Learning: Conduct and Community in the Republic of Letters, 1680–1750*. New Haven, CT: Yale University Press.

Golinski, Jan. 1998. *Making Natural Knowledge: Constructivism and the History of Science*. Cambridge: Cambridge University Press.

Gómez, Pablo F. 2017. *The Experiential Caribbean: Creating Knowledge and Healing in the Early Modern Atlantic*. Chapel Hill: University of North Carolina Press.

Goodman, Dena. 1994. *The Republic of Letters: A Cultural History of the French Enlightenment*. Ithaca, NY: Cornell University Press.

Goodman, Eleanor. 2001. "Royal Piety: Faith, Religious Politics, and the Experience of Art at the Convent of the Descalzas Reales in Madrid." PhD diss., Institute of Fine Arts, New York University.

Goulding, Robert. 2009. "Pythagoras in Paris: Petrus Ramus Imagines the Prehistory of Mathematics." *Configurations* 17 (1–2): 51–86.

———. 2010. *Defending Hypatia: Ramus, Savile, and the Renaissance Rediscovery of Mathematical History*. Dordrecht, Netherlands: Springer.

Gouwens, Kenneth. 1998. *Remembering the Renaissance: Humanist Narratives of the Sack of Rome*. Leiden, Netherlands: Brill.

———. 2012. "Humanists, Historians, and the Fullness of Time in Renaissance Rome." In *Rethinking the High Renaissance*, edited by Jill Burke, 95–110. Burlington, VT: Ashgate.

Graevius, Johannes. 1681. *Oratio de Cometis*. Utrecht, Netherlands: R. Zyll.

Grafton, Anthony. 1975. "J. J. Scaliger's Indices to J. Gruter's *Inscriptiones Antiquae*: A Note on Leiden University Library MS Scal. 11." *Lias* 2: 109–13.

———. (1977) 1991. "On the Scholarship of Politian and Its Context." *Journal of the Warburg and Courtauld Institutes* 40: 150–88. Reprinted in *Defenders of the Text: The Traditions of Scholarship in an Age of Science, 1450–1800*, 47–75. Cambridge, MA: Harvard University Press.

———. 1983. "Protestant versus Prophet: Isaac Casaubon on Hermes Trismegistus." *Journal of the Warburg and Courtauld Institutes* 46: 78–93.

———. 1983–93. *Joseph Scaliger: A Study in the History of Classical Scholarship*. 2 vols. Oxford: Oxford University Press.

———. 1985. "Renaissance Readers and Ancient Texts: Comments on Some Commentaries." *Renaissance Quarterly* 38: 615–49.

———. (1987) 2001. "Portrait of Justus Lipsius." *American Scholar* 56: 382–90. Reprinted in *Bring Out Your Dead: The Past as Revelation*, 227–43. Cambridge, MA: Harvard University Press.

———. 1988. "The Availability of Ancient Works." In *Cambridge History of Renaissance Philosophy*, edited by Charles B. Schmitt, Quentin Skinner, Eckhard Kessler, and Jill Kraye, 767–91. Cambridge: Cambridge University Press.

———. 1990a. *Forgers and Critics: Creativity and Duplicity in Western Scholarship*. Princeton, NJ: Princeton University Press.

———. 1990b. "Humanism, Magic, and Science." In *The Impact of Humanism in Western Europe*, edited by Anthony Goodman and Angus MacKay, 99–117. London: Longman.

———. 1991a. *Defenders of the Text: The Traditions of Scholarship in an Age of Science, 1450–1800*. Cambridge, MA: Harvard University Press.

———. 1991b. "Humanism and Science in Rudolphine Prague: Kepler in Context." In *The Literary Culture in the Holy Roman Empire*, edited by J. A. Parente Jr., George C. Schoolfield, and Richard Erich Schade, 19–45. Chapel Hill: University of North Carolina Press.

———. 1993. *Rome Reborn: The Vatican Library and Renaissance Culture*. Washington, DC: Library of Congress.

———. 1996. "The New Science and the Traditions of Humanism." In *The Cambridge Companion to Renaissance Humanism*, edited by Jill Kraye, 203–23. Cambridge: Cambridge University Press.

———. 1997a. *The Footnote: A Curious History*. Cambridge, MA: Harvard University Press.

———. 1997b. "Giovanni Pico della Mirandola: Trials and Triumphs of an Omnivore." In *Commerce with the Classics: Ancient Books and Renaissance Readers*, 93–134. Ann Arbor: University of Michigan Press.

———. 1997c. "Johannes Kepler: The New Astronomer Reads Ancient Texts." In *Commerce with the Classics: Ancient Books and Renaissance Readers*, 185–224. Ann Arbor: University of Michigan Press.

———. 1997d. "The Rest vs. the West." *New York Review of Books* (April 10).

———. 1998. "The Revival of Antiquity: A Fan's Notes on Recent Work." *American Historical Review* 103: 118–21.

———. 1999. *Cardano's Cosmos: The Worlds and Works of a Renaissance Astrologer*. Cambridge, MA: Harvard University Press.

———. 2000. *Leon Battista Alberti, Master Builder of the Italian Renaissance*. New York: Hill & Wang.

———. (2001) 2009. "Where Was Salomon's House? Ecclesiastical History and the Intellectual Origins of Bacon's *New Atlantis*." In *Die europäische Gelehrtenrepublik im Zeitalter des Konfessionalismus*, edited by Herbert Jaumann, 21–38. Wiesbaden: Harrassowitz. Reprinted in *Worlds Made by Words: Scholarship and Community in the Modern West*, 98–113. Cambridge, MA: Harvard University Press.

———. 2002. *Bring Out Your Dead: The Past as Revelation*. Cambridge, MA: Harvard University Press.

———. 2007a. "Say Anything: What the Renaissance Teaches Us about Torture." *New Republic* (Nov. 5): 22–24.

———. 2007b. *What Was History? The Art of History in Early Modern Europe*. Cambridge: Cambridge University Press.

———. 2009. "Kindled: Reading Electronically and the Other Way." *New Republic* (18 Nov.): 33–38.

————. 2010a. "Britain: The Disgrace of the Universities." *New York Review of Books* (March 9).

————. 2010b. "Humanities and Inhumanities." *New Republic* (March 11): 32–36.

————. 2011a. "Beyond Comparison: Sarah Palin and the Blood Libel." *New Republic* (Feb. 17): 7.

————. 2011b. *The Culture of Correction in Renaissance Europe.* London: British Library.

————. 2011c. "The Most Charming Pagan." *New York Review of Books* (Dec. 8).

————. 2011d. "The Public Intellectual and the American University: Robert Morss Lovett Revisited." *American Scholar* 70 (Autumn): 53.

————. 2011e. "Roy Rosenzweig: Scholarship as Community." In *Clio Wired: The Future of the Past in the Digital Age,* by Roy Rosenzweig, ix–xxiv. New York: Columbia University Press.

————. 2011f. "Scholars of the World Unite!" *National Interest* (Jan.–Feb.): 75–80.

————. 2011g. "Wisconsin: The Cronon Affair." *New Yorker News Blog* (March 28).

————. 2014a. "Arnaldo Momigliano and the Tradition of Ecclesiastical History." In *The Legacy of Arnaldo Momigliano,* edited by Tim Cornell and Oswyn Murray, 53–76. London: Warburg Institute.

————. 2014b. "The Enclosure of the American Mind." *New York Times* (Aug. 24): BR20.

————. 2016. "Christianity's Jewish Origins Rediscovered: The Roles of Comparison in Early Modern Ecclesiastical Scholarship." *Erudition and the Republic of Letters* 1: 13–42.

————. 2017. "Spinoza's Hermeneutics: Some Heretical Thoughts." In *Scriptural Authority and Biblical Criticism in the Dutch Golden Age: God's Word Questioned,* edited by Dirk van Miert, Henk J. M. Nellen, Piet Steenbakkers, and Jetze Touber, 177–96. Oxford: Oxford University Press.

Grafton, Anthony, and James Grossman. 2011. "No More Plan B: A Very Modest Proposal for Graduate Programs in History." *Perspectives on History* (Oct.).

Grafton, Anthony, and Lisa Jardine. 1986. *From Humanism to the Humanities: Education and the Liberal Arts in Fifteenth- and Sixteenth-Century Europe.* Cambridge, MA: Harvard University Press.

Grafton, Anthony, and Glenn W. Most, eds. 2016a. *Canonical Texts and Scholarly Practices: A Global Comparative Approach.* Cambridge: Cambridge University Press.

————. 2016b. "How to Do Things with Texts." In *Canonical Texts and Scholarly Practices: A Global Comparative Approach,* edited by Anthony Grafton and Glenn W. Most, 1–13. Cambridge: Cambridge University Press.

Grafton, Anthony, and William Newman. 2006. *Secrets of Nature: Astrology and Alchemy in Early Modern Europe.* Cambridge, MA: MIT Press.

Grafton, Anthony, and Nancy Siraisi, eds. 1999. *Natural Particulars: Nature and the Disciplines in Early Modern Europe.* Cambridge, MA: MIT Press.

Grafton, Anthony, and Joanna Weinberg, with Alastair Hamilton. 2011. *"I Have Always Loved the Holy Tongue": Isaac Casaubon, the Jews, and a Forgotten Chapter in Renaissance Scholarship.* Cambridge, MA: Belknap Press of Harvard University Press.

Grafton, Anthony, and Megan Williams. 2006. *Christianity and the Transformation of the Book: Origen, Eusebius, and the Library of Caesarea.* Cambridge, MA: Belknap Press of Harvard University Press.

Grafton, Anthony, with April Shelford and Nancy Siraisi. 1992. *New Worlds, Ancient Texts: The Power of Tradition and the Shock of Discovery.* Cambridge, MA: Harvard University Press.

Greenblatt, Stephen. 1980. *Renaissance Self-Fashioning: From More to Shakespeare.* Chicago: University of Chicago Press.

———. 1990. *Marvelous Possessions: The Wonder of the New World.* Chicago: University of Chicago Press.

———. 2011. *The Swerve: How the World Became Modern.* New York: W. W. Norton. Published in the United Kingdom as *The Swerve: How the Renaissance Began.* London: Bodley Head.

Greene, Molly. 2010. *Catholic Pirates and Greek Merchants: A Maritime History of the Mediterranean.* Princeton, NJ: Princeton University Press.

Grell, Chantall. 1993. *L'Histoire entre érudition et philosophie.* Paris: PUF.

Grendi, Edoardo. 1977. "Micro-analisi e storia sociale." *Quaderni storici* 35: 506–20.

Grendler, Paul. 2002. *The Universities of the Italian Renaissance.* Baltimore: Johns Hopkins University Press.

Griggs, Tamara. 2007. "Universal History from Counter-Reformation to Enlightenment." *Modern Intellectual History* 4: 219–47.

Grosrichard, Alain. (1979) 1998. *Structure du sérail: la fiction du despotisme oriental dans l'Occident classique.* Paris: Seuil. Translated as: *The Sultan's Court: European Fantasies of the East.* Translated by L. Heron. New York: Verso.

Gruterus, Ianus. 1603. *Inscriptiones antiquae totius orbis Romani.* Heidelberg: ex officina Commeliniana.

Gruzinski, Serge. 2010. *What Time Is It There? America and Islam at the Dawn of Modern Times.* Translated by Jean Birrell. Cambridge: Polity Press.

———. 2015. *The Eagle and the Dragon: Globalization and European Dreams of Conquest in China and America in the Sixteenth Century.* Translated by Jean Birrell. Cambridge: Polity Press.

Guenée, Bernard. 1980. *Histoire et culture historique dans l'Occident médiéval.* Paris: Aubier Montaigne.

Guerrini, Luigi. 2009. "The Accademia dei Lincei, the New World, and the *Libellus de Medicinalibus Indorum Herbis.*" In *Flora: The Aztec Herbal,* edited by Martin Clayton, Luigi Guerrini, and Alejandro de Ávila Blomberg, 20–44. Series B, Natural History, pt. 8 of *The Paper Museum of Cassiano dal Pozzo.* London: Royal Collection; Harvey Miller.

Guibovich Pérez, Pedro. 2001. "The Printing Press in Colonial Peru: Production Process and Literary Categories in Lima, 1584–1699." *Colonial Latin American Review* 10 (2): 167–88.

Guicciardini, Niccolò. 2009. *Isaac Newton on Mathematical Certainty and Method.* Cambridge: MA: MIT Press.

Guldi, Jo, and David Armitage. 2014. *The History Manifesto.* Cambridge: Cambridge University Press.

Hall, Marie Boas. 1962. *The Scientific Renaissance, 1450–1630.* New York: Harper.

Hamadeh, Shirine. 2004. "Ottoman Expressions of Early Modernity and the 'Inevitable' Question of Westernization." *Journal of the Society of Architectural Historians* 63: 32–51.

———. 2007. *The City's Pleasures: Istanbul in the Eighteenth Century.* Seattle: University of Washington Press.

Hämäläinen, Pekka, and Samuel Truett. 2011. "On Borderlands." *Journal of American History* 98: 338–61.

Hamilton, Alastair. 1985. *William Bedwell, the Arabist, 1563–1632*. Leiden, Netherlands: Brill.

———. 1994. "An Egyptian Traveller in the Republic of Letters: Josephus Barbatus or Abudacnus the Copt." *Journal of the Warburg and Courtauld Institutes* 57: 123–50.

———, ed. 2005. *The Republic of Letters and the Levant*. Leiden, Netherlands: Brill.

———. 2006. *The Copts and the West, 1439–1822: The European Discovery of the Egyptian Church*. Oxford: Oxford University Press.

Hamilton, Alastair, and Francis Richard. 2004. *André du Ryer and Oriental Studies in Seventeenth-Century France*. London: Arcadian Library.

Hamling, Tara, and Catherine Richardson, eds. 2016. *Everyday Objects: Medieval and Early Modern Material Culture and Its Meanings*. London: Routledge.

Hankins, James. 1990. *Plato in the Italian Renaissance*. 2 vols. Leiden, Netherlands: Brill.

———, ed. 2000. *Renaissance Civic Humanism: Reappraisals and Reflections*. Cambridge: Cambridge University Press.

———. 2013. "Ficino's Critique of Lucretius." In *The Rebirth of Platonic Theology: Proceedings of a Conference . . . (Florence, 26–27 April 2007)*, edited by James Hankins and Fabrizio Meroi, 137–54. Florence: Olschki.

Hardwick, Lorna, and Christopher Stray, eds. 2008. *A Companion to Classical Receptions*. Malden, MA: Blackwell.

Hardy, Nicholas. 2017. *Criticism and Confession: The Bible in the Seventeenth-Century Republic of Letters*. Oxford: Oxford University Press.

Hardy, Nicholas and Dmitri Levitin, eds. 2020. *Confessionalisation and Erudition in Early Modern Europe: An Episode in the History of the Humanities*. Oxford: Oxford University Press.

Harkness, Deborah. 2007. *The Jewel House: Elizabethan London and the Scientific Revolution*. New Haven, CT: Yale University Press.

Harman, Oren and Peter L. Galison. 2008. "Epistemic Virtues and Leibnizian Dreams: On the Shifting Boundaries between Science, Humanities, and Faith." *The European Legacy: Toward New Paradigms* 13: 551–75.

Harris, A. Katie. 2007. *From Muslim to Christian Grenada: Inventing a City's Past in Early Modern Spain*. Baltimore: Johns Hopkins University Press.

Hartog, François. 2003. *Régimes d'historicité: Présentisme et expériences du temps*. Paris: Seuil.

Hasse, Dag Nikolaus. 2016. *Success and Suppression: Arabic Sciences and Philosophy in the Renaissance*. Cambridge, MA: Harvard University Press.

Haugen, Kristine. 2011. *Richard Bentley: Poetry and Enlightenment*. Cambridge, MA: Harvard University Press.

———. 2012. "Thomas Lydiat's Scholarship in Prison: Discovery and Disaster in the Seventeenth Century." *Bodleian Library Record* 25: 183–216.

Hausse, Heidi. 2019. "The Locksmith, the Surgeon, and the Mechanical Hand: Communicating Technical Knowledge in Early Modern Europe." *Technology and Culture* 60:1, pp. 34–64.

Hazard, Paul. (1935) 2013. *The Crisis of the European Mind, 1680–1715*. Translated by J. Lewis May, introduction by Anthony Grafton. Reprinted, New York: New York Review Books.

Head, Randolph C. 2019. *Making Archives in Early Modern Europe: Proof, Information, and Political Record-Keeping, 1400–1700*. Cambridge: Cambridge University Press.

Hernández, Francisco. 1790. *Opera, cum edita, tum inedita, ad autographi fidem et integritatem expressa*. Edited by Casimiro Gómez Ortega. Madrid: Ibarra.

Heyberger, Bernard. 1994. *Les chrétiens du Proche-Orient au temps de la réforme catholique: Syrie, Liban, Palestine, XVIIe–XVIIIè siècles*. Rome: École française de Rome.

————. 1995. "La carrière manquée d'un ecclésiastique oriental en Italie: Timothée Karnûsh, archevêque syrien catholique de Mardîn." *Bulletin de la faculté des lettres de Mulhouse* 19: 31–47.

————. 2010a. "L'islam et les Arabes chez un érudit maronite au service de l'Eglise catholique (Abraham Ecchellensis)." *Al-Qantara* 31: 481–512.

————, ed. 2010b. *Orientalisme, science et controverse: Abraham Ecchellensis (1605–1664)*. Turnhout, Belgium: Brepols.

Hill, Christopher. 1972. *The World Turned Upside Down: Radical Ideas during the English Revolution*. London: Temple Smith; New York: Viking Press.

Hobbins, Daniel. 2009. *Authorship and Publicity before Print: Jean Gerson and the Transformation of Late Medieval Learning*. Philadelphia: University of Pennsylvania Press.

Hobsbawm, Eric. 1972. "The Social Function of the Past: Some Questions." *Past and Present* 55: 3–17.

Hochstrasser, Timothy. 2000. *Natural Law Theories in the Early Enlightenment*. Cambridge: Cambridge University Press.

Hodgen, Margaret. 1964. *Early Anthropology in the Sixteenth and Seventeenth Centuries*. Philadelphia: University of Pennsylvania Press.

Hoffman, George. 2005. "The Investigation of Nature." In *The Cambridge Companion to Montaigne*, edited by Ullrich Langer, 163–82. Cambridge: Cambridge University Press.

Hogarth, Rana. 2017. *Medicalizing Blackness: Making Racial Differences in the Atlantic World, 1780–1840*. Chapel Hill: University of North Carolina Press.

Höpfl, Harro. 2004. *Jesuit Political Thought: The Society of Jesus and the State, c. 1540–1640*. Cambridge: Cambridge University Press.

Hopkins, A. G., ed. 2002. *Globalization in World History*. London: Pimlico.

Hsia, R. Po-chia, ed. 2007. *Reform and Expansion, 1500–1660*. Vol. 6 of *The Cambridge History of Christianity*. Cambridge: Cambridge University Press.

Huff, Toby. 2011. *Intellectual Curiosity and the Scientific Revolution*. Cambridge: Cambridge University Press.

Hunt, Lynn. 2014. *Writing History in the Global Era*. New York: W. W. Norton.

————. 2018. *History: Why It Matters*. Cambridge: Polity Press.

Hunter, Ian. 2001. *Rival Enlightenments: Civil and Metaphysical Philosophy in Early Modern Germany*. Cambridge: Cambridge University Press.

————. 2007. *The Secularisation of the Confessional State: The Political Thought of Christian Thomasius*. Cambridge: Cambridge University Press.

Hunter, Michael. 1998. *Archives of the Scientific Revolution: The Formation and Exchange of Ideas in Seventeenth-Century Europe*. Woodbridge, UK: Boydell Press.

Huppert, George. 1970. *The Idea of Perfect History: Historical Erudition and Historical Philosophy in Renaissance France*. Urbana: University of Illinois Press.

Israel, Jonathan. 2001. *Radical Enlightenment: Philosophy and the Making of Modernity, 1650–1750*. Oxford: Oxford University Press.

————. 2006. *Enlightenment Contested: Philosophy, Modernity, and the Emancipation of Man, 1670–1752*. Oxford: Oxford University Press.

Iversen, Erik. 1993. *The Myth of Egypt and Its Hieroglyphs in European Tradition.* Copenhagen: Gad, 1961. Reprinted, Princeton, NJ: Princeton University Press.

Izbicki, Thomas. 1999. "Cajetan's Attack on Parallels between Church and State." *Cristianesimo nella storia* 20 (1): 81–89.

———. 2006. "Representation in Nicholas of Cusa." In *Repraesentatio: Mapping a Keyword for Churches and Governance; Proceedings of the San Miniato International Workshop, October 13–16, 2004*, edited by Massimo Faggioli and Alberto Meloni, 61–78. Münster: LIT.

Jacob, Christian, ed. 2007–11. *Lieux de Savoir.* 2 vols. Paris: Albin Michel.

Jardine, Lisa. 1974. *Francis Bacon: Discovery and the Art of Discourse.* Cambridge: Cambridge University Press.

———. 1993. *Erasmus, Man of Letters: The Construction of Charisma in Print.* Princeton, NJ: Princeton University Press.

———. 1996. *Worldly Goods: A New History of the Renaissance.* New York: Nan A. Talese.

Jardine, Lisa, and Jerry Brotton. 2000. *Global Interests: Renaissance Art between East and West.* London: Reaktion.

Jardine, Lisa, and Anthony Grafton. 1990. "'Studied for Action': How Gabriel Harvey Read His Livy." *Past & Present* 129: 30–78.

Jenkins, Allan, and Patrick Preston. 2007. *Biblical Scholarship and the Church: A Sixteenth-Century Crisis of Authority.* Aldershot, UK: Ashgate.

Jenkins, Eugenia Zuroski. 2013. *A Taste for China: English Subjectivity and the Prehistory of Orientalism.* Oxford: Oxford University Press.

Johannesson, Kurt. 1991. *The Renaissance of the Goths in Sixteenth-Century Sweden: Johannes and Olaus Magnus as Politicians and Historians.* Translated by James Larson. Berkeley: University of California Press.

Johns, Adrian. 2010. *Piracy: The Intellectual Property Wars from Gutenberg to Gates.* Chicago: University of Chicago Press.

Jones, C. W. 1934. "Polemius Silvius, Bede, and the Names of the Months." *Speculum* 9 (1): 50–56.

Jones, Roger, and Nicholas Penny. 1983. *Raphael.* New Haven, CT: Yale University Press.

Jorink, Eric. 2010. *Reading the Book of Nature in the Dutch Golden Age.* Translated by Peter Mason. Leiden, Netherlands: Brill.

Joy, Lynn Sumida. 1987. *Gassendi the Atomist: Advocate of History in an Age of Science.* Cambridge: Cambridge University Press.

Jurdjevic, Mark. 2014. *A Great and Wretched City: Promise and Failure in Machiavelli's Florentine Political Thought.* Cambridge, MA: Harvard University Press.

Jütte, Daniel. (2011) 2015. *The Age of Secrecy: Jews, Christians, and the Economy of Secrets, 1400–1800.* Translated by J. Riemer. New Haven, CT: Yale University Press.

———. 2013. "Interfaith Encounters between Jews and Christians in the Early Modern Period and Beyond: Toward a Framework." *American Historical Review* 118: 378–400.

Kagan, Richard L. 2000. *Urban Images of the Hispanic World, 1493–1793.* New Haven, CT: Yale University Press.

———. 2012. "'The Spanish Turn': The Discovery of Spanish Art in the United States, 1887–1920." In *Collecting Spanish Art: Spain's Golden Age and America's Gilded Age*, edited by Inge Reist and José Luis Colomer, 21–41. New York: The Frick Collection.

————. 2019. *The Spanish Craze: America's Fascination with the Hispanic World, 1779–1939.* Lincoln: University of Nebraska Press.

Kahn, Victoria. 1994. *Machiavellian Rhetoric: From the Counter-Reformation to Milton.* Princeton, NJ: Princeton University Press.

Kane, William Terence. 1940. *Jean Garnier: Librarian.* Chicago: University of Chicago Press.

Kantorowicz, Ernst. 1981. *The King's Two Bodies: A Study in Medieval Political Theology.* Princeton, NJ: Princeton University Press.

Kaplan, Benjamin J. 2010. *Divided by Faith: Religious Conflict and the Practice of Toleration in Early Modern Europe.* Cambridge, MA: Harvard University Press.

Kaplan, Yosef. 2000. *An Alternative Path to Modernity: The Sephardi Diaspora in Western Europe.* Leiden, Netherlands: Brill.

Kasl, Ronda, ed. 2009. *Sacred Spain: Art and Belief in the Spanish World.* Indianapolis: Indianapolis Museum of Art; distributed by Yale University Press.

Kassell, Lauren. 2005. *Medicine and Magic in Elizabethan London: Simon Forman; Astrologer, Alchemist, and Physician.* Oxford: Oxford University Press.

Keller, Vera. 2012. "Mining Tacitus: Secrets of Empire, Nature, and Art in the Reason of State." *British Journal for the History of Science* 45: 189–212.

————. 2015. *Knowledge and the Public Interest, 1575–1725.* New York: Cambridge University Press.

Keller, Vera, Anna Marie Roos, and Elizabeth Yale, eds. 2018. *Archival Afterlives: Life, Death, and Knowledge-Making in Early Modern British Scientific and Medical Archives.* Leiden: Brill.

Keller-Rahbé, Edwige, ed. 2010. *Les arrière-boutiques de la littérature: auteurs et imprimeurs-libraires aux XVIe et XVIIe siècles.* Toulouse: Presses Universitaires du Mirail.

Kelley, Donald R. 1964. "*Historia Integra*: François Baudouin and His Conception of History." *Journal of the History of Ideas* 25: 35–57.

————. 1970. *Foundations of Modern Historical Scholarship: Language, Law, and History in the French Renaissance.* New York: Columbia University Press.

————, ed. 1997. *History and the Disciplines: The Reclassification of Knowledge in Early Modern Europe.* Rochester, NY: University of Rochester Press.

————. 1998. *Faces of History: Historical Inquiry from Herodotus to Herder.* New Haven, CT: Yale University Press.

————. 1999. "Writing Cultural History in Early Modern Europe: Christophe Milieu and His Project." *Renaissance Quarterly* 52: 342–65.

Kelly, Joan. 1977. "Did Women Have a Renaissance?" In *Becoming Visible: Women in European History*, edited by Renate Bridenthal and Claudia Koonz, 174–201. Boston: Houghton Mifflin.

Kelly, Samantha. 2015. "The Curious Case of Ethiopian Chaldean: Fraud, Philology, and Cultural (Mis)Understanding in European Conceptions of Ethiopia." *Renaissance Quarterly* 68: 1227–64.

Kemp, Martin. 1990. *The Science of Art: Optical Themes in Western Art from Brunelleschi to Seurat.* New Haven, CT: Yale University Press.

Kepler, Johannes. 1997. *The Harmony of the World.* Translated by Eric Aiton, A. M. Duncan, and J. V. Field. Philadelphia: American Philosophical Society.

Kerrigan, William, and Gordon Braden. 1989. *The Idea of the Renaissance.* Baltimore: Johns Hopkins University Press.

Klein, Ursula, and E. C. Spary. 2009. *Materials and Expertise in Early Modern Europe: Between Market and Laboratory*. Chicago: University of Chicago Press.

Knapp, Jeffrey. 2005. "What Is a Co-author?" *Representations* 89: 1–29.

Köhler, Johann David. 1728. *Sylloge aliquot scriptorum de bene ordinanda et ornanda bibliotheca*. Frankfurt: Joannes Stein.

Koselleck, Reinhart. 1984. "Historia Magistra Vita: Über die Auflösung des Topos im Horizont neuzeitlich bewegter Geschichte." In *Vergangene Zukunft: Zur Semantik geschichtlicher Zeiten*, 38–66. Frankfurt am Main: Suhrkamp.

———. 1988. *Critique and Crisis: Enlightenment and the Pathogenesis of Modern Society*. Cambridge, MA: MIT Press.

Krajewski, Markus. 2018. *The Server: A History of Assistance from the Present to the Baroque*. Translated by Ilinca Iurascu. New Haven, CT: Yale University Press.

Krämer, Fabian. 2014. *Ein Zentaur in London: Lektüre und Beobachtung in der frühneuzeitlichen Naturforschung*. Affalterbach, Germany: Didymos-Verlag.

Kraye, Jill. 2001. "L'interprétation platonicienne de l'*Enchirdion* d'Épictète proposée par Politien: Philologie et philosophie dans la Florence du XVᵉ siècle, à la fin des années 70." In *Penser entre les lignes: Philologie et philosophie au Quattrocento*, edited by Fosca Mariani Zini, 161–77. Villeneuve d'Ascq, France: Presses Universitaire du Septentrion.

———. 2007a. "From Medieval to Early Modern Stoicism." In *Continuities and Disruptions between the Middle Ages and the Renaissance: Proceedings of the Colloquium Held at the Warburg Institute, 15–16 June 2007*, edited by Charles Burnett, 1–23. Louvain-la-Neuve, Belgium: FIDEM.

———. 2007b. "The Revival of Hellenistic Philosophies." In *The Cambridge Companion to Renaissance Philosophy*, edited by James Hankins, 97–112. Cambridge: Cambridge University Press.

———. 2008. "Pico on the Relationship of Rhetoric and Philosophy." In *Pico della Mirandola: New Essays*, edited by Michael V. Dougherty, 13–36. Cambridge: Cambridge University Press.

———. 2014a. "Epicureanism and Other Hellenistic Philosophies." In *Brill's Encyclopaedia of the Neo-Latin World*, vol. 1, edited by Philip Ford, Jan Bloemendal, Charles Fantazzi, and Craig Kallendorf, 619–29. Leiden, Netherlands: Brill.

———. 2014b. "Lucretius—Editions and Commentaries." In *Brill's Encyclopaedia of the Neo-Latin World*, vol. 2, edited by Philip Ford, Jan Bloemendal, Charles Fantazzi, and Craig Kallendorf, 1038–40. Leiden, Netherlands: Brill.

———. 2016. "Beyond Moral Philosophy: Renaissance Humanism and the Philosophical Canon." *Rinascimento* 56: 3–22.

———. 2017. "The Early Modern Turn from Roman to Greek Stoicism." In *For a Skeptical Peripatetic: Festschrift in Honour of John Glucker*, edited by Yosef Z. Liebersohn, Ivor Ludlam, and Amos Edelheit, 293–305. Sankt Augustin, Germany: Academia Verlag.

Kraynack, Robert P. 1990. *History and Modernity in the Thought of Thomas Hobbes*. Ithaca, NY: Cornell University Press.

Krazek, Rafal. 2011. *Montaigne et la philosophie du plaisir: Pour une lecture épicurienne des Essais*. Paris: Classiques Garnier.

Kristeller, Paul Oskar. 1944–45. "Humanism and Scholasticism in the Italian Renaissance." *Byzantion* 17: 346–74.

———. 1961. "Humanism and Scholasticism in the Italian Renaissance." Reprinted in

Renaissance Thought: The Classic, Scholastic, and Humanist Strains, 92–119. New York: Harper & Row.

Kushner, Eva. 2018. *Pontus de Tyard et son oeuvre poétique*. Paris: Classiques Garnier.

Kusukawa, Sachiko. 2012. *Picturing the Book of Nature: Image, Text, and Argument in Sixteenth-Century Human Anatomy and Medical Botany*. Chicago: University of Chicago Press.

Lachaud, Frédérique. 2015. "Filiation and Context: The Medieval Afterlife of the *Policraticus*." In *A Companion to John of Salisbury*, edited by C. Grellard and F. Lachaud, 377–438. Leiden, Netherlands: Brill.

Lagrée, Jaqueline. 2016. "Justus Lipsius." In *The Routledge Handbook of the Stoic Tradition*, edited by John Sellars, 160–73. London: Routledge.

Laplanche, François. 1986. *L'écriture, le sacré et l'histoire: Érudits et politiques protestants devant la Bible en France au XVIIe siècle*. Amsterdam: APA-Holland University Press.

Lässig, Simone. 2016. "The History of Knowledge and the Expansion of the Historical Research Agenda." *Bulletin of the German Historical Institute* 59: 29–58.

Latour, Bruno. 1993. *We Have Never Been Modern*. Translated by Catherine Porter. Cambridge, MA: Harvard University Press.

Lehmann, Paul. 1911. *Johannes Sichardus und die von ihm benutzten Bibliotheken und Handschriften*. Munich: Beck.

Lennox, Jeffers. 2017. *Homelands and Empires: Indigenous Spaces, Imperial Fictions, and Competition for Territory in Northeastern North America, 1690–1763*. Toronto: University of Toronto Press.

Leijenhorst, Cees. 2002. *The Mechanisation of Aristotelianism: The Late Aristotelian Setting of Thomas Hobbes' Natural Philosophy*. Leiden, Netherlands: Brill.

Leinkauf, Thomas. 2017. *Grundriss Philosophie des Humanismus und der Renaissance (1350–1600)*. 2 vols. Hamburg: Felix Meiner Verlag.

León Portilla, Miguel. 2002. *Bernardino de Sahagún: First Anthropologist*. Norman: University of Oklahoma Press.

Léry, Jean de. 1990. *History of a Voyage to the Land of Brazil, Otherwise Called America*. Translated by Janet Whatley. Berkeley: University of California Press.

Levi della Vida, Giorgio. 1948. *Documenti intorno alle relazioni delle chiese orientali con la S. Sede durante il pontificato di Gregorio XIII*. Vatican City: Biblioteca apostolica vaticana.

Levine, Joseph M. 1977. *Dr. Woodward's Shield: History, Science, and Satire in Augustan England*. Berkeley: University of California Press.

———. 1991. *The Battle of the Books: History and Literature in the Augustan Age*. Ithaca, NY: Cornell University Press.

———. 1999. *The Autonomy of History: Truth and Method from Erasmus to Gibbon*. Chicago: University of Chicago Press.

Levitin, Dmitri. 2012. "From Sacred History to the History of Religion: Paganism, Judaism, and Christianity in European Historiography from Reformation to 'Enlightenment.'" *Historical Journal* 55: 1117–60.

———. 2015. *Ancient Wisdom in the Age of the New Science: Histories of Philosophy in England, c. 1640–1700*. Cambridge: Cambridge University Press.

Levy, F. J. 1967. *Tudor Historical Thought*. San Marino, CA: Huntington Library.

Lewis, Bernard. 1982. *The Muslim Discovery of Europe*. New York: W. W. Norton.

Lewis, Rhodri. 2018. "Shakespeare, Olaus Magnus, and Monsters of the Deep." *Notes and Queries* 65: 76–81.

Liang, Yuen-Gen, and Touba Ghadessi. 2013. "The Interdisciplinary Humanities: A Platform for Experiential Learning of Workplace Skills." *Perspectives on History* (April).

Ligota, Christopher R. 1987. "Annius of Viterbo and Historical Method." *Journal of the Warburg and Courtauld Institutes* 50: 44–56.

Lindebergius, Petrus. 1591. *De praecipuorum tam in sacris quam in Ethnicis scriptis numerorum Nobilitate, mysterio, et eminentia.* Rostock, Germany: Myliander.

Lipsius, Justus. 1604. *Manuductionis ad Stoicam philosophiam libri tres. . . .* Antwerp: Officina Plantiniana.

Locke, John. (1689) 2010. *A Letter concerning Toleration and Other Writings.* Edited by Mark Goldie. Indianapolis: Liberty Fund.

Lockhart, James. 1992. *The Nahuas after the Conquest: A Social and Cultural History of the Indians of Central Mexico, Sixteenth through Eighteenth Centuries.* Stanford, CA: Stanford University Press.

———. 1993. *We People Here: Nahuatl Accounts of the Conquest of Mexico.* Vol. 1 of *Repertorium Columbianum.* Berkeley: University of Calfornia Press.

Long, A. A. 2003. "Stoicism in the Philosophical Tradition: Spinoza, Lipsius, Butler." In *Hellenistic and Early Modern Philosophy*, edited by Jon Miller and Brad Inwood, 7–29. Cambridge: Cambridge University Press.

Long, Pamela O. 2001. *Openness, Secrecy, Authorship: Technical Arts and the Culture of Knowledge from Antiquity to the Renaissance.* Baltimore: Johns Hopkins University Press.

———. 2011. *Artisan/Practitioners and the Rise of the New Sciences, 1400–1600.* Corvallis: Oregon State University Press.

———. 2015. "Trading Zones in Early Modern Europe." *Isis* 106: 840–47.

Loop, Jan. 2013. *Johan Heninrich Hottinger: Arabic and Islamic Studies in the Seventeenth Century.* Oxford: Oxford University Press.

López Fadul, Valeria A. Escauriaza. 2015. "Languages, Knowledge, and Empire in the Early Modern Iberian World, 1492–1650." PhD diss., Princeton University.

Lotito, Mark A. 2019. *The Reformation of Historical Thought.* Leiden, Netherlands: Brill.

Love, Harold. 2002. *Attributing Authorship.* Cambridge: Cambridge University Press.

Lüthy, Christoph. 2000. "The Fourfold Democritus on the Stage of Early Modern Science." *Isis* 93: 443–79.

Luttrell, Anthony. 1975. *Medieval Malta: Studies on Malta before the Knights.* London: British School at Rome.

Lyon, Gregory B. 2003. "Baudouin, Flacius, and the Plan for the Magdeburg Centuries." *Journal of the History of Ideas* 64 (2): 253–72.

MacCormack, Sabine. 1991. *Religion in the Andes: Vision and Imagination in Early Colonial Peru.* Princeton, NJ: Princeton University Press.

Macfie, Alexander Lyon, ed. 2001. *Orientalism: A Reader.* New York: New York University Press.

Mack, Rosamond E. 2002. *Bazaar to Piazza: Islamic Trade and Italian Art, 1300–1600.* Berkeley: University of California Press.

MacLean, Gerald, ed. 2005. *Re-Orienting the Renaissance: Cultural Exchanges with the East.* Basingstoke, UK: Palgrave Macmillan.

Maclean, Ian. 2006. "The 'Sceptical Crisis' Reconsidered: Galen, Rational Medicine, and the *Libertas philosophandi.*" *Early Science and Medicine* 11: 247–74.

MacPhail, Eric. 2000. "Montaigne's New Epicureanism." *Montaigne Studies* 12: 91–103.

Magaloni Kerpel, Diana. 2014. *The Colors of the New World: Artists, Materials, and the Creation of the "Florentine Codex."* Los Angeles: Getty Research Institute.

Magnus, Olaus. 1555. *Historia de gentibus septentrionalibus.* Rome: Giovanni Maria di Viotti.

Maillard, Jean-Francois, Judit Kecskeméti, Catherine Magnien, and Monique Portalier, eds. 1999. *La France des Humanistes: Hellénistes I.* Turnhout, Belgium: Brepols.

Malcolm, Noel. 2002. *Aspects of Hobbes.* Oxford: Oxford University Press.

———. 2010. *Reason of State, Propaganda, and the Thirty Years' War: An Unknown Translation by Thomas Hobbes.* Oxford: Oxford University Press.

———. 2019. *Useful Enemies: Islam and the Ottoman Empire in Western Political Thought, 1450–1750.* Oxford: Oxford University Press.

Malusa, Luciano. 1993. "Thomas Stanley's *History of Philosophy.*" In *Models of the History of Philosophy: From Its Origins in the Renaissance to the "Historia Philosophica,"* edited by Giovanni Santinello, Francesco Bottin, Gregorio Piaia, and Constance Blackwell, 163–203. Dordrecht, Netherlands: Kluwer.

Marcaida López, José Ramón. 2014. *Arte y ciencia en el Barroco español: Historia natural, coleccionismo y cultura visual.* Seville: Fundación Focus-Abengoa; and Madrid: Marcial Pons.

Marchand, Suzanne L. 2009. *German Orientalism in the Age of Empire: Religion, Race, and Scholarship.* Cambridge: Cambridge University Press.

———. 2013. "Has the History of Disciplines Had Its Day?" In *Rethinking Modern European Intellectual History,* edited by Darrin McMahon and Samuel Moyn, 131–52. Oxford: Oxford University Press.

———. 2019. "How Much Knowledge Is Worth Knowing? An American Intellectual Historian's Thoughts on the *Geschichte des Wissens.*" *Berichte zur Wissenschaftsgeschichte* 42, Nos. 2–3 (September): 126–49.

Margócsy, Dániel. 2014. *Commercial Visions: Science, Trade, and Visual Culture in the Dutch Golden Age.* Chicago: University of Chicago Press.

Markey, Lia. 2011. "Istoria della terra chiamata la Nuova Spagna: The History and Reception of Sahagún's Codex at the Medici Court." In *Colors between Two Worlds: The Florentine Codex of Bernardino de Sahagún,* edited by Gerhard Wolf and Joseph Connors, with Louis A. Waldman, 198–218. Florence: Kunsthistorisches Institut; Villa I Tatti.

Marr, Alexander. 2011. *Between Raphael and Galileo: Mutio Oddi and the Mathematical Culture of Late Renaissance Italy.* Chicago: University of Chicago Press.

Marshall, John. 2006. *John Locke, Toleration, and Early Enlightenment Culture: Religious Intolerance and Arguments for Religious Toleration in Early Modern and "Early Enlightenment" Europe.* Cambridge: Cambridge University Press.

Martin, Craig. 2014. *Subverting Aristotle: Religion, History, and Philosophy in Early Modern Science.* Baltimore: Johns Hopkins University Press.

Martin, Ged. 2004. *Past Futures: The Impossible Necessity of History.* Toronto: University of Toronto Press.

Martin, John J. 2004. *Myths of Renaissance Individualism.* New York: Palgrave Macmillan.

Martindale, Charles, and Richard F. Thomas, ed. 2006. *Classics and the Uses of Reception.* Malden, MA: Blackwell.

Martinelli, Lucia Cesarini. 1980. "Sesto Empirico e una dispera enciclopedia delle arti e delle scienze di Angelo Poliziano." *Rinascimento* 20: 327–58.

Mason, Peter. 1990. *Deconstructing America: Representations of the Other.* New York: Routledge.

Mathes, W. Michael. 1985. *The America's First Academic Library: Santa Cruz de Tlatelolco.* Sacramento: California State Library Foundation.

Mattioli, Emilio. 1965. "Luciano tra Pico e Poliziano." In *L'Opera e il pensiero di Giovanni Pico della Mirandola nella storia dell'umanesimo: Convegno internazione (Mirandola: 15–18 settembre 1963),* 189–95. Vol. 2. Florence: Istituto Nazionale di Studi sul Rinascimento.

Matytsin, Anton. 2016. *The Specter of Skepticism in the Age of Enlightenment.* Baltimore: Johns Hopkins University Press.

Maxson, Brian 2013. *The Humanist World of Renaissance Florence.* Cambridge: Cambridge University Press.

Maxwell, Julie. 2004. "Counter-Reformation Versions of Saxo: A New Source for 'Hamlet'?" *Renaissance Quarterly* 57 (2): 518–60.

Mayer, August L. 1923. "Retratos españoles en el extranjero." *Boletín de la Sociedad Española de Excursiones* 31 (2): 123–25.

McCahill, Elizabeth. 2013. *Reviving the Eternal City: Rome and the Papal Court 1420–1447.* Cambridge, MA: Harvard University Press.

McCormick, Ted. 2009. *William Petty and the Ambitions of Political Arithmetic.* Oxford: Oxford University Press.

McDonough, Kelly S. 2014. *The Learned Ones: Nahua Intellectuals in Postconquest Mexico.* Tucson: University of Arizona Press.

McGuire, J. E., and P. M. Rattansi. 1966. "Newton and the Pipes of Pan." *Notes and Records of the Royal Society of London* 21: 108–43.

McKendrick, Neil, John Brewer, and J. H. Plumb. 1982. *The Birth of a Consumer Society: The Commercialization of Eighteenth-Century England.* London: Europa.

McKim-Smith, Gridley. 1974. *The Passion Scenes of Gregorio Fernández.* PhD diss., Harvard University.

McKitterick, David. 2003. *Print, Manuscript, and the Search for Order, 1450–1830.* Cambridge: Cambridge University Press.

McManamon, John, S.J. 2016. *Caligula's Barges and the Renaissance Origins of Nautical Archaeology under Water.* College Station: Texas A & M University Press.

McManus, Stuart. 2016. "The Global Lettered City: Humanism and Empire in Colonial Latin America and the Early Modern World." PhD diss., Harvard University.

McNeill, William. 1963. *The Rise of the West: A History of the Human Community.* Chicago: University of Chicago Press.

———. 1989. *Age of Gunpowder Empires, 1450–1800.* Washington, DC: American Historical Association.

Meissner, A. A. W., ed. 1847. *Vollständiges Englisch-Deutsches und Deutsch-Englisches Wörterbuch.* Leipzig: A. Liebeskind.

Melton, James Van Horn. 2001. *The Rise of the Public in Enlightenment Europe.* Cambridge: Cambridge University Press.

Mendelsohn, Andrew, with V. Hess. 2010. "Case and Series: Medical Knowledge and Paper Technology, 1600–1900." *History of Science* 48: 287–314.

Mercier, Stéphane. 2006. "Un ange au secours du Portique: Autour de la traduction du *Manuel* d'Épictète par Politien." *Folia Electronica Classica* 11 (Jan.–June).

Meserve, Margaret. 2008. *Empires of Islam in Renaissance Historical Thought.* Cambridge, MA: Harvard University Press.

Miller, Jon, and Brad Inwood, ed. 2003. *Hellenistic and Early Modern Philosophy.* Cambridge: Cambridge University Press.

Miller, Peter N. 2001. "Taking Paganism Seriously: Anthropology and Antiquarianism in Early Seventeenth-Century Histories of Religion." *Archiv für Religionsgeschichte* 3: 183–209.

———, ed. 2007. *Momigliano and Antiquarianism: Foundations of the Modern Cultural Sciences.* Toronto: University of Toronto.

———. 2015. *Peiresc's Mediterranean World.* Cambridge, MA: Harvard University Press.

Miller, Peter N., and François Louis, eds. 2012. *Antiquarianism and Intellectual Life in Europe and China, 1500–1800.* Ann Arbor: University of Michigan Press.

Millstone, Noah. 2014. "Seeing like a Statesman in Early Stuart England." *Past & Present* 223: 77–127.

———. 2016. *Manuscript Circulation and the Invention of Politics in Early Stuart England.* Cambridge: Cambridge University Press.

Minnich, Nelson. 1969. "Concepts of Reform Proposed at the Fifth Lateran Council." *Archivum Historiae Pontificiae* 7: 163–251.

Mintz, Sidney. 1985. *Sweetness and Power: The Place of Sugar in Modern History.* New York: Penguin.

Mitter, Partha. 1977. *Much Maligned Monsters: History of European Reactions to Indian Art.* Oxford: Clarendon Press.

Modigliani, Anna. 2013. *Congiurare all'antica: Stefano Porcari, Niccolò V, Roma 1453.* Rome: Roma nel Rinascimento.

Molho, Anthony. 1998. "The Italian Renaissance, Made in the USA." In *Imagined Histories: American Historians Interpret the Past,* edited by Anthony Molho and Gordon Wood, 263–94. Princeton, NJ: Princeton University Press.

Momigliano, Arnaldo. 1950. "Ancient History and the Antiquarian." *Journal of the Warburg and Courtauld Institutes* 13: 285–315.

———. 1990. *The Classical Foundations of Modern Historiography.* Berkeley: University of California Press.

Mommsen, Theodore E. 1942. "Petrarch's Conception of the 'Dark Ages.'" *Speculum* 17: 226–42.

Monfasani, John. 1999. "The Pseudo-Aristotelian *Problemata* and Aristotle's *De animalibus* in the Renaissance." In *Natural Particulars: Nature and the Disciplines in Renaissance Europe,* edited by Anthony Grafton and Nancy Siraisi, 205–47. Cambridge, MA: MIT Press.

———. 2006a. *Kristeller Reconsidered: Essays on His Life and Scholarship.* New York: Italica.

———. 2006b. "The Renaissance as the Concluding Phase of the Middle Ages." *Bullettino dell'Istituto Storico Italiano per Il Medio Evo* 108: 165–85.

———. 2014. Review of Rocco Rubini, *The Other Renaissance: Italian Humanism between Hegel & Heidegger. Rassegna Europea di Letteratura Italiana* 43: 131–36.

Monnas, Lisa. 2008. *Merchants, Princes, and Painters: Silk Fabrics in Italian and Northern Paintings, 1300–1550*. New Haven, CT: Yale University Press.

———. 2012. *Renaissance Velvets*. London: V&A.

Montebello, Philippe de. 1995–96. "Director's Foreword: Antonio Ratti Textile Center." *Metropolitan Museum of Art Bulletin* 53 (3): 5–9.

Montero Sobrevilla, Iris. 2015. "Transatlantic Hum: Natural History and the Itineraries of the Torpid Hummingbird, ca. 1521–1790." PhD diss., Cambridge University.

Morford, Mark. 1991. *Stoics and Neostoics: Rubens and the Circle of Lipsius*. Princeton, NJ: Princeton University Press.

Moyn, Samuel, and Andrew Sartori, eds. 2013. *Global Intellectual History*. New York: Columbia University Press.

Muller, Richard A. 2003. *Holy Scripture: The Cognitive Foundation of Theology*. Vol. 2 of *Post-Reformation Reformed Dogmatics: The Rise and Development of Reformed Orthodoxy, ca. 1520–1725*. 2nd ed. Grand Rapids, MI: Baker Academic.

Mulryne, J. R., Helen Watanabe-O'Kelly, and Margaret Shewring, eds. 2004. *Europa Triumphans: Court and Civic Festivals in Early Modern Europe*. Aldershot, UK: Ashgate.

Mulsow, Martin. 2005. "Antiquarianism and Idolatry: The *Historia* of Religions in the Seventeenth Century." In *Historia: Empiricism and Erudition in Early Modern Europe*, edited by Gianna Pomata and Nancy G. Siraisi, 181–210. Cambridge, MA: MIT Press.

———. 2007. *Die unanständige Gelehrtenrepublik: Wissen, Libertinage und Kommunikation in der Frühen Neuzeit*. Stuttgart: Metzler.

———. 2012. *Prekäres Wissen: Eine andere Ideengeschichte der Frühen Neuzeit*. Berlin: Suhrkamp.

———. 2019. "History of Knowledge." In *Debating New Approaches to History*, edited by Marek Tamm and Peter Burke, 159–88. London: Bloomsbury Academic.

Mundy, Barbara E. 1996. *The Mapping of New Spain: Indigenous Cartography and the Maps of the Relaciones Geográficas*. Chicago: University of Chicago Press.

Muratori, Cecilia, and Gianni Paganini, eds. 2016. *Early Modern Philosophers and the Renaissance Legacy*. Cham, Germany: Springer.

Murphy, Kathleen S. 2011. "Translating the Vernacular: Indigenous and African Knowledge in the Eighteenth-Century British Atlantic." *Atlantic Studies* 8: 29–48.

Murphy, Mekado. 2017. "Waiting for the Credits to End? Movies Are Naming More Names." *New York Times* (May 26).

Muslow, Martin. 2019. "History of Knowledge." In *Debating New Approaches to History*, ed. Marek Tamm and Peter Burke, 159–88. London: Bloomsbury Academic.

Muteau, C., and J. Garnier. 1858–60. *Galerie bourguignonne*. 3 vols. Paris: A. Durand.

Myers, Kathleen Ann. 1993. "The Representations of New World Phenomena: Visual Epistemology and Gonzalo Fernández de Oviedo's Illustrations." In *Early Images of the Americas: Transfer and Invention*, edited by Jerry M. Williams and Earl Lewis, 183–213. Tucson: University of Arizona Press.

———. 2007. *Fernández de Oviedo's Chronicle of America: A New History for a New World*. Austin: University of Texas Press.

Nadel, George. 1964. "Philosophy of History before Historicism." *History and Theory* 3: 291–315.

Nagel, Alexander, and Christopher S. Wood. 2010. *Anachronic Renaissance*. Cambridge, MA: MIT Press.

Nahrendorf, Carsten. 2019. "Antike Universalgeschichte und Säkularisierung im Melanchthonkreis: Georg Majors Edition des Justinus (1526–37) und die *Chronica Carionis* (1532)." *Daphnis* 47: 407–46.

Naquin, Nicholas. 2013. "On the Shoulders of Hercules: Erasmus, the Froben Press, and the 1516 Jerome Edition in Context." PhD diss., Princeton University.

Nauert, Charles G., Jr. 1979. "Humanists, Scientists, and Pliny: Changing Approaches to a Classical Author." *American Historical Review* 84: 72–85.

Naya, Emmanuel. 2001. "Traduire les *Hypotyposes pyrrhoniennes*: Henri Estienne entre la fièvre quarte et la folie chrétienne." In *Le scepticisme au XVIᵉ et au XVIIᵉ siècle*, edited by Pierre-François Moreau, 48–101. Paris: Albin Michel.

Nederman, Cary. 2009. *Lineages of European Political Thought: Explorations along the Medieval Modern Divide from John of Salisbury to Hegel*. Washington, DC: Catholic University of America Press.

Needham, Joseph. 1954–. *Science and Civilisation in China*. Cambridge: Cambridge University Press.

Nellen, Henk J. M., and Piet Steenbakkers. 2017. "Biblical Philology in the Long Seventeenth Century: New Orientations." In *Scriptural Authority and Biblical Criticism in the Dutch Golden Age: God's Word Questioned*, edited by Dirk van Miert, Henk J. M. Nellen, Piet Steenbakkers, and Jetze Touber. Oxford: Oxford University Press.

Nelles, Paul. 1997. "The Library as an Instrument of Discovery: Gabriel Naudé and the Uses of History." In *History and the Disciplines: The Reclassification of Knowledge in Early Modern Europe*, edited by Donald R. Kelley, 41–57. Rochester, NY: University of Rochester Press.

———. 2000. "*Historia litteraria* and Morhof: Private Teaching and Professorial Libraries at the University of Kiel." In *Mapping the World of Learning: The* Polyhistor *of Daniel Georg Morhof*, edited by Françoise Waquet, 31–56. Wiesbaden: Harrassowitz.

Neugebauer, Otto. 1951. *The Exact Sciences in Antiquity*. Princeton, NJ: Princeton University Press.

Newman, William R. 2004. *Promethean Ambitions: Alchemy and the Quest to Perfect Nature*. Chicago: University of Chicago Press.

———. 2006. *Atoms and Alchemy: Chymistry and the Experimental Origins of the Scientific Revolution*. Chicago: University of Chicago Press.

Nicéron, Jean-Pierre. 1745. *Mémoires pour servir à l'histoire des hommes illustres dans la république des lettres, avec un catalogue raisonné de leurs ouvrages*. Vol. 43. Paris: Briasson.

Nicholas of Cusa. 1991. *The Catholic Concordance*. Translated by P. E. Sigmund. Cambridge: Cambridge University Press.

Nielsen, Michael. 2011. *Reinventing Discovery: The New Era of Networked Science*. Princeton, NJ: Princeton University Press.

Norbrook, David, Stephen Harrison, and Philip Hardie, eds. 2016. *Lucretius and the Early Modern*. Classical Presences. Oxford: Oxford University Press.

Northeast, Catherine. 1991. *The Parisian Jesuits and the Enlightenment, 1700–1762*. Oxford: Voltaire Foundation.

Norton, Marcy. 2006. "Tasting Empire: Chocolate and the European Internalization of Mesoamerican Aesthetics." *American Historical Review* 111: 660–91.

———. 2008. *Sacred Gifts, Profane Pleasures: A History of Tobacco and Chocolate in the Atlantic World*. Ithaca, NY: Cornell University Press.

Nothaft, C. Philipp E. 2012. *Dating the Passion: The Life of Jesus and the Emergence of Scientific Chronology (200–1600)*. Leiden, Netherlands: Brill.

———. 2018. *Scandalous Error: Calendar Reform and Calendrical Astronomy in Medieval Europe*. Oxford: Oxford University Press.

Novick, Peter. 1988. *That Noble Dream: The "Objectivity Question" and the American Historical Profession*. Cambridge: Cambridge University Press.

O'Brien, Emily. 2015. *The Commentaries of Pope Pius II and the Crisis of the Fifteenth-Century Papacy*. Toronto: University of Toronto Press.

O'Brien, Patrick. 2006. "Historiographical Traditions and Modern Imperatives for the Restoration of Global History." *Journal of Global History* (March): 3–39.

Ogborn, Miles. 2008. *Global Lives: Britain and the World, 1550 to 1800*. Cambridge: Cambridge University Press.

———. 2013. "'It's Not What You Know . . .': Encounters, Go-Betweens, and the Geography of Knowledge." *Modern Intellectual History* 10: 163–75.

Ogilvie, Brian. 2003. "The Many Books of Nature: Renaissance Naturalists and Information Overload." *Journal of the History of Ideas* 64 (1): 29–40.

———. 2006. *The Science of Describing: Natural History in Renaissance Europe*. Chicago: University of Chicago Press.

Oldfather, W. A. 1927. *Contribution towards a Bibliography of Epictetus*. Urbana: University of Illinois.

Olds, Katrina. 2015. *Forging the Past: Invented Histories in Counter-Reformation Spain*. New Haven, CT: Yale University Press.

Ortolano, Guy. 2009. *The Two Cultures Controversy: Science, Literature, and Cultural Politics in Postwar Britain*. Cambridge: Cambridge University Press.

Osler, Margaret J., ed. 1991. *Atoms, Pneuma, and Tranquility: Epicurean and Stoic Themes in European Thought*. Cambridge: Cambridge University Press.

———. 2010. "Becoming an Outsider: Gassendi in the History of Philosophy." In *Insiders and Outsiders in Seventeenth-Century Philosophy*, edited by G. A. J. Rogers, Tom Sorell, and Jill Kraye, 23–42. New York: Routledge.

Osterhammel, Jürgen. (1998) 2018. *Unfabling the East: The Enlightenment's Encounter with Asia*. Translated by R. Savage. Princeton, NJ: Princeton University Press, 2018.

———. 2005. *Globalization: A Short History*. Princeton, NJ: Princeton University Press.

———. 2011. "Globalizations." In *Oxford Handbook of World History*, edited by Jerry H. Bentley, 89–104. Oxford: Oxford University Press.

Otten, Willemein. 2001. "Medieval Scholasticism: Past, Present, and Future." *Dutch Review of Church History* 81: 275–89.

Ouvry-Vial, Brigitte, and Anne Réach-Ngô, eds. 2010. *L'acte éditorial: Publier à la Renaissance et aujourd'hui*. Paris: Classiques Garnier.

Overfield, James. 1984. *Humanism and Scholasticism in Late Medieval Germany*. Princeton, NJ: Princeton University Press.

Pabel, Hilmar M. 2008. *Herculean Labours: Erasmus and the Editing of St. Jerome's Letters in the Renaissance*. Leiden, Netherlands: Brill.

Pade, Marianne, ed. 2005. *On Renaissance Commentaries*. Hildesheim, Germany: Georg Olms.

Paganini, Gianni. 2007. "Hobbes's Critique of the Doctrine of Essences and Its Sources."

In *The Cambridge Companion to Hobbes's* Leviathan, edited by Patricia Springborg, 337–57. Cambridge: Cambridge University Press.

Pagden, Anthony. 1991. "*Ius et Factum*: Text and Experience in the Writings of Bartolomé de Las Casas." *Representations* 33: 147–62.

Palma, Juan de la. 1636. *Vida de la serenísima infanta sor Margarita de la Cruz*. Madrid: Imprenta Real.

Palmer, Ada. 2014. *Reading Lucretius in the Renaissance*. Cambridge, MA: Harvard University Press.

Palumbo, Margherita. 2007. "Marcello, Cristoforo." *Dizionario biografico degli Italiani*. Vol. 69. Rome: Istituto della Enciclopedia Italiana.

Panofsky, Erwin. 1944. "Renaissance and Renascences." *Kenyon Review* 6: 201–36.

———. 1960. *Renaissance and Renascences in Western Art*. New York: Harper & Row.

Panofsky, Erwin, and Fritz Saxl. 1933. "Classical Mythology in Mediaeval Art." *Metropolitan Museum Studies* 4: 228–80.

Pardo Tomás, José. 2013. "Conversion Medicine: Communication and Circulation of Knowledge in the Franciscan Convent College of Tlatelolco, 1527–1577." *Quaderni Storici* 48 (1): 21–42.

———. 2016. "Making Natural History in New Spain, 1525–1590." In *The Globalization of Knowledge in the Iberian Colonial World*, edited by Helge Wendt, 29–51. Max Planck Research Library for the History and Development of Knowledge, Proceedings, Vol. 10. Berlin: Edition Open Access.

Park, Katharine, and Lorraine Daston, eds. 2006. *Early Modern Science*. Vol. 3 of *The Cambridge History of Science*. Cambridge: Cambridge University Press.

Parrish, Susan Scott. 2006. *American Curiosity: Cultures of Natural History in the Colonial British Atlantic World*. Chapel Hill: University of North Carolina Press.

Parry, Richard. 2020. "Episteme and Techne." In *Stanford Encyclopedia of Philosophy*, edited by Edward N. Zalta. Stanford University, 1997–. Article published April 11, 2003; last modified March 27, 2020. https://plato.stanford.edu/entries/episteme-techne/.

Passannante, Gerard. 2011. *The Lucretian Renaissance: Philology and the Afterlife of Tradition*. Chicago: University of Chicago Press.

Peck, Amelia, ed. 2013. *Interwoven Globe: The Worldwide Textile Trade, 1500–1800*. New Haven, CT: Yale University Press.

Pérez Mínguez, Fidel. 1923. "Notas de un peregrino: Desconocidos descubiertos." *Boletín de la Sociedad Española de Excursiones* 31 (3): 206–12.

Pérez-Ramos, Antonio. 1988. *Francis Bacon's Idea of Science and the Maker's Knowledge Tradition*. Oxford: Clarendon Press.

Perler, Dominik. 2004. "Was There a 'Pyrrhonian Crisis' in Early Modern Philosophy? A Critical Notice of Richard H. Popkin." *Archiv für Geschichte der Philosophie* 85: 209–20.

Perrone, Benigno F. 1973. "Il 'De re publica christiana' nel pensiero filosofico e politico di Pietro Galatino." In *Studi di storia pugliese in onore di Giuseppe Chiarelli*, edited by Michele Paone, vol. 2., pp. 499–632. Galatina: Congedo.

Petrarch. 1975. *Rerum familiarum libri, I–VIII*. Edited and translated by Aldo S. Bernardo. Albany: State University of New York Press.

———. 2017. *Francesco Petrarca: Selected Letters*. Vol. 1. Translated by Elaine Fantham. Cambridge, MA: Harvard University Press.

Pettegree, Andrew. 2010. *The Book in the Renaissance*. New Haven, CT: Yale University Press.

Pickering, Andrew. 1995. *The Mangle of Practice: Time, Agency, and Science*. Chicago: University of Chicago Press.

Pickstone, John. 2000. *Ways of Knowing: A New History of Science, Technology, and Medicine*. Chicago: University of Chicago.

Pocock, J. G. A. 1957. *The Ancient Constitution and the Feudal Law: A Study of English Historical Thought in the Seventeenth Century*. 1st ed. Cambridge: Cambridge University Press.

———. 1975. *The Machiavellian Moment: Florentine Political Thought and the Atlantic Republican Tradition*. Princeton: Princeton University Press.

———. 1987. *The Ancient Constitution and the Feudal Law: A Study of English Historical Thought in the Seventeenth Century*. 2nd ed. Cambridge: Cambridge University Press.

———. 1999–2015. *Barbarism and Religion*. 6 vols. Cambridge: Cambridge University Press.

Poliziano, Angelo. 1986. *Lamia: Praelectio in Priora Aristotelis analytica*, edited and commentary by Ari Wesseling. Leiden: Brill.

———. 2010. *Lamia: Text, Translation, and Introductory Studies*. Edited by Christopher S. Celenza. Leiden, Netherlands: Brill.

Pollock, Sheldon. 2015. "Introduction." In *World Philology*, edited by Sheldon Pollock, Benjamin A. Elman, and Ku-ming Kevin Chang. Cambridge, MA: Harvard University Press.

Pollock, Sheldon, Benjamin A. Elman, and Ku-ming Kevin Chang, eds. 2015. *World Philology*. Cambridge, MA: Harvard University Press.

Pomata, Gianna. 2005. "*Praxis Historialis*: The Uses of *Historia* in Early Modern Medicine." In *Historia: Empiricism and Erudition in Early Modern Europe*, edited by Gianna Pomata and Nancy G. Siraisi, 105–46. Cambridge, MA: MIT Press.

Pomata, Gianna, and Nancy G. Siraisi, eds. 2005. *Historia: Empiricism and Erudition in Early Modern Europe*. Cambridge, MA: MIT Press.

Pomeranz, Kenneth. 2000. *The Great Divergence: China, Europe, and the Making of the Modern World Economy*. Princeton, NJ: Princeton University Press.

Pomponazzi, Pietro. 1970. *Corso inediti dell'insegnamento padovano*. Edited by Antonino Poppi. 2 vols. Padua: Antenore.

Pont, Robert. 1599. *A Newe Treatise on the Right Reckoning of Yeares*. Edinburgh: R. Waldegrave.

Poole, William. 2010. *The World Makers: Scientists of the Restoration and the Search for the Origins of the Earth*. Oxford: Peter Lang.

Popkin, Richard H. 1960. *The History of Scepticism from Erasmus to Descartes*. Assen, Netherlands: Van Gorcum.

———. 1979. *The History of Scepticism from Erasmus to Spinoza*. Berkeley: University of California Press.

———. 2003. *The History of Scepticism from Savonarola to Bayle*. New York: Oxford University Press.

Popper, Nicholas. 2006. "'Abraham, Planter of Mathematics': Histories of Mathematics and Astrology in Early Modern Europe." *Journal of the History of Ideas* 67: 87–106.

———. 2011. "An Ocean of Lies: The Problem of Historical Evidence in the Sixteenth Century." *Huntington Library Quarterly* 74 (3): 375–400.

————. 2012. *Walter Ralegh's* History of the World *and the Historical Culture of the Late Renaissance*. Chicago: University of Chicago Press.

————. 2016. "The Sudden Death of the Burning Salamander: Reading Experiment and the Transformation of Natural Historical Practice in Early Modern Europe." *Erudition and the Republic of Letters* 1 (4): 464–90.

Post, Gaines, Kimon Giocarinis, and Richard Kay. 1955. "The Medieval Heritage of a Humanistic Ideal: 'Scientia donum Dei est, unde non vendi potest.'" *Traditio* 11: 195–234.

Pratt, Mary-Louise. 1992. *Imperial Eyes: Studies in Travel Writing and Transculturation*. London: Routledge.

Premo, Bianca. 2017. *Enlightenment on Trial: Ordinary Litigants and Colonialism in the Spanish Empire*. New York: Oxford University Press.

Price, David. 2011. *Johannes Reuchlin and the Campaign to Destroy Jewish Books*. Oxford: Oxford University Press.

Primavesi, Oliver. 2011. "Henri II Estienne über philosophische Dichtung: Eine Fragmentsammlung als Beitrag zu einer poetologischen Kontroverse." In *The Presocratics from the Latin Middle Ages to Hermann Diels: Akten der 9. Tagung der Karl und Gertrud Abel-Stiftung vom 5.–7. Oktober 2006 in München*, edited by Oliver Primavesi and Katharina Luchner, 157–95. Stuttgart: Franz Steiner.

Principe, Lawrence, Hjalmar Fors, and H. Otto Sibum. 2016. "From the Library to the Laboratory and Back Again: Experiment as a Tool for Historians of Science." *Ambix* 63: 85–97.

Proske, Beatrice Gilman. 1967. *Juan Martínez Montañés: Sevillian Sculptor*. New York: Hispanic Society of America.

Prown, Jules David. 1982. "Mind in Matter: An Introduction to Material Culture Theory and Method." *Winterthur Portfolio* 17 (1): 1–19.

Pumfrey, Stephen. 1991. "The History of Science and the Renaissance Science of History." In *Science, Culture, and Popular Belief in Renaissance Europe*, edited by Stephen Pumfrey, Paolo L. Rossi, and Maurice Slawinski, 48–70. Manchester: Manchester University Press.

Quantin, Jean-Louis. 2009. *The Church of England and Christian Antiquity: The Construction of a Confessional Identity in the 17th Century*. Oxford: Oxford University Press.

Quaquarelli, Leonardo, and Zita Zanardi. 2005. *Pichiana: Bibliografia delle edizioni e degli studi*. Florence: Olschki.

Quillen, Carol. 1998. *Rereading the Renaissance: Petrarch, Augustine, and the Language of Humanism*. Ann Arbor: University of Michigan Press.

Quint, David. 1985. "Humanism and Modernity: A Reconsideration of Bruni *Dialogues*." *Renaissance Quarterly* 38: 423–45.

Quirós Rosado, Roberto. 2008, 2009. "Nobleza, iglesia y comercio indiano: El caso de Cristóbal García de Segovia (1622–1692)." *Hidalguía* 331: 839–60; 332: 43–83.

Radway, Robyn Dora. 2017. "Vernacular Diplomacy in Central Europe: Statesmen and Soldiers between the Habsburg and Ottoman Empires, 1543–1593." PhD diss., Princeton University.

Raggio, Olga. 1965. "A Neapolitan Christmas Crib." *Metropolitan Museum of Art Bulletin* 24: 151–58.

Raj, Kapil. 2007. *Relocating Modern Science: Circulation and the Construction of Knowledge in South Asia and Europe, 1650–1900*. Basingstoke, UK: Palgrave Macmillan.

Ralegh, Walter. 1614. *The History of the World*. London: Walter Burre.

Ramachandran, Ayesha. 2015. *The Worldmakers: Global Imagining in Early Modern Europe*. Chicago: University of Chicago Press.

Rampling, Jennifer. 2014. "Transmuting Sericon: Alchemy as 'Practical Exegesis' in Early Modern England." *Osiris* 29: 19–34.

Rankin, Alisha. 2013. *Panaceia's Daughters: Noblewomen as Healers in Early Modern Germany*. Chicago: University of Chicago.

Raphael, Renée. 2017. *Reading Galileo: Scribal Technologies and the* Two New Sciences. Baltimore: Johns Hopkins University Press.

Raven, Diederick, and Wolfgang Krohn. 2000. "Introduction." In *The Social Origins of Modern Science*, by Edgar Zilsel, edited by Diederick Raven, Wolfgang Krohn, and Robert S. Cohen. Dordrecht, Netherlands: Kluwer.

Reeves, Marjorie. 1993. *The Influence of Prophecy in the Later Middle Ages: A Study in Joachimism*. Notre Dame, IN: University of Notre Dame Press.

Reinert, Sophus. 2011. *Translating Empire: Emulation and the Origins of Political Economy*. Cambridge, MA: Harvard University Press.

Reist, Inge, and José Luis Colomer, eds. 2012. *Collecting Spanish Art: Spain's Golden Age and America's Gilded Age*. New York: The Frick Collection, in association with the Centro de Estudios Europa Hispánica and the Center for Spain in America.

Restall, Matthew, and Florine Asselbergs, eds. 2007. *Invading Guatemala: Spanish, Nahua, and Maya Accounts of the Conquest Wars*. University Park: Pennsylvania State University Press.

Revel, Jacques. 1995. "Microanalysis and the Construction of the Social." In *Histories: French Constructions of the Past*, edited by Jacques Revel and Lynn Hunt, 492–502. New York: New Press, 1995.

Rey Bueno, Mar. 2004. "Junta de herbolarios y tertulias espagíricas: El círculo cortesano de Diego de Cortavila (1597–1657)." *Dynamis* 24: 243–67.

Rheinberger, Hans-Jörg. 2010. *On Historicizing Epistemology: An Essay*. Translated by David Fernbach. Stanford, CA: Stanford University Press.

———. 2016. "Culture and Nature in the Prism of Knowledge." *History of the Humanities* 1: 166–81.

Richardson, Catherine, Tara Hamling, and David Gaimster, eds. 2016. *The Routledge Handbook of Material Culture in Early Modern Europe*. New York: Routledge.

Richardson, Heather. 2017. "I Don't Like to Talk about Politics . . ." Facebook post (Jan. 28). https://www.facebook.com/heather.richardson.986/posts/654265404770041.

Roberts, Lissa, Simon Schaffer, and Peter Dear, eds. 2007. *The Mindful Hand*. Amsterdam: Koninklijke Nederlandse Akademie van Wetenschappen.

Roberts, Meghan K. 2016. *Sentimental Savants: Philosophical Families in Enlightenment France*. Chicago: University of Chicago Press.

Roberts, Sean. 2013. *Printing a Mediterranean World: Florence, Constantinople, and the Renaissance of Geography*. Cambridge, MA: Harvard University Press.

Robertson, John. 2005. *The Case for the Enlightenment: Scotland and Naples, 1680–1760*. Cambridge: Cambridge University Press.

Robichaud, Denis. 2018. *Plato's Persona: Marsilio Ficino, Renaissance Humanism, and Platonic Traditions*. Philadelphia: University of Pennsylvania Press.

Robinson, Chase F. 2003. *Islamic Historiography*. Cambridge: Cambridge University Press.

Rockefeller, Stuart Alexander. 2011. "Flow." *Current Anthropology* 52: 557–78.

Rodgers, Daniel T. 2014. "Cultures in Motion: An Introduction." In *Cultures in Motion*, edited by Daniel T. Rodgers, Bhavani Raman, and Helmut Reimitz, 1–19. Princeton, NJ: Princeton University Press.

Romani, V. 1988. *Il "Syntagma de arte typographica" di Juan Caramuel ed altri testi secenteschi sulla tipografia e l'edizione*. Rome: Vecchiarelli.

Roos, Anna Marie. 2019. *Martin Lister and His Remarkable Daughters: The Art of Science in the Seventeenth Century*. Oxford: Bodleian Library.

Rosaleny, Vicente Raga. 2009. "The Current Debate about Montaigne's Skepticism." In *Skepticism in the Modern Age: Building on the Work of Richard Popkin*, edited by José R. Maia Neto, Gianni Paganini, and John Christian Laursen, 55–70. Leiden, Netherlands: Brill.

Rosenberg, Daniel. 2003. "Early Modern Information Overload." *Journal of the History of Ideas* 64: 1–9.

Rosenberg, Daniel, and Anthony Grafton. 2010. *Cartographies of Time*. New York: Princeton Architectural Press.

Rosenthal, Jean-Laurent, and R. Bin Wong. 2011. *Before and Beyond Divergence: The Politics of Economic Change in China and Europe*. Cambridge, MA: Harvard University Press.

Rosenwald, Michael S. 2017. "In Divided America, History Is Weaponized to Praise or Condemn Trump." *Washington Post* (Feb. 6).

Ross, Alan. 2015. *Daum's Boys: Schools and the Republic of Letters in Early Modern Germany*. Manchester: Manchester University Press.

Ross, Sarah. 2009. *The Birth of Feminism: Women as Intellect in Renaissance Italy and England*. Cambridge, MA: Harvard University Press.

———. 2016. *Everyday Renaissances: The Quest for Cultural Legitimacy in Venice*. Cambridge, MA: Harvard University Press.

Rossi, Paolo. 1970. *Philosophy, Technology, and the Arts in the Early Modern Era*. Translated by Salvator Attanasio. New York: Harper & Row.

Rothman, Aviva. 2017. *The Pursuit of Harmony: Kepler on Cosmos, Confession, and Community*. Chicago: University of Chicago Press.

Rothman, E. Natalie. 2006. "Between Venice and Istanbul: Trans-imperial Subjects and Cultural Mediation in the Early Modern Mediterranean." PhD diss., University of Michigan.

———. 2009. "Interpreting Dragomans: Boundaries and Crossings in the Early Modern Mediterranean." *Comparative Studies in Society and History* 51: 771–800.

———. 2012. *Brokering Empire: Trans-imperial Subjects between Venice and Istanbul*. Ithaca, NY: Cornell University Press.

———. Forthcoming. *The Dragoman Renaissance: Diplomatic Interpreters and the Routes of Orientalism*. Ithaca, NY: Cornell University Press.

Rothschild, Emma. 2011. *The Inner Life of Empires: An Eighteenth-Century History*. Princeton, NJ: Princeton University Press.

Rowland, Ingrid. 1998. *The Culture of the High Renaissance: Ancients and Moderns in Sixteenth-Century Rome*. Cambridge: Cambridge University Press.

———. 2000. *The Ecstatic Journey: Athanasius Kircher in Baroque Rome*. Chicago: University of Chicago Press.

Rousseau, G. S., and Roy Porter, eds. 1990. *Exoticism in the Enlightenment.* Manchester: Manchester University Press.

Rozen, Minna. 2002. *A History of the Jewish Community in Istanbul: The Formative Years, 1453–1566.* Leiden, Netherlands: Brill.

Rubiés, Joan-Pau. 2000. *Travel and Ethnology in the Renaissance: South India through European Eyes, 1250–1625.* Cambridge: Cambridge University Press.

———. 2006. "Theology, Ethnography, and the Historicization of Idolatry." *Journal of the History of Ideas* 67 (4): 571–96.

Rublack, Ulinka. 2016. "Renaissance Dress, Cultures of Making, and the Period Eye." *West 86th: A Journal of Decorative Arts, Design History, and Material Culture* 23 (1): 6–34.

Ruderman, David. 2010. *Early Modern Jewry: A New Cultural History.* Princeton, NJ: Princeton University Press.

Ruehl, Martin. 2015. *The Italian Renaissance in the German Historical Imagination, 1860–1930.* Cambridge: Cambridge University Press.

Rummel, Erika. 1995. *The Humanist-Scholastic Debate in the Renaissance and Reformation.* Cambridge, MA: Harvard University Press.

———. 2000. *The Confessionalization of Humanism in Reformation Germany.* Oxford: Oxford University Press.

Rusconi, Roberto. 1992. "An Angelic Pope before the Sack of Rome." In *Prophetic Rome in the High Renaissance Period*, edited by Marjorie Reeves, 157–87. Oxford: Clarendon Press.

Rüstem, Ünver. 2019. *Ottoman Baroque: The Architectural Refashioning of Eighteenth-Century Istanbul.* Princeton, NJ: Princeton University Press.

Safier, Neil. 2008. *Measuring the New World: Enlightenment Science and South America.* Chicago: University of Chicago Press.

Sahagún, Bernardino de. 1950–1982. *Florentine Codex: General History of the Things of New Spain.* Translated by Arthur J. O. Anderson and Charles E. Dibble. 12 vols. Santa Fe, NM: School of American Research.

Said, Edward W. 1978. *Orientalism.* New York: Pantheon.

———. 1993. *Culture and Imperialism.* New York: Vintage.

Saliba, George. 2007. *Islamic Science and the Making of the European Renaissance.* Cambridge: MIT Press.

Salomon, Xavier F. 2017. "'An Arsenal of Passions': The Frick Self-Portrait from Seville to Manhattan." In *Murillo: The Self-Portraits*, edited by Xavier F. Salomon, Letizia Treves, and Maria Álvarez-Garcillian, 69–85. New York: The Frick Collection; New Haven, CT: Yale University Press.

Salomoni, Mario. 1955. *De principatu.* Milan: A. Giuffrè.

Santana-Acuña, Alvaro. 2016. "The End of the Traditional Art Gallery? Part 2: Dark Intermediaries, 1% Collectors, and Global Galleries." *Books and Ideas* (May 9). http://www.booksandideas.net/The-End-of-the-Traditional-Art-Gallery-Part-Two.html.

Sardar, Marika. 2013. "Silk along the Seas: Ottoman Turkey and Safavid Iran in the Global Textile Trade." In *Interwoven Globe: The Worldwide Textile Trade, 1500–1800*, edited by Amelia Peck, 66–81. New Haven, CT: Yale University Press.

Scaliger, Joseph Justus. 1583. *Opus novum de emendatione temporum.* Paris: Patisson.

———. 1627. *Epistolae.* Leiden, Netherlands: Elzevir.

Schaffer, Simon, Lissa Roberts, Kapil Raj, and James Delbourgo, eds. 2009. *The Brokered World: Go-Betweens and Global Intelligence, 1770–1820.* Sagamore Beach, MA: Science History.

Schama, Simon. 1987. *The Embarrassment of Riches: An Interpretation of Dutch Culture in the Golden Age.* New York: Knopf.

Schiebinger, Londa L. 2007. *Plants and Empire: Colonial Bioprospecting in the Atlantic World.* Cambridge, MA: Harvard University Press.

———. 2017. *Secret Cures of Slaves: People, Plants, and Medicine in the Eighteenth-Century Atlantic World.* Stanford: Stanford University Press.

Schiebinger, Londa, and Claudia Swan, eds. 2005. *Colonial Botany: Science, Commerce, and Politics in the Early Modern World.* Philadelphia: University of Pennsylvania Press.

Schiffman, Zachary Sayre. 2011. *The Birth of the Past.* Baltimore: Johns Hopkins University Press.

Schlanger, Zoë. 2017. "Rogue Scientists Race to Save Climate Data from Trump." *Wired* (Jan. 19).

Schmidt-Biggemann, Wilhelm. 1983. *Topica Universalis: Eine Modellgeschichte humanistischer und barocker Wissenschaft.* Hamburg: Felix Meiner.

Schmitt, Charles B. 1972. *Cicero Scepticus: A Study of the Influence of the Academica in the Renaissance.* The Hague: Martinus Nijhoff.

———. 1983a. *Aristotle and the Renaissance.* Cambridge, MA: Harvard University Press.

———. 1983b. "The Rediscovery of Ancient Skepticism in Modern Times." In *The Skeptical Tradition,* edited by Miles Burnyeat, 225–51. Berkeley: University of California Press.

Schnapp, Alain, ed. 2013. *World Antiquarianism: Comparative Perspectives.* Los Angeles: Getty Research Institute.

Schorske, Carl E. 1979. *Fin-de-Siècle Vienna: Politics and Culture.* New York: Knopf.

Screech, M. A. 1998. *Montaigne's Annotated Copy of Lucretius: A Transcription and Study of the Manuscript, Notes, and Pen-Marks.* Geneva: Droz.

Seifert, Arno. 1976. *Cognitio Historica: Die Geschichte als Namengeberin der frühneutzlichen Empirie.* Berlin: Duncker & Humblot.

Semmelhack, Elizabeth. 2009. *On a Pedestal: Renaissance Chopines to Baroque Heels.* Toronto: Bata Shoe Museum.

———. 2015. *Standing Tall: The Curious History of Men in Heels.* Toronto: Bata Shoe Museum.

Senellart, Michel. 1989. *Machiavélisme et raison d'état.* Paris: PUF.

Serjeantson, Richard. 1998. "Testimony, Authority, and Proof in Seventeenth-Century England." PhD diss., Cambridge University.

———. 2006. "Proof and Persuasion." in *The Cambridge History of Science,* vol. 3: *Early Modern Science,* edited by Katherine Park and Lorraine Daston, 132–75. Cambridge: Cambridge University Press.

Serrera Contreras, Juan M. 1983. *Hernando de Esturmio.* Seville: Diputación Provincial de Sevilla.

Seth, Suman. 2018. *Difference and Disease: Medicine, Race, and the Eighteenth-Century British Empire.* Cambridge: Cambridge University Press.

Sewell, William H. 2005. *Logics of History: Social Theory and Social Transformation.* Chicago: University of Chicago Press.

———. 2010. "The Empire of Fashion and the Rise of Capitalism in Eighteenth-Century France." *Past & Present* 206: 81–120.

Shalev, Zur. 2012. *Sacred Words and Worlds: Geography, Religion, and Scholarship, 1550–1700*. Leiden, Netherlands: Brill.

Shapin, Steven. 1984. "Pump and Circumstance: Robert Boyle's Literary Technology." *Social Studies of Science* 14: 481–519.

———. 1989. "The Invisible Technician." *American Scientist* 77: 554–63.

———. 1994. *A Social History of Truth: Civility and Science in Seventeenth-Century England*. Chicago: University of Chicago Press.

———. 1996. *The Scientific Revolution*. Chicago: University of Chicago Press.

Shapin, Steven, and Simon Schaffer. 1985. *Leviathan and the Air-Pump: Hobbes, Boyle, and the Experimental Life*. Princeton, NJ: Princeton University Press.

Shapiro, Barbara. 1979. "History and Natural History in Sixteenth- and Seventeenth-Century England: An Essay on the Relationship between Humanism and Science." In *English Scientific Virtuosi in the Sixteenth and Seventeenth Centuries*, edited by Barbara Shapiro and Robert Frank, 1–55. Los Angeles: Clark Library.

———. 1991. "Early Modern Intellectual Life: Humanism, Religion, and Science in Seventeenth Century England." *History of Science* 29 (1): 45–71.

———. 1993. *Probability and Certainty in Seventeenth-Century England: A Study of the Relationships between Natural Science, Religon, History, Law, and Literature*. Princeton, NJ: Princeton University Press.

———. 2000. *A Culture of Fact: England, 1550–1720*. Ithaca, NY: Cornell University Press.

Sheehan, Jonathan. 2003. "From Philology to Fossils: The Biblical Encyclopedia in Early Modern Europe." *Journal of the History of Ideas* 64 (1): 41–60.

———. 2005. *The Enlightenment Bible: Translation, Scholarship, Culture*. Princeton, NJ: Princeton University Press.

Shelford, April G. 2007. *Transforming the Republic of Letters: Pierre-Daniel Huet and European Intellectual Life, 1650–1720*. Rochester, NY: Boydell & Brewer.

Sherman, William H. 2008. *Used Books: Marking Readers in Renaissance England*. Philadelphia: University of Pennsylvania Press.

Shuger, Debora K. 1994. *The Renaissance Bible: Scholarship, Sacrifice, and Subjectivity*. Berkeley: University of California Press.

SilverMoon. 2007. "The Imperial College of Tlatelolco and the Emergence of a New Nahua Intellectual Elite in New Spain (1500–1760)." PhD diss., Duke University.

Siraisi, Nancy. 1997. *The Clock and the Mirror: Girolamo Cardano and Renaissance Medicine*. Princeton, NJ: Princeton University Press.

———. 2007. *History, Medicine, and the Traditions of Renaissance Learning*. Ann Arbor: University of Michigan Press.

———. 2012. "Medicine, 1450–1620, and the History of Science." *Isis* 103 (3): 491–514.

———. 2013. *Communities of Learned Experience: Epistolary Medicine in the Renaissance*. Baltimore: Johns Hopkins University Press.

Sire, H. J. A. 1994. *The Knights of Malta*. New Haven, CT: Yale University Press.

Siskin, Clifford, and William Warner, ed. 2010. *This Is Enlightenment*. Chicago: University of Chicago Press.

Sixtus Senensis. 1566. *Bibliotheca Sancta*. Venice: Franciscus Franciscius Senensis.

Skinner, Quentin. 1978. *The Foundations of Modern Political Thought.* 2 vols. Cambridge: Cambridge University Press.

Smail, Daniel Lord. 2016. *Legal Plunder: Household Debt Collection in Late Medieval Europe.* Cambridge, MA: Harvard University Press.

Smentek, Kristel. 2009. "Looking East: Jean-Étienne Liotard, the Turkish Painter." In *Globalizing Cultures: Art and Mobility in the Eighteenth Century*, edited by Nebahat Avcioğlu and Finbarr Barry Flood, 84–112. Ars orientalis 39. Washington, DC: Smithsonian Institution.

Smith, Pamela H. 2004. *The Body of the Artisan: Art and Experience in the Scientific Revolution.* Chicago: University of Chicago Press.

———. 2009. "Science on the Move: Recent Trends in the History of Early Modern Science." *Renaissance Quarterly* 62: 345–75.

———. 2013. "The History of Science as a Cultural History of the Material World." In *Cultural Histories of the Material World*, edited by Peter Miller, 210–25. Ann Arbor: University of Michigan Press.

———. 2014. "Knowledge in Motion: Following Itineraries of Matter in the Early Modern World." In *Cultures in Motion*, edited by Daniel T. Rodgers, Bhavani Raman, and Helmut Reimitz, 109–33. Princeton, NJ: Princeton University Press.

Smith, Pamela H., and Paula Findlen, ed. 2002. *Merchants and Marvels: Commerce, Science, and Art in Early Modern Europe.* New York: Routledge.

Smith, Pamela H., Amy Meyers, and Harold Cook, eds. 2014. *Ways of Making and Knowing: The Material Culture of Empirical Knowledge.* Ann Arbor: University of Michigan Press.

Smith, Pamela H., and Benjamin Schmidt, eds. 2007. *Making Knowledge in Early Modern Europe: Practices, Objects, and Texts, 1400–1800.* Chicago: University of Chicago Press.

Smoller, Laura. 1994. *History, Prophecy, and the Stars: The Christian Astrology of Pierre d'Ailly, 1350–1420.* Princeton, NJ: Princeton University Press.

Smyth, Marina. 1996. *Understanding the Universe in Seventh-Century Ireland.* Woodbridge, UK: Boydell.

Snow, C. P. 1959. *The Two Cultures and the Scientific Revolution.* Cambridge: Cambridge University Press.

Snyder, Jon R. 2009. *Dissimulation and the Culture of Secrecy in Early Modern Europe.* Berkeley: University of California Press.

Snyder, Timothy. 2017. "Donald Trump and the New Dawn of Tyranny." *Time* (March 3).

Solano, Francisco de. 1988. *Cuestionarios para la formación de las relaciones geográficas de Indias: Siglos XVI/XIX.* Madrid: CSIC.

Soler, Léna, Sjoerd Zwart, Michael Lynch, and Vincent Israel-Jost, eds. 2014. *Science after the Practice Turn in the Philosophy, History, and Social Studies of Science.* New York: Routledge.

Soll, Jacob. 2005. *Publishing the Prince: History, Reading, and the Birth of Political Criticism.* Ann Arbor: University of Michigan Press.

———. 2009. *The Information Master: Jean-Baptiste Colbert's Secret State Intelligence System.* Ann Arbor: University of Michigan Press.

———. 2014. *The Reckoning: Financial Accountability and the Rise and Fall of Nations.* New York: Basic Books.

———. 2016. "The Grafton Method, or the Science of Tradition." In *For the Sake of Learn-*

ing: Essays in Honor of Anthony Grafton, vol. 2, edited by Ann Blair and Anja-Silvia Goeing, 1018–31. Leiden, Netherlands: Brill.

Sommervogel, Carlos, ed. 1890–1932. *Bibliothèque de la Compagnie de Jésus*. 11 vols. Paris: Alphonse Picard.

Sorell, Tom. 2000. "Hobbes's Uses of the History of Philosophy." In *Hobbes and History*, edited by G. A. J. Rogers and Tom Sorell, 81–95. London: Routledge.

Sorkin, David. 2008. *The Religious Enlightenment: Protestants, Jews, and Catholics from London to Vienna*. Princeton, NJ: Princeton University Press.

Sotheby's. 1994. *European Works of Art, Arms, and Armour; Tapestries and Furniture*. Sale 6521 (Jan. 11).

Southern, R. W. 1995–2001. *Scholastic Humanism and the Foundations of Europe*. 2 vols. Oxford: Wiley-Blackwell.

Spence, Jonathan D. 1984. *The Memory Palace of Matteo Ricci*. New York: Penguin.

———. 1988. *The Question of Hu*. New York: Knopf.

Spinoza, Benedict de. 2007. *Theological-Political Treatise*. Edited by Jonathan Israel and translated by Michael Silverthorne. Cambridge: Cambridge University Press.

Stagl, Justin. 1995. *A History of Curiosity: The Theory of Travel, 1550–1800*. London: Routledge.

Staley, David. 2007. *History and Future: Using Historical Thinking to Imagine the Future*. Lanham, MD: Lexington Books.

Stanley, Thomas. 1659. *The History of Philosophy, the Fourth Part, Containing the Sceptick Sect*. London: H. Moseley and T. Dring.

Starn, Randolph. 1998. "Renaissance Redux." *American Historical Review* 103: 122–24.

Stearns, Peter N. 2016. *Globalization in World History*. 2nd ed. New York: Routledge.

Stephens, Walter. 1979. "Berosus Chaldaeus: Counterfeit and Fictive Editors of the Early Sixteenth Century." PhD diss., Cornell University.

Stigler, Stephen. 1986. "John Craig and he Probability of History: From the Death of Christ to the Birth of Laplace." *Journal of the American Statistical Association* 81 (396): 879–87.

Stinger, Charles. 1998. *The Renaissance in Rome*. Bloomington: Indiana University Press.

Stols, Alexandre M. 1964. "Descripción del códice." In *Libellus de medicinalibus Indorum herbis: Manuscrito azteca de 1552, según traducción latina de Juan Badiano; Versión española con estudios y comentarios por diversos autores*, by Martín de la Cruz and Juan Badiano, 229–36. Mexico City: IMSS.

Stolzenberg, Daniel. 2013. *Egyptian Oedipus: Athanasius Kircher and the Secrets of Antiquity*. Chicago: University of Chicago Press.

Stone, Lawrence. 1979. "The Revival of Narrative: Reflections on a New Old History." *Past & Present* 85: 3–24.

Stout, Jeffrey. 2004. *Democracy and Tradition*. Princeton, NJ: Princeton University Press.

Strang, Cameron. 2018. *Frontiers of Science: Imperialism and Natural Knowledge in the Gulf South Borderlands, 1500–1850*. Chapel Hill: University of North Carolina Press for Omohundro Institute of Early American History and Culture.

Strassler, Robert B., ed. 2007. *The Landmark Herodotus: The Histories*. New York: Pantheon.

Stroumsa, Guy G. 2010. *A New Science: The Discovery of Religion in the Age of Reason*. Cambridge, MA: Harvard University Press.

Subrahmanyam, Sanjay. 1997. "Connected Histories: Notes towards a Reconfiguration of Early Modern Eurasia." *Modern Asian Studies* 31: 735–62.

———. 2003. "Turning the Stones Over: Sixteenth-Century Millenarianism from the Tagus to the Ganges." *Indian Economic and Social History Review* 40: 129–61.

———. 2005. "On World Historians in the Sixteenth Century." *Representations* 91: 26–57.

———. 2009. "Between a Rock and a Hard Place: Some Afterthoughts." In *The Brokered World: Go-Betweens and Global Intelligence, 1770–1820*, edited by Simon Schaffer, Lissa Roberts, Kapil Raj, and James Delbourgo, 429–40. Sagamore Beach, MA: Science History.

———. 2011. *Three Ways to Be Alien: Travails and Encounters in the Early Modern World.* Waltham, MA: Brandeis University Press.

———. 2012. *Courtly Encounters: Translating Courtliness and Violence in Early Modern Eurasia.* Cambridge, MA: Harvard University Press.

———. 2015. "On Early Modern Historiography." In *The Construction of a Global World, 1400–1800 CE*, part 2, *Patterns of Change*, edited by Jerry H. Bentley, Sanjay Subrahmanyam, and Merry E. Wiesner-Hanks, 425–45. Vol. 6 of *The Cambridge World History.* Cambridge: Cambridge University Press.

———. 2017. *Europe's India: Words, People, Empires, 1500–1800.* Cambridge, MA: Harvard University Press.

Swan, Claudia. 2015. "Exotica on the Move: Birds of Paradise in Early Modern Holland." *Art History* 38 (4): 620–35.

Sweet, James H. 2013. *Domingos Álvares, African Healing, and the Intellectual History of the Atlantic World.* Chapel Hill: University of North Carolina Press.

Swerdlow, Noel, and Otto Neugebauer. 1984. *Mathematical Astronomy in Copernicus' De Revolutionibus.* New York: Springer.

Syson, Luke, Sheena Wagstaff, Emerson Bowyer, and Brinda Kumar, eds. 2018. *Like Life: Sculpture, Color, and the Body.* New Haven, CT: Yale University Press.

Taylor, Charles. 2007. *A Secular Age.* Cambridge, MA: Harvard University Press.

Taylor, E. G. R. 1954. *The Mathematical Practitioners of Tudor and Stuart England.* Cambridge: Cambridge University Press, for the Institute of Navigation.

———. 1966. *The Mathematical Practitioners of Hanoverian England, 1714–80.* Cambridge: Cambridge University Press, for the Institute of Navigation.

Taylor, Jean Gelman. 2009. *The Social World of Batavia: Europeans and Eurasians in Colonial Indonesia.* 2nd ed. Madison: University of Wisconsin Press.

Temkin, Moshik. 2017. "Historians Shouldn't Be Pundits." Editorial. *New York Times* (June 26).

Térey, Gabriel von. 1923. "Die Murillo-Bilder der Sammlung Eugen Boross in Larchmond (New York)." *Der Cicerone* 15: 769–74.

Terraciano, Kevin. 2010. "Three Texts in One: Book XII of the *Florentine Codex*." *Ethnohistory* 57: 51–72.

Thomas, Keith. 1971. *Religion and the Decline of Magic: Studies of Popular Beliefs in Sixteenth- and Seventeenth-Century England.* London: Weidenfeld and Nicolson.

Thomas, Nicholas. 1991. *Entangled Objects: Exchange, Material Culture, and Colonialism in the Pacific.* Cambridge, Mass.: Harvard University Press.

Thomson, Ann. 1987. *Barbary and Enlightenment: European Attitudes towards the Maghreb in the 18th Century.* Leiden, Netherlands: Brill.

Tiffany, Tanya J. 2016. "'Little Idols': Royal Children and the Infant Jesus in the Devotional Practice of Sor Margarita de la Cruz (1567–1633)." *The Early Modern Child in Art and History*, edited by Matthew Knox Averett, 35–48. London: Routledge.

Tinguely, Frédéric. 2000. *L'Écriture du Levant à la Renaissance: Enquête sur les Voyageurs Français dans l'Empire de Soliman le Magnifique*. Geneva: Droz.

Tommasino, Pier Mattia. 2013. *L'Alcorano di Macometto: Storia di un libro del Cinquecento europeo*. Bologna: Il Mulino.

Toomer, G. J. 1996. *Eastern Wisedome and Learning: The Study of Arabic in Seventeenth-Century England*. Oxford: Clarendon Press.

Touber, Jetze. 2016. "God's Word in the Dutch Republic." In *God in the Enlightenment*, edited by William J. Bulman and Robert G. Ingram, 157–81. Oxford: Oxford University Press.

———. 2017. "Biblical Philology and Hermeneutical Debate in the Dutch Republic in the Second Half of the Seventeenth Century." In *Scriptural Authority and Biblical Criticism in the Dutch Golden Age: God's Word Questioned*, edited by Dirk van Miert, Henk J. M. Nellen, Piet Steenbakkers, and Jetze Touber, 325–47. Oxford: Oxford University Press.

———. 2018. *Spinoza and Biblical Philology in the Dutch Republic, 1660–1710*. Oxford: Oxford University Press.

Townsend, Camille. 2006. *Malintzin's Choices: An Indian Woman in the Conquest of Mexico*. Albuquerque: University of New Mexico Press.

Travers, Robert. 2015. "The Connected Worlds of Haji Mustapha (c. 1730–91): A Eurasian Cosmopolitan in Eighteenth-Century Bengal." *Indian Economic and Social History Review* 52: 297–333.

Trentmann, Frank. 2016. *Empire of Things: How We Became a World of Consumers, from the Fifteenth Century to the Twenty-First*. New York: HarperCollins.

Trivellato, Francesca. 2011. "Is There a Future for Italian Microhistory in the Age of Global History?" *California Italian Studies* 2. https://escholarship.org/uc/item/0z94n9hq.

———. 2012. *Familiarity of Strangers: The Sephardic Diaspora, Livorno, and Cross-Cultural Trade in the Early Modern Period*. New Haven, CT: Yale University Press.

Trivellato, Francesca, Leor Halevi, and Catia Antunes, eds. 2014. *Religion and Trade: Cross-Cultural Exchanges in World History, 1000–1900*. New York: Oxford University Press.

Tsing, Anna Lowenhaupt. 2005. *Friction: An Ethnography of Global Connection*. Princeton, NJ: Princeton University Press.

Tuck, Richard. 2012. "Hobbes and Tacitus." In *Hobbes and History*, edited by G. A. J. Rogers and Thomas Sorell, 99–111. London: Routledge.

Turner, James C. 2014. *Philology: The Forgotten Origins of the Modern Humanities*. Princeton, NJ: Princeton University Press.

Tyard, Pontus de. 1603. *De recta nominum impositione*. Lyon, France: Roussin.

———. 2007. *La droite imposition des noms*. Edited by Jean Céard and Jean-Claude Margolin. Vol. 7 of *Oeuvres completes*. Paris: Champion.

Ulrich, Laurel Thatcher. 2015. *Tangible Things: Making History through Objects*. Oxford: Oxford University Press.

Union académique internationale. 1960–. *Catalogus translationum et commentariorum: Mediaeval and Renaissance Translations and Commentaries, Annotated Lists, and Guides*. Washington, DC: Catholic University of America Press.

U.S. Department of State. n.d. "Policy Planning Committee." Accessed March 14, 2017. https://www.state.gov/s/p/.

Valensi, Lucette. 2008. *Mardochée Naggiar: enquête sur un inconnu.* Paris: Stock.

Vanautgaerden, Alexandre. 2012. *Erasme typographe.* Geneva: Droz.

Van Liere, Katherine. 2003. "After Nebrija: Academic Reformers and the Teaching of Latin in Sixteenth-Century Salamanca." *Sixteenth Century Journal* 23: 1065–1105.

Van Liere, Katherine, Simon Ditchfield, and Howard Louthan, eds. 2012. *Sacred History: Uses of the Christian Past in the Renaissance World.* Oxford: Oxford University Press.

Van Miert, Dirk. 2018. *The Emancipation of Biblical Philology in the Dutch Republic, 1590–1670.* Oxford: Oxford University Press.

Van Miert, Dirk, Henk J. M. Nellen, Piet Steenbakkers, and Jetze Touber, eds. 2017. *Scriptural Authority and Biblical Criticism in the Dutch Golden Age: God's Word Questioned.* Oxford: Oxford University Press.

Varey, Simon, ed. 2000. *The Mexican Treasury: The Writings of Dr. Francisco Hernández.* Translated by Rafael Chabrán, Cynthia L. Chamberlin, and Simon Varey. Stanford, CA: Stanford University Press.

Varey, Simon, Rafael Chabrán, and Dora B. Weiner, eds. 2000. *Searching for the Secrets of Nature: The Life and Works of Dr. Francisco Hernández.* Stanford, CA: Stanford University Press.

Veeser, Harold Aram, ed. 1989. *The New Historicism.* New York: Routledge.

Vessey, Mark. 1994. "Erasmus' Jerome: The Publishing of a Christian Author." *Erasmus Studies* 14: 62–99.

———. 2017. "A More Radical Renaissance." *Erasmus Studies* 37: 23–44.

Viesca Treviño, Carlos. 1995. ". . . Y Martín de la Cruz, autor del *Códice de la Cruz Badiano*, era un médico tlatelolca de carne y hueso." *Estudios de Cultura Náhuatl* 25: 479–98.

Villey, Pierre. 1933. *Les Sources et l'évolution des Essais de Montaigne.* 2 vols. Paris: Librairie Hachette.

Viroli, Maurizio. 2005. *From Politics to Reason of State: The Acquisition and Transformation of the Language of Politics, 1250–1600.* Cambridge: Cambridge University Press.

Visceglia, Maria. 2009. *Riti di corte e simboli della regalità: I regni d'Europa e del Mediterraneo dal Medioevo all'età moderna.* Rome: Salerno.

Visser, Arnoud S. Q. 2011. *Reading Augustine in the Reformation: The Flexibility of Intellectual Authority in Europe, 1500–1620.* New York: Oxford University Press.

Viti, Paolo. 1994a. "Pico e Poliziano." In *Pico, Poliziano e l'Umanesimo di fine Quattrocento: Biblioteca Medicea Laurenziana, 4 novembre–31 dicembre 1994: Catalogo,* edited by Paolo Viti, 103–25. Florence: Olschki.

———, ed. 1994b. *Pico, Poliziano e l'Umanesimo di fine Quattrocento: Biblioteca Medicea Laurenziana, 4 novembre–31 dicembre 1994: Catalogo.* Florence: Olschki.

———. 2015. "Note sulla traduzione di Angelo Poliziano del *Manuale* di Epitteto." In *Il ritorno dei classici nell'umanesimo: Studi in memoria di Gianvito Resta,* edited by Gabriella Albanese, Claudio Ciociola, Mariarosa Cortesi, and Claudia Villa, 621–33. Florence: Edizioni del Galluzzo.

Vives, Juan Luis. 1531. *De tradendis disciplinis.* Antwerp: Excudebat M. Hillenius.

———. 1913. *Vives: On Education.* Translated by Foster Watson. Cambridge: Cambridge University Press.

Völkel, Markus. 1987. *"Pyrrhonismus historicus" und "fides historica": Die Entwicklung der deutschen historischen Methodologie unter dem Gesichtspunkt der historischen Skepsis.* Frankfurt: Lang.

Voltaire, 2000. *Candide and Other Writings.* Translated by D. Wootton. Indianapolis: Hackett.

Vrolijk, Arnoud. 2010. "The Prince of Arabists and His Many Errors: Thomas Erpenius's Image of Joseph Scaliger and the Edition of the *Proverbia Arabica* (1614)." *Journal of the Warburg and Courtauld Institutes* 73: 297–325.

Walker, D. P. 1972. *The Ancient Theology: Studies in Christian Platonism from the Fifteenth to the Eighteenth Century.* London: Duckworth.

Wallace, William E. 1994. *Michelangelo at San Lorenzo.* Cambridge: Cambridge University Press.

Waquet, François. 1989. "Qu'est-ce que la République les Lettres? Essai de sémantique historique." *Bibliothèque de l'École des Chartes* 147: 473–502.

———. 1993. "Les éditions de correspondances savantes et les idéaux de la République des Lettres." *XVIIe Siècle* 45: 99–118.

Warsh, Molly. 2018. *American Baroque: Pearls and the Nature of Empire, 1492–1700.* Chapel Hill: University of North Carolina Press for Omohundro Institute of Early American History and Culture.

Weber, Max. 1930. *The Protestant Ethic and the Spirit of Capitalism.* New York: Scribner.

Webster, Susan Verdi. 1998. *Art and Ritual in Golden-Age Spain: Sevillian Confraternities and the Processional Sculpture of Holy Week.* Princeton, NJ: Princeton University Press.

Weinberg, Bernard. 1961. *A History of Literary Criticism in the Italian Renaissance.* 2 vols. Chicago: University of Chicago Press.

Welch, Evelyn, ed. 2017. *Fashioning the Early Modern: Dress, Textiles, and Innovation in Europe, 1500–1800.* Oxford: Oxford University Press.

Westfall, Carol William. 1974. *In This Most Perfect Paradise: Alberti, Nicholas V, and the Invention of Conscious Urban Planning in Rome, 1447–55.* University Park: Pennsylvania State University Press.

White, Joseph Blanco. 1825. *Letters from Spain.* 2nd ed. London: Henry Colburn.

Whitney, Elspeth. 1990. *Paradise Restored: The Mechanical Arts from Antiquity through the Thirteenth Century.* Philadelphia: Transactions of the American Philosophical Society.

Wilentz, Sean. 2017. "Trump Ushers in the American Age of Paranoia: Like Goldwater, He Promises a Return to a Lost Era of National Greatness." *Newsweek* (Jan. 19).

Wilkinson, Robert. 2007. *Orientalism, Aramaic, and Kabbalah in the Catholic Reformation: The First Printing of the Syriac New Testament.* Leiden, Netherlands: Brill.

Williams, Wes. 2016. "'Well Said/Well Thought': How Montaigne Read His Lucretius." In *Lucretius and the Early Modern*, edited by David Norbrook, Stephen Harrison, and Philip Hardie, 135–60. Oxford: Oxford University Press.

Wills, John E., Jr. 2015. "The First Global Dialogues: Inter-cultural Relations, 1400–1800." In *The Construction of a Global World, 1400–1800 CE*, part 2, *Patterns of Change*, edited by Jerry H. Bentley, Sanjay Subrahmanyam, and Merry E. Wiesner-Hanks, 50–79. Vol. 6 of *The Cambridge World History.* Cambridge: Cambridge University Press.

Witt, Ronald. 2003. *In the Footsteps of the Ancients: The Origins of Humanism from Lovato to Bruni.* Boston: Brill.

————. 2011. *The Two Latin Cultures and the Foundation of Renaissance Humanism in Medieval Italy*. Cambridge: Cambridge University Press.

Witt, Ronald, John Najemy, Craig Kallendorf, and Werner Gundersheimer. 1996. "AHR Forum: Hans Baron's Renaissance Humanism." *American Historical Review* 101: 107–44.

Wolf, Gerhard, and Joseph Connors, with Louis A. Waldman, eds. 2011. *Colors between Two Worlds: The Florentine Codex of Bernardino de Sahagún*. Florence: Kunsthistorisches Institut; Villa I Tatti.

Wolff, É. 1999. "L'utilisation du texte de Lucrèce par Gassendi dans le *Philosophiae Epicuri syntagma*." In *Présence de Lucrèce: Actes du Colloque tenu à Tours (3–5 décembre 1988)*, edited by Rémy Poignault, 327–36. Tours, France: Centre de Recherches A. Piganiol.

Wootton, David. 2015. *The Invention of Science: A New History of the Scientific Revolution*. London: Penguin.

Woudhuysen, Henry. 1996. *Sir Philip Sidney and the Circulation of Manuscripts, 1558–1640*. Oxford: Clarendon Press.

Wunder, Amanda. 2017. *Baroque Seville: Sacred Art in a Century in Crisis*. University Park: Pennsylvania State University Press.

Yale, Elizabeth. 2016. *Sociable Knowledge: Natural History and the Nation in Early Modern Britain*. Philadelphia: University of Pennsylvania Press.

————. 2019. "A Letter Is a Paper House: Home, Family, and Natural Knowledge." In *Working with Paper: Gendered Practices in the History of Knowledge*, edited by Carla Bittel, Elaine Leong, and Christine van Oertzen. Pittsburgh: University of Pittsburgh Press.

Yates, Frances. 1964. *Giordano Bruno and the Hermetic Tradition*. London: Routledge.

Yeazell, Ruth Bernard. 2000. *Harems of the Mind: Passages of Western Art and Literature*. New Haven, CT: Yale University Press.

Yeo, Richard R. 2003. "A Solution to the Multitude of Books: Ephraim Chambers's *Cyclopaedia* (1728) as 'the Best Book in the Universe.'" *Journal of the History of Ideas* 64 (1): 61–72.

————. 2014. *Notebooks, English Virtuosi, and Early Modern Science*. Chicago: University of Chicago Press.

Yonan, Michael. 2011. "Toward a Fusion of Art History and Material Culture Studies." *West 86th: A Journal of Decorative Arts, Design History, and Material Culture* 18 (2): 232–48.

Yoran, Hanan. 2007. "Florentine Civic Humanism and the Emergence of Modern Ideology." *History and Theory* 46: 326–44.

Zaken, Avner Ben. 2010. *Cross-Cultural Scientific Exchanges in the Eastern Mediterranean, 1560–1660*. Baltimore: Johns Hopkins University Press.

Zedelmaier, Helmut. 1992. *Bibliotheca Universalis und Bibliotheca Selecta: Das Problem der Ordnung des gelehrten Wissens in der frühen Neuzeit*. Cologne: Böhlau.

Zetina, Sandra, Tatiana Falcón, Elsa Arroyo, and José Luis Ruvalcaba. 2011. "The Encoded Language of Herbs: Material Insights into the *de la Cruz-Badiano Codex*." In *Colors between Two Worlds: The Florentine Codex of Bernardino de Sahagún*, edited by Gerhard Wolf and Joseph Connors, with Louis A. Waldman, 220–55. Florence: Kunsthistorisches Institut; Villa I Tatti.

Zilsel, Edgar. 1945. "The Genesis of the Idea of Scientific Progress." *Journal of the History of Ideas* 6: 325–49.

Zwinger, Theodor. 1565. *Theatrum vitae humanae*. Basel: Oporinus.

———. 1571. *Theatrum vitae humanae*. Basel: Froben.

———. 1586. *Theatrum humanae vitae*. Basel: Episcopius.